# Photogrammetry and Remote Sensing

# Photogrammetry and Remote Sensing

Edited by **Matt Weilberg**

SYRAWOOD
PUBLISHING HOUSE

New York

Published by Syrawood Publishing House,
750 Third Avenue, 9th Floor,
New York, NY 10017, USA
www.syrawoodpublishinghouse.com

**Photogrammetry and Remote Sensing**
Edited by Matt Weilberg

© 2016 Syrawood Publishing House

International Standard Book Number: 978-1-68286-107-3 (Hardback)

Printed in the United States of America.

# Contents

# Preface

The science of taking measurements using photographs is called photogrammetry and it complements the discipline of remote sensing. This book attempts to assist those with a goal of delving into the field of photogrammetry and remote sensing by covering photogrammetric methods and their applications in diverse fields such as architecture, engineering, geology, meteorology, etc. From theories to research to practical applications, case studies related to all contemporary topics of relevance to this field have been included in this book. It is highly recommended for students, academicians and researchers of this field.

Significant researches are present in this book. Intensive efforts have been employed by authors to make this book an outstanding discourse. This book contains the enlightening chapters which have been written on the basis of significant researches done by the experts.

Finally, I would also like to thank all the members involved in this book for being a team and meeting all the deadlines for the submission of their respective works. I would also like to thank my friends and family for being supportive in my efforts.

**Editor**

# PRECISE INDOOR LOCALIZATION FOR MOBILE LASER SCANNER

R. Kaijaluoto*, A. Kukko, J. Hyyppä

Finnish Geospatial Research Institute (FGI) - Center of Excellence in Laser Scanning Research, FI-02430 Masala, Finland
(risto.kaijaluoto@nls.fi, antero.kukko@nls.fi, juha.coelasr@gmail.com)

**KEY WORDS:** Mobile Laser Scanning, Indoor localization, Simultaneous Localization and Mapping, Point Cloud

**ABSTRACT:**

Accurate 3D data is of high importance for indoor modeling for various applications in construction, engineering and cultural heritage documentation. For the lack of GNSS signals hampers use of kinematic platforms indoors, TLS is currently the most accurate and precise method for collecting such a data. Due to its static single view point data collection, excessive time and data redundancy are needed for integrity and coverage of data. However, localization methods with affordable scanners are used for solving mobile platform pose problem. The aim of this study was to investigate what level of trajectory accuracies can be achieved with high quality sensors and freely available state of the art planar SLAM algorithms, and how well this trajectory translates to a point cloud collected with a secondary scanner.

In this study high precision laser scanners were used with a novel way to combine the strengths of two SLAM algorithms into functional method for precise localization. We collected five datasets using Slammer platform with two laser scanners, and processed them with altogether 20 different parameter sets. The results were validated against TLS reference. The results show increasing scan frequency improves the trajectory, reaching 20 mm RMSE levels for the best performing parameter sets. Further analysis of the 3D point cloud showed good agreement with TLS reference with 17 mm positional RMSE. With precision scanners the obtained point cloud allows for high level of detail data for indoor modeling with accuracies close to TLS at best with vastly improved data collection efficiency.

## 1 INTRODUCTION

Demand for digital 3D models of building indoors has been growing as the cost of producing one has been getting smaller. They can be used for variety of purposes from creating virtual worlds to monitoring building condition. In building monitoring periodically taken 3D measurements can be used to asses the structural integrity of a building by for example measuring of supporting beams or walls have bulged or moved. 3D model of a building can also help with planning renovations and can also be used for assessing the result. Virtual models of important cultural and historical sites can be used as marketing for the lesser known ones and for the popular ones it can serve as a way to visit it alone as crowds of tourists can greatly affect the atmosphere. While virtual visit to a site is not the same as an actual visit, having the option enables people from all over the world without the resources or time for actual visit to have the experience. Some sites can also be too delicate for actual visits by tourists so a virtual model can be the only way to provide access for wider public.

To create point clouds and models of building interiors, terrestial laser scanners (TLS) are commonly employed. While this provides good quality point clouds, data collection requires planning and is time consuming. This is especially true in cluttered spaces with low visibility where the amount of reference targets and scanning locations required for a occlusion free point cloud can quickly grow to a large number yielding also to an excessive data redundancy.

With Mobile Laser Scanner (MLS) continuously taking measurements, large area can be covered quickly and occlusions are much less of a problem (Kukko, 2013). To achieve this, the trajectory of the MLS platform must be known with high precision. Outdoors, the combination of inertial measurement unit (IMU) and global navigation satellite system (GNSS) can provide good estimate (Kaartinen et al., 2012) but indoors in the absence of GNSS signals other methods must be employed. While there are some interesting experimental localization results (Lehtola et al., 2015), horizontally mounted laser scanners are widely used for localization and mapping purposes indoors. Scanners typically provide with accurate spatial information about the world with little noise and when combined to a Simultaneous Localization and Mapping (SLAM) algorithms they can provide trajectory of a platform in unknown environment (Thrun et al., 2005, p.153, p.309).

SLAM problem is difficult because there is a circular dependency between a need for a map for localization and a desire to know location for building the map. Fortunately this problem can be solved by iteratively localizing the pose relative to the starting pose and by using this information to construct the map (Thrun et al., 2005, p. 309). A SLAM algorithm can be divided to a frontend and a backend. The frontend deals with sensory input doing scan matching and calculating spatial relations between subsequent scans. The backend works with collected information and trying to optimize poses to keep the resulting trajectory coherent in relation to reality (Konolige et al., 2010). In practise a SLAM frontend is enough to solve the SLAM problem, but as the small errors in the scan matching will inevitably accumulate reducing accuracy, in complex environments with longer trajectories and many loop closures (returning to an area visited before) a backend can greatly improve the result.

Aim of this research is to investigate what level of trajectory accuracies can be achieved with high quality sensors and freely available state of the art planar SLAM algorithms, and how well this trajectory translates to point cloud collected with a secondary scanner.

## 2 METHODS

The proposed approach for the indoor mapping problem is to combine high quality sensors to state of the art 2D SLAM algorithms. Slammer indoor MLS platform consists of a NovAtel SPAN Flexpak6 GNSS receiver with tactical grade IMU (UIMU-LCI) and two Faro Focus 3D (120S and X330) high precision laser scanners mounted on a wheeled cart as seen in Figure 1.

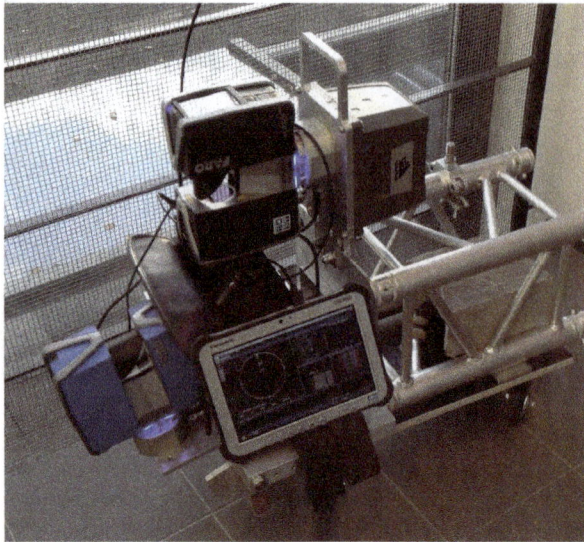

Figure 1: SLAMmer platform consists of IMU (NovAtel UIMU LCI), horizontally mounted scannes for SLAM (FARO Focus 3D120S), and secondary scanner for 3D point cloud generation (FARO Focus 3DX330). The scanners are interchangeable. Tablet computer is used for IMU and timing data recording.

In the setting the 120S laser scanner is mounted horizontally for measuring the platform movement, while the X330 in front of the system is tilted 10 degrees for producing 3D point cloud of the scene (Figure 5) (angle is adjustable at 10 degree steps for optimizing the configuration).

For data processing Robotic Operating System (ROS) framework is used (ROS, 2014). It contains among other things, numerous open source libraries for solving SLAM and localization problems and tools for accompanying data processing. As all of the algorithms can be used within the common framework it provides a straightforward way for trying, comparing and combining different algorithms.

As an original approach we combine two SLAM algorithms with different strengths, Karto Open library (SRI International, 2014) and Hector SLAM (Kohlbrecher et al., 2011). Hector SLAM algorithm utilizes the full scan rate of modern high frequency laser scanners, by using every subsequent scan the search space for rigid-body transformation between scans stays small and the transformation can be found in real time. The correct transformation is found by optimizing the alignment of scan endpoints with the map learned so far with Gauss-Newton method (Kohlbrecher et al., 2011).

Hill climbing optimization methods, are prone to getting stuck at local optima so to ensure convergence to global minimum, the algorithm maintains a pyramid of multiple occupancy grids with each having half the resolution of the preceding one. Scan matching is started with the coarsest resolution occupancy grid and then repeated on finer grids using the result of scan matching done at the previous resolution as the starting guess. To reach precision greater than available from discrete sized grid cells bilinear filtering is employed. Even though the high quality scan matcher gives the algorithm good accuracy, during closing a loop the lack of optimizing backend shows up as a discrete jump in the trajectory when the pose recovers.

Karto (SRI International, 2014) uses correlative scan matching algorithm by (Olson, 2009) in its frontend and Sparse Pose Adjustment (SPA) (Konolige et al., 2010) as its backend.

The scan matcher requires an estimate of movement as its first guess (from example odometer, in our case from Hector SLAM) and then performs a search around it at discrete steps over a search window. To narrow down the search over large 3D space $(x, y, \theta)$ of possible transformations the area is first evaluated at a coarser scale to find areas of interest for a finer search. Many transformation candidates around the correct one are evaluated and provide the algorithm with reliable estimate for the covariance of the found transformation (Olson, 2009), this value is important for the backend as a weight for the constraint created from the scan match. The downside of discrete steps is the limit on accuracy they give. This is especially noticeable if every scan from the scanner is input to the algorithm, the displacement between scans can be less than the search resolution which causes the localization to become unreliable. As we are processing the data offline the resolution of the search grid could be set to a value small enough, in theory this should only result in a longer processing time but unfortunately the algorithm became unstable at around 2 mm resolution, which still causes jitter to the trajectory.

The SPA based backend of the Karto uses a pose graph structure where scanner poses form the nodes and the edges between them are the scan matching results from the frontend. When returning to an explored area and closing a loop, the current scan is matched with a chain of previously processed ones and if the match is sufficiently good, an additional constraint between those poses is created. During optimization the poses are moved around to minimize the error caused by the constraints and this reduces the errors accumulated before the loop closure. The Karto with its SPA backend has been shown to give excellent results when compared to other algorithms (Vincent et al., 2010).

The Karto library is mainly aimed for mobile robots and online use, but as we do the processing offline with all of the data available some changes can be made. In addition to using parameters which provide higher accuracy and processing load too great for online computation, a modification to the loop closing function was made. In its original form a loop is closed immediately when a match between current scan and a chain of nearby processed scans gives a response value greater than a predetermined threshold. In the studied approach matching is tried between the chain of old, processed scans and all of the "current" scans which are located close enough to be possible loop closure candidates. After trying matching all of them, the one with the highest response value is selected and used for closing the loop. This is advantageous as a better match leads to less error and the match generally gets better as the scanning locations get closer together. In original form, if the threshold is small a suboptimal loop closure is made when the current location is still far away from the chain of scans but if the threshold is too high, some not as good but still useful loop closures might be missed altogether. With our approach more loop closures can be made while ensuring that they are as good as possible.

The movement of the platform and scanner mounted on it causes distortion to the resulting scans, which adds error to scan matching. If this movement is known, it can be compensated by transforming each measurement of the scan according to the scanner pose at the time of measurement. The IMU deployed is of sufficient quality so the IMU measurements could be used for estimating this movement but to keep the estimates from drifting away, fusion of the IMU and SLAM estimates with help of for example extended Kalman filter would be required.

As an intermediate measure we only consider the rotational component of the movement during a scan. This is more straightforward as the IMU provides angular velocity measurements and

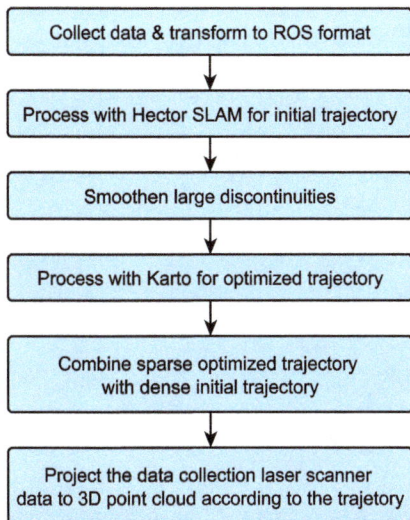

Figure 2: Pipeline from data collection to finalized point cloud.

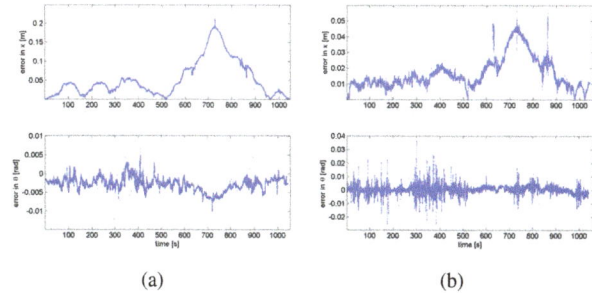

Figure 3: Errors of test run X2 as time series. (a) trajectory from Hector SLAM before processing with Karto, (b) the result of the enhancement with Karto. Top row is the position error and bottom row is for the rotational error. Without optimizing backend errors can grow unboundedly.

orientation estimates based on them. Even though the absolute heading (rotation around z-axis) value based on the IMU measurements drifts away from the truth, it can be used for estimating relative rotation over short periods of time accurately. Hector SLAM algorithm was augmented with this feature. Currently only the scanner start and end poses within a scan are used while the intermediate poses corresponding to each single scan point measurement are interpolated. Unfortunately Karto algorithm makes assumptions of undistorted scan when doing scan matching, which makes implementing this kind of movement compensation more difficult, and it is a task of future experimenting to be fully exploited.

In our approach we use the Karto library as an ad-hoc backend for the Hector SLAM by using the trajectory from Hector as the initial guess for Karto. This overcomes the problems caused by the lack of backend in Hector while enabling us to use its high quality scan matcher.

The proposed pipeline can be seen in Figure 2. First the data is transformed to ROS applicable format and then input to Hector SLAM for calculating the initial trajectory. Before its use in Karto, the large discontinuities caused by loop closures in the Hector trajectory are smoothed away. This is accomplished by checking each transformation between two poses and comparing it to the preceding ones, if there is large change in transformation length or direction from the preceding ones, the transformation is replaced by an average of the preceding transformations.

The smoothed trajectory is then input to Karto and processed. By setting the Karto parameter which penalizes transformations differing from initial guess to a sufficiently large value it can be made to largely follow the initial trajectory. To speed up the processing only a subset of the scans are processed by Karto, in our case three per second. The sparse and optimized trajectory is then combined with the dense Hector trajectory. This is accomplished by performing an affine transformation to each subtrajectory between two optimized poses which align the subtrajectory endpoints with the optimized trajectory. Headings of the poses between optimized endpoints are interpolated.

Finally the complete trajectory is used to transform the scans made by the tilted laser scanner to form 3D point cloud of the whole area.

## 3 EXPERIMENTS

To test our approach we made 5 test runs in the FGI office, which has both long narrow corridors and cluttered spaces with short visibility. The test runs took different routes, details of them are in Table 1. All tests were made with same IMU logging settings, acceleration and angular velocity measurements were logged at 200 Hz and attitude calculated by SPAN at 20 Hz. Horizontal laser scanner maximum range was set to 41 meters which is close to the the length of the longest corridor and enables the scanner to detect everything in its field of view.

| | Data acquisition settings and trajectory length | | | | |
|---|---|---|---|---|---|
| Test | Rot. Rate | Ang. Res. | p/scan | Length | Speed |
| | Hz | mrad | | m | m/s |
| X1 | 59.7 | 0.767 | 6828 | 256.9 | 0.27 |
| X2 | 59.7 | 0.767 | 6828 | 265.6 | 0.25 |
| S1 | 47.7 | 0.613 | 8536 | 268.5 | 0.52 |
| S2 | 47.7 | 0.613 | 8536 | 312.1 | 0.52 |
| S3 | 47.7 | 0.613 | 8536 | 232.6 | 0.49 |

Table 1: Test runs starting with X were made with X330 scanner as the horizontal scanner and ones starting with S with 120S. Length is the length of the trajectory and speed is the average movement speed.

Faro X330 laser scanner was used in TLS mode to generate a point cloud of the study area for geometric reference. Altogether 31 TLS scans were taken, which were then registered together into a single point cloud by using Faro Scene software and target spheres 199 mm and 145 mm in diameter placed around the area.

For extracting reference trajectories for the test runs, a 50 mm thick slice around the elevation of the horizontal laser scanner from the floor level (about 50 cm) was extracted from the reference point cloud and the scans collected during each of the test runs were matched to it with the Hector SLAM algorithm. The use of Hector SLAM instead of using Monte Carlo Localization or just matching the scans to the reference map is advantageous as there are places (under desks etc.), which were occluded and not seen in the TLS based reference map and the slam algorithm can generate a map for those locations and use it for localization. The parameters of the Hector SLAM were set to only update the underlying map with very low probability so the original geometry of the TLS based reference map was preserved and used for matching when available.

After being generated by Hector SLAM the reference trajectories were interactively inspected by checking that each scan correctly

| Test | Position error | | | Heading error | | |
|------|------|------|------|------|------|------|
| | RMSE | SD | Max | RMSE | SD | Max |
| | mm | mm | mm | deg | deg | deg |
| X1h | 48.7 | 22.5 | 124.1 | 0.11 | 0.06 | 1.04 |
| X1a | 42.7 | 30.5 | 117.7 | 0.16 | 0.11 | 2.04 |
| X1b | 16.4 | 7.3 | 40.38 | 0.13 | 0.10 | 1.86 |
| X1d | 37.3 | 22.2 | 97.6 | 0.19 | 0.12 | 2.00 |
| X2h | 71.7 | 47.9 | 213.3 | 0.18 | 0.09 | 0.69 |
| X2a | 20.8 | 14.3 | 68.7 | 0.16 | 0.12 | 2.00 |
| X2b | 28.7 | 20.9 | 83.4 | 0.16 | 0.12 | 2.02 |
| X2d | 19.4 | 10.1 | 58.8 | 0.16 | 0.12 | 2.07 |
| S1h | 192.9 | 130.0 | 520,7 | 0.56 | 0.28 | 7.9 |
| S1a | 104.8 | 70.4 | 280.2 | 0.35 | 0.20 | 1.78 |
| S1b | 283.5 | 143.4 | 582.5 | 1.37 | 0.72 | 3.05 |
| S1c | 148.8 | 87.3 | 376.3 | 0.53 | 0.33 | 2.47 |
| S1d | 75.1 | 45.8 | 204.1 | 0.35 | 0.22 | 1.80 |
| S2h | 128.9 | 62.1 | 282.5 | 0.45 | 0.26 | 6.06 |
| S2a | 33.6 | 14.3 | 205.8 | 0.32 | 0.22 | 3.45 |
| S2b | 35.6 | 11.2 | 233.4 | 0.29 | 0.21 | 3.59 |
| S2c | 33.5 | 21.5 | 202.9 | 0.32 | 0.22 | 3.54 |
| S2d | 87.1 | 50.7 | 177.0 | 0.37 | 0.24 | 3.40 |
| S3h | 230.6 | 157.4 | 593.0 | 0.51 | 0.24 | 7.76 |
| S3a | 72.8 | 60.7 | 258.0 | 0.33 | 0.23 | 1.74 |
| S3b | 73.3 | 56.4 | 239.1 | 0.33 | 0.21 | 1.75 |
| S3c | 58.5 | 29.5 | 142.4 | 0.29 | 0.20 | 1.82 |
| S3d | 81.7 | 45.9 | 202.3 | 0.31 | 0.20 | 1.87 |

Table 2: Trajectory errors for the test runs. h = initial trajectory from Hector SLAM, others are processed by Karto. Letters a to d represent different Karto parameter sets. a,b = 5 mm matching resolution. c,d = 2.5 mm matching resolution. a,c have smaller penalty multiplier for matches differing from initial guess while b,d have larger one.

matches the slice of the reference point cloud. A trajectory modifier tool, which enables easy inspection and modification if necessary, was created and used for this task. This inspection was necessary as the matching by Hector SLAM occasionally fails and the trajectory can deviate more than 50 mm from the correct one. With the scans distorted by movement, precise manual correction was extremely difficult and some small (under 10mm) errors remain in the reference trajectories.

Results of processing with SLAM algorithms can be seen in Table 2 and an example trajectory can be seen in Figure 4. While each test run was different, the main difference was the order in which the different corridors of the library (left hand side of the trajectory) were visited. The error metrics and figures were calculated with the help of Rawseeds metrics computation toolkit (Andrea Bonarini and Tardos, 2006). The trajectories were tied to the reference ones and through them to the TLS point cloud only by setting the starting pose to the reference starting pose. While this is not optimal, as the poses in the pose graph of Karto are all anchored to the static starting pose this proved not to cause any problems.

The computed trajectories were combined with the data scanner for generating 3D point clouds of the FGI study area. The point cloud data was scanned with 30 Hz scan frequency and 244,000 points a second, maximum range was set to 6.0 m. The point clouds were validated for geometric accuracy against the TLS reference point cloud also used for the SLAM trajectory evaluation. An example of a point cloud data collected in Test 2 is illustrated in Figure 5. The point cloud in Test 1 includes 31 million points, while the Test 2 cloud has about 40 million points in total.

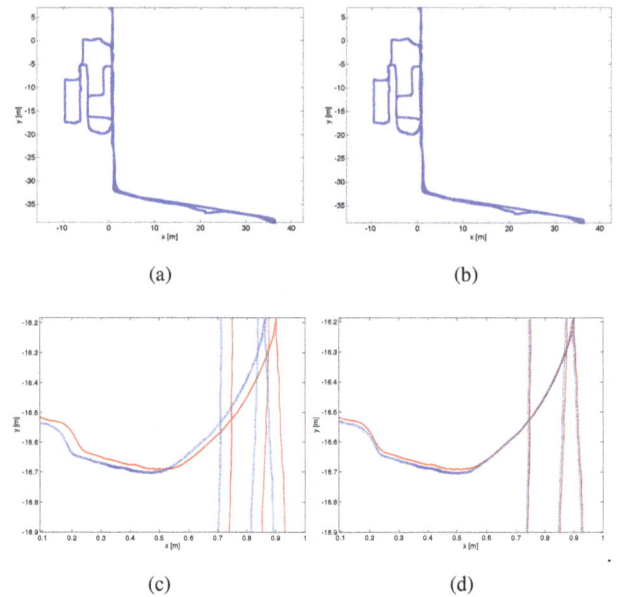

(a)    (b)

(c)    (d)

Figure 4: Trajectories of test run X2, start and end in the topmost part, red represents reference. (c) and (d) are detail from the whole trajectory. (a), (c) are initial trajectories from Hector SLAM. (b), (d) are from Karto. Detection of loop closures by Karto's backend keeps the position from drifting far away.

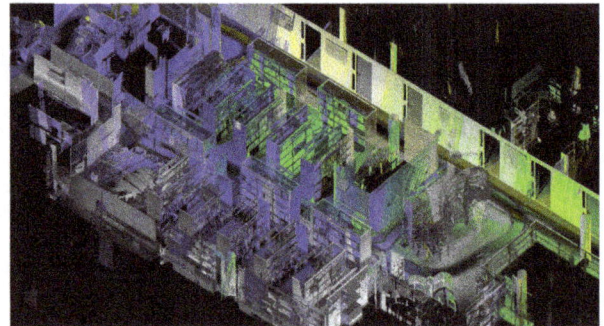

Figure 5: 3D point cloud collected with the Slammer system from test X2d. The precision and accuracy of the data regardless of the complex scene is well represented. Point coloring is based on the laser intensity and data file number to illustrate the progress of the data acquisition.

The geometric error of the point clouds were analyzed on eight pylons in the study area. Two of them are visible on bottom right in Figure 5. The center of the pylon was measured from the TLS and Slammer point cloud interactively. The resulting errors are seen in Table 3. The maximum errors for both cases were found from the lower corridor. The error grows steadily starting from the bend.

To get the scale of the error over the whole building the lengths of the main corridors were measured. From the TLS cloud the upper (library) corridor length was determined to be 43.498 m, Test 1 shows 43.492 m, and Test 2 43.501 m length correspondingly. The lower corridor length from the TLS cloud was 44.039 m, while it was found to be 44.056 m for the Test 1 data, and 44.068 m for the Test 3. This results about an error of 0.1-0.5%. However, the horizontal location error for the data point cloud varies from place to place. This variation is summarized in Table 3.

| | Test X1b | Test X2d |
|---|---|---|
| n | 18 | 19 |
| | m | m |
| Mean | 0.021 | 0.014 |
| SD | 0.018 | 0.010 |
| RMSE | 0.026 | 0.017 |
| Minimum | 0.004 | 0.004 |
| Maximum | 0.083 | 0.036 |

Table 3: Planar errors in the point cloud. n is the number of locations compared.

Figure 6: Details of the building wall, doors and interior features captured by Slammer. Points are colored by laser intensity.

## 4 DISCUSSION

The majority of the trajectory error for each test run was caused by a slight error in the estimated orientation when visiting the over 40 m long lower corridor of the building. This can be seen as the large rise in error in Figure 3, while the magnitude of error differs, all test runs showed similar tendency. The corridor is quite feature poor and as such difficult for scan matching and as extra challenge the lobby which connects the two corridors contains glass and metal surfaces on multiple sides.

Most noticeable thing about the results is the difference in position accuracy between the test runs with different laser scanners. While the X330 is upgraded version of 120S and should provide better performance, the more likely culprit is the almost halved movement speed of the X1 and X2 trajectories. Slower movement speed and higher scanner rotation rate lead to less of distortion to the scans which helps scan matching.

Another source of error is the combining of the dense initial trajectory to the optimized sparse one. Currently when a subtrajectory between two optimized poses is transformed to fit between the endpoints, it is scaled same amount in both X and Y directions. When the scanner is almost stationary, this scaling can magnify the small random jitter in the localization to centimetre range. This effect could be reduced by scaling only along the line between the endpoints. As an example, the maximum error spike in Figure 3(b) is caused by this.

Out of the test runs taken with 120S, S2 stands out with its lower errors. The S2 test run was the longest and also the most convoluted with the trajectory crossing itself numerous times. These provide the Karto algorithm with many chances to perform loop closures which is its main strength, each loop closure ties the trajectory together and helps reduce accumulated error. As an opposite example in the S1b test the cost the backend tries to minimize somehow became NaN and as a result, no pose optimization was performed. Without the backend Karto library provides no advantages over Hector SLAM.

As can be seen in the Table 2 no parameter combination which would have provided best results for each test was found. In total the Karto library has tens of different parameters which interact with each other so finding the optimal combination is quite difficult. To ensure robust results, movement should be stable and the path should be planned in a way which leads to many loop closures.

## 5 CONCLUSIONS

In this study we investigated use and effect of high-end laser scanners on providing localization data for SLAM algorithms. The proposed method combines two SLAM algorithms for computing accurate trajectory for 3D point cloud generation. The trajectory estimation was first carried out using frontend Hector SLAM with a high quality scan matcher for providing initial odometry for Karto algorithm. Karto has a backend for detecting loop closures and is used to optimize poses drifting due to cumulating error of the scan matching.

We collected five different trajectories, and processed them using different parameter sets. The resulting trajectories were evaluated against TLS reference. The results show, that the scanner version used for SLAM might have a role in in the performance of the algorithm, but most probably is a feat of increased scan frequency and slower data acquisition. The analysis of the 3D point cloud generated from the secondary scanner data based on the resulting SLAM trajectories show good agreement with the TLS reference. The point cloud generated using the best accuracy trajectory show 17 mm RMSE and 36 mm maximum error for horizontal position. The dimensions of the building show 0.1-0.5% error.

While the point cloud quality achievable by TLS scanning might not be reached just yet, the centimetre range localization accuracy provided by the proposed combination approach is enough to produce coherent and visually pleasing point clouds with vastly faster data collection. With the modular sensor suite of Slammer adding synchronous RGB or infrared cameras to the platform is also possible.

These results are also only preliminary and can expected to be improved by further modifying the SLAM algorithm, by integrating the IMU measurements more robustly and by improving the data collection practises. Mounting a second, cheaper laser scanner horizontally should also be tried to make comparison to see how much of the good performance is caused by the expensive laser scanner and how much by the SLAM algorithms.

### REFERENCES

Andrea Bonarini, Wolfram Burgard, G. F. M. M. D. G. S. and Tardos, J. D., 2006. Rawseeds: Robotics advancement through web-publishing of sensorial and elaborated extensive data sets. In: In proceedings of IROS'06 Workshop on Benchmarks in Robotics Research.

Kaartinen, H., Hyyppä, J., Kukko, A., Jaakkola, A. and Hyyppä, H., 2012. Benchmarking the performance of mobile laser scanning systems using a permanent test field. Sensors 12(9), pp. 12814–12835.

Kohlbrecher, S., Meyer, J., von Stryk, O. and Klingauf, U., 2011. A flexible and scalable slam system with full 3d motion estimation. In: Proc. IEEE International Symposium on Safety, Security and Rescue Robotics (SSRR), IEEE.

Konolige, K., Grisetti, G., Kummerle, R., Burgard, W., Limketkai, B. and Vincent, R., 2010. Efficient sparse pose adjustment for 2d mapping. In: Intelligent Robots and Systems (IROS), 2010 IEEE/RSJ International Conference on, IEEE, pp. 22–29.

Kukko, A., 2013. Mobile Laser Scanning–System development, performance and applications. PhD thesis.

Lehtola, V. V., Virtanen, J.-P., Kukko, A., Kaartinen, H. and Hyyppä, H., 2015. Localization of mobile laser scanner using classical mechanics. ISPRS Journal of Photogrammetry and Remote Sensing 99, pp. 25–29.

Olson, E. B., 2009. Real-time correlative scan matching. In: Robotics and Automation, 2009. ICRA'09. IEEE International Conference on, IEEE, pp. 4387–4393.

ROS, 2014. Robotic Operating System. http://www.ros.org. Accessed 2.11.2014.

SRI International, 2014. Karto. http://www.kartorobotics.com/. Accessed 2.11.2014.

Thrun, S., Burgard, W. and Fox, D., 2005. Probabilistic Robotics (Intelligent Robotics and Autonomous Agents). The MIT Press.

Vincent, R., Limketkai, B. and Eriksen, M., 2010. Comparison of indoor robot localization techniques in the absence of gps. In: SPIE Defense, Security, and Sensing, International Society for Optics and Photonics, pp. 76641Z–76641Z.

# 2

# IMPROVEMENT OF 3D MONTE CARLO LOCALIZATION
# USING A DEPTH CAMERA AND TERRESTRIAL LASER SCANNER

S. Kanai [a],*, R. Hatakeyama [a], H. Date [a]


[a] Graduate School of Information Science and Technology, Hokkaido University, Kita-ku, Sapporo 060-0814, Japan
- {kanai, hdate}@ssi.ist.hokudai.ac.jp, r_hatakeyama@sdm.ssi.ist.hokudai.ac.jp

**Commission IV/WG7 & V/4**

**KEY WORDS:** Monte Carlo Localization, Depth Camera, Terrestrial Laser Scanner, GPGPU, IMU, Scene Simulation

**ABSTRACT:**

Effective and accurate localization method in three-dimensional indoor environments is a key requirement for indoor navigation and lifelong robotic assistance. So far, Monte Carlo Localization (MCL) has given one of the promising solutions for the indoor localization methods. Previous work of MCL has been mostly limited to 2D motion estimation in a planar map, and a few 3D MCL approaches have been recently proposed. However, their localization accuracy and efficiency still remain at an unsatisfactory level (a few hundreds millimetre error at up to a few FPS) or is not fully verified with the precise ground truth. Therefore, the purpose of this study is to improve an accuracy and efficiency of 6DOF motion estimation in 3D MCL for indoor localization. Firstly, a terrestrial laser scanner is used for creating a precise 3D mesh model as an environment map, and a professional-level depth camera is installed as an outer sensor. GPU scene simulation is also introduced to upgrade the speed of prediction phase in MCL. Moreover, for further improvement, GPGPU programming is implemented to realize further speed up of the likelihood estimation phase, and anisotropic particle propagation is introduced into MCL based on the observations from an inertia sensor. Improvements in the localization accuracy and efficiency are verified by the comparison with a previous MCL method. As a result, it was confirmed that GPGPU-based algorithm was effective in increasing the computational efficiency to 10-50 FPS when the number of particles remain below a few hundreds. On the other hand, inertia sensor-based algorithm reduced the localization error to a median of 47mm even with less number of particles. The results showed that our proposed 3D MCL method outperforms the previous one in accuracy and efficiency.

## 1. INTRODUCTION

With recent interest in indoor navigation and lifelong robotic assistance for human life support, there is an increased need for more effective and accurate localization method in three dimensional indoor environments. The *localization* is the ability to determine the robot's position and orientation in the environment. So far, three typical methods have been proposed for the indoor localization; (1) based only on internal sensors such as an odometry or an inertial navigation, (2) utilizing observations of the environment from outer sensors in an a priori or previously learned map such as Monte Carlo Localization (MCL), and (3) relying on infrastructures previously-installed in the environments such as distinct landmarks such as bar-codes, WiFi access points or surveillance camera networks (Borenstein et al., 1997).

Among them, MCL gives one of the promising solutions for the localization when previously created environment map is available and the system installation should be realized at a low cost. The MCL is a kind of probabilistic state estimation methods (Thrun et al., 2005) which can provide a comprehensive and real-time solution to the localization problem. However, previous work of MCL has been mostly limited to 2D motion estimation in a planar map using 2D laser scanners (Dellaert et al., 1999; Thrun et al., 2001). Recently, a few 3D MCL approaches have been proposed where rough 3D models and consumer-level depth cameras are used as the environment maps and outer sensors (Fallon et al., 2012;

Hornung et al., 2014; Jeong et al., 2013). However, their localization accuracy and efficiency still remain at an unsatisfactory level (a few hundreds millimetre error at up to a few FPS) (Fallon et al., 2012; Hornung et al., 2014), or the accuracy is not fully verified using the precise ground truth (Jeong et al., 2013).

Therefore, the purpose of this study is to improve an accuracy and efficiency of 6DOF motion estimation in 3D Monte Carlo Localization (MCL) for indoor localization. To this end, firstly, a terrestrial laser scanner is used for creating a precise 3D mesh model as an environment map, and a professional-level depth camera is installed in the system as an outer sensor. GPU scene simulation is also introduced to upgrade the speed of prediction phase in MCL. Moreover, GPGPU programming is implemented to realize further speed up of the likelihood estimation phase, and anisotropic particle propagation is introduced based on the observations from an inertia sensor in MCL. The improvements in the localization accuracy and efficiency are verified by the comparison with a previous 3D MCL method (Fallon et al., 2012).

## 2. 3D MONTE CARLO LOCALIZATION

Monte Carlo Localization (MCL) is one of probabilistic state estimation methods (Thrun et al., 2005) using observation from outer sensor. The position and orientation of a system to be estimated is expressed by a state variable. A probability density function of the state variables is represented approximately by a finite set of "particles" each of which expresses a discrete instance of the state variables, and a progress of the probability

*Corresponding author:
Satoshi Kanai ( kanai@ssi.ist.hokudai.ac.jp )

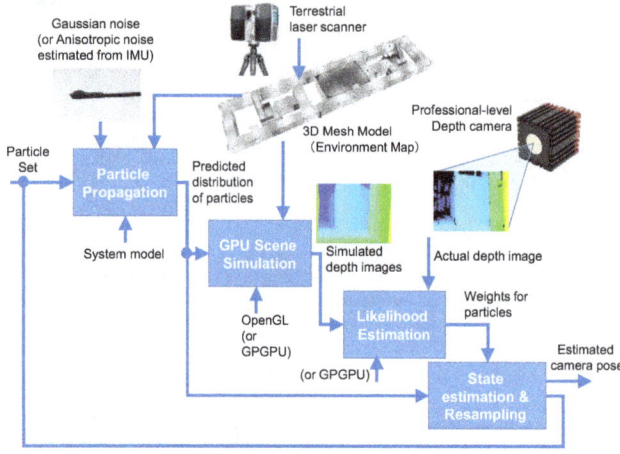

Figure 1. Our 3D MCL baseline algorithm

distribution is estimated by repeating propagation, likelihood evaluation and weighting based on the observation, and resampling of the particles.

The 3D MCL algorithms of our study are extensions of the previous 3D MCL approach which made use of GPU scene simulation (Fallon et al., 2012). As our extensions, in order to increase the accuracy of a map and an outer sensor, a terrestrial laser scanner is introduced for creating a precise 3D mesh model as an environment map, and a professional-level depth camera is installed as an outer sensor. Moreover, in our second extensions, a GPGPU programming is implemented to realize a speed up of the localization process, and an anisotropic particle propagation based on the observations from an inertia sensor is introduced in order to increase the localization accuracy.

Figure 1 shows a flow of our baseline 3D MCL algorithm. The state variable $x_t = [x_t, y_t, z_t, \varphi_t, \theta_t, \psi_t]$ denotes a 6-DOF pose of the depth camera at a time step $t$, and a probability distribution of $x_t$ is expressed by two different set of particles $\mathcal{X}_{t|t-1}$ and $\mathcal{X}_{t|t}$ as Eqs (1) and (2).

$$\mathcal{X}_{t|t-1} \equiv \left\{ x_{t|t-1}^{(i)} \right\} \qquad (1)$$

$$\mathcal{X}_{t|t} \equiv \left\{ x_{t|t}^{(i)} \right\} \qquad (2)$$

where, $\mathcal{X}_{t|t-1}$ is called a predicted distribution, $\mathcal{X}_{t|t}$ is called a filter distribution, and $x_{t|t-1}^{(i)}$ and $x_{t|t}^{(i)}$ represent an $i$-th particle of each distribution respectively.

The depth camera pose $x_t$ can be estimated by the following steps;

1) **Initialization**: An initial filter distribution $\mathcal{X}_{0|0}$ at $t = 0$ is created by assigning a same state $x_{0|0}$ into all of the particles in the distribution as Eq(3).

$$\mathcal{X}_{0|0} = \{x_{0|0}, x_{0|0}, \ldots\ldots, x_{0|0}\} \qquad (3)$$

2) **Propagation**: A predicted distribution $\mathcal{X}_{t+1|t}$ of the next time step $t + 1$ is generated from a current filtered distribution $\mathcal{X}_{t|t}$ by applying a system model of 6-DOF motion of the depth camera. In the baseline algorithm, the system model is given by a Gaussian noise model as Eq(4).

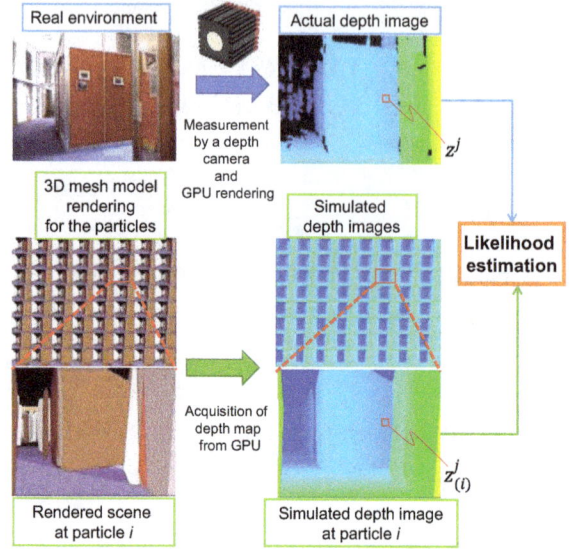

Figure 2. GPU scene simulation

$$x_{t+1|t}^{(i)} = x_{t|t}^{(i)} + \aleph(0, \sigma^2) \qquad (4)$$

where, $\aleph(0, \sigma^2)$ is a 6D normal random numbers with variances $\sigma^2 = [\sigma_x^2, \sigma_y^2, \sigma_z^2, \sigma_\varphi^2, \sigma_\theta^2, \sigma_\psi^2]$. By substituting $x_{t|t-1}^{(i)}$ with $x_{t+1|t}^{(i)}$ in $\mathcal{X}_{t|t-1}$, a predicated distribution at the next time step is updated to $\mathcal{X}_{t+1|t}$.

3) **GPU scene simulation**: As shown in Figure 2, a set of simulated depth images is generated in a premade a 3D mesh model of an environment map. Each simulated depth image $Z_{(i)}^G$ is easily obtained by GPU rendering whose viewpoint coincides with a camera pose expressed by an $i$-th particle $x_{t+1|t}^{(i)} \in \mathcal{X}_{t+1|t}$ in the predicted distribution. In this study, an OpenGL function $glReadPixels()$ is used to quickly obtain an two-dimennsional array of normalized depth values of $Z_{(i)}^G$ from a depth buffer of GPU in one go.

4) **Likelihood estimation and weighting**: A raw depth image at time step $t + 1$ is captured from the depth camera, and then a simple moving average filter among consecutive frames is applied to the image to suppress noises of depth values. This filtered depth image is also rendered by GPU to generate a rendered actual depth image $Z^D$ which has a same image format as a simulated image $Z_{(i)}^G$. Then $Z^D$ is compared with each of the simulated image $Z_{(i)}^G$. As a result of the comparison, a likelihood of $i$-th particle $l_{(i)} \in [0, 1]$ is evaluated by Eqs(5) and (6).

$$\tilde{l}_{(i)} = \sum_{j=1}^{M} 1 - \frac{\left| z_D^j - z_{(i)}^j \right|}{0.5 + \left| z_D^j - z_{(i)}^j \right|} \qquad (5)$$

$$l_{(i)} = \frac{\tilde{l}_{(i)} - \tilde{l}_{min}}{\tilde{l}_{max} - \tilde{l}_{min}} \qquad (6)$$

where, $z_D^j$ is a depth value at a pixel $j$ in the rendered actual depth image $Z^D$, $z_{(i)}^j$ is a depth value at a pixel $j$ in the simulated image $Z_{(i)}^G$ corresponding to an $i$-th particle, and $M$ is the number of pixels in a depth image. $\tilde{l}_{max}$ and $\tilde{l}_{min}$ are the maximum and minimum likelihood values respectively among elements of a set $\{\tilde{l}_{(i)}\}$ each of which is evaluated by Eq(5) corresponding to an $i$-th particle. Using Eq(6), a normalized likelihood value $l_{(i)}(\in [0,1])$ for an $i$-th particle can be obtained.

Once a normalized likelihood value $l_{(i)}$ is obtained, a normalized weight $w_{(i)}(\in [0,1])$ is assigned to the $i$-th particle in the predicted distribution $\mathcal{X}_{t+1|t}$ based on $l_{(i)}$ as Eqs(7) and (8).

$$\tilde{w}_{(i)} = \exp\left[-\frac{(\tilde{l}_{(i)}-1)^2}{0.05^2}\right] \qquad (7)$$

$$w_{(i)} = \frac{\tilde{w}_{(i)}}{\sum_{i=1}^{N} \tilde{w}_{(i)}} \qquad (8)$$

where, $N$ is the number of particles.

5) **State estimation and resampling**: The most probable estimate of the depth camera pose $x_{t+1}$ is obtained as a weighted sum of the particles in $\mathcal{X}_{t+1|t}$ as Eq(9).

$$x_{t+1} = \sum_{x_{t+1|t}^{(i)} \in \mathcal{X}_{t+1|t}} w_{(i)} x_{t+1|t}^{(i)} \qquad (9)$$

Finally, a new filtered distribution $\mathcal{X}_{t+1|t+1}$ is recreated by resampling particles in $\mathcal{X}_{t+1|t}$ so that an existence probability of $i$-th particle approximately equals to $w_{(i)}$. Roulette wheel selection method (Doucet et al, 2001) is used for the resampling.

By repeating the above from step 2) to 5), the particle distribution can be updated according to the observation from the depth camera, and state estimation of the camera pose $x_t$ is sequentially updated. GPU rendering is introduced in step 2) for generating $N$ simulated depth images in real time.

### 3. EFFICIENCY IMPROVEMENT OF THE LOCALIZAITON BASED ON GPGPU PROGRAMMING

As shown in Figure 3-(a), in our baseline MCL algorithm described in section 2, processing other than GPU scene simulation (step 2)) are executed by CPU, and every simulated depth image $Z_{(i)}^G$ which is rendered in GPU has to be uploaded directly to CPU right after the rendering using an OpenGL function *glReadPixels()*. As a result, $(N \times M \times D)$ byte data has to be uploaded in all from GPU to CPU in every time step, where $N$ means the number of total particles, $M$ is the number of pixels in a depth image, and $D$ is the number of bytes of one pixel data. In our setting, a professional-level depth camera (SR4000) is used, and it generates a depth image of $176\times144$ pixels in every frame. When a float-type variable (4 bytes) is used per pixel and 100 particles are included in a distribution, around 10Mbyte data has to be uploaded from GPU to CPU in every frame. Due to relatively slow execution speed of *glReadPixels()*, its transmission delay is not negligible, and it was observed that the delay caused a considerable bottleneck of the baseline MCL algorithm.

To reduce this delay, General Purpose computing on GPU (GPGPU) (Eberly, 2014) is introduced in the algorithm so that the GPU takes both GPU scene simulation (Step 2)) and likelihood estimation (Step 3)) as shown in Figure 3-(b). The

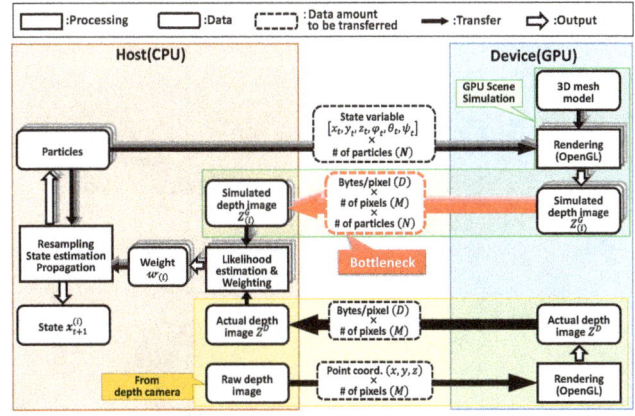

(a) Baseline algorithm

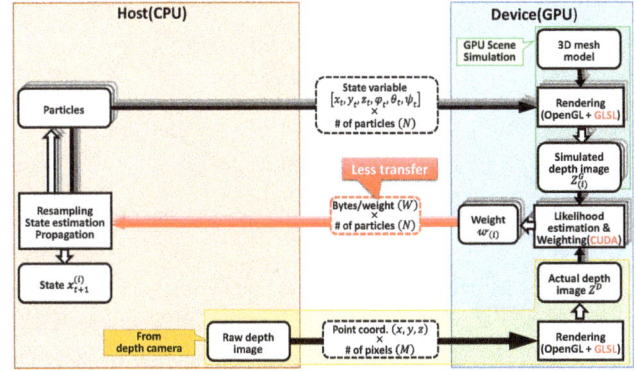

(b) GPGPU-based algorithm

Figure 3. Difference in MCL between the baseline and GPGPU-based algorithm

parallel processing of likelihood estimations for all particles can be handily implemented in GPGPU. In the GPGPU implementation, the simulated depth image $Z_{(i)}^G$ is directly rendered using GLSL (OpenGL Shading Language). And using CUDA (Sanders et al., 2010), a resulting rendered image can be compared with the rendered actual depth image $Z^D$, and likelihood estimation and weighting can be processed on the GPU without transferring any simulated depth image data to the CPU. Finally, only a set of weights of the particles $\{w_{(i)}\}$ only has to be transferred from GPU to CPU. Moreover, a likelihood value at a pixel $j$ in the right hand side of Eq.(5) can also be evaluated independently of the other pixel. So we parallelized this pixel-wise likelihood calculation using CUDA.

As a result of this implementation, a set of weights $\{w_{(i)}\}$ whose data amount is around $(N \times W)$ byte ($W$: a number of bytes per one weight) in total only has to be uploaded from GPU to CPU per every frame. This can significantly reduce the data amount to be uploaded to about 400Byte. The effect of the implementation on our localization performance is explained in 5.2.

### 4. ACCURACY IMPROVEMENT OF THE LOCALIZAITON WITH THE AID OF AN INERTIA SENSOR

As the other approach for the accuracy improvement in the localization, a small 6-DOF inertial measurement unit (IMU) is installed in the system. With the aid of this IMU, the

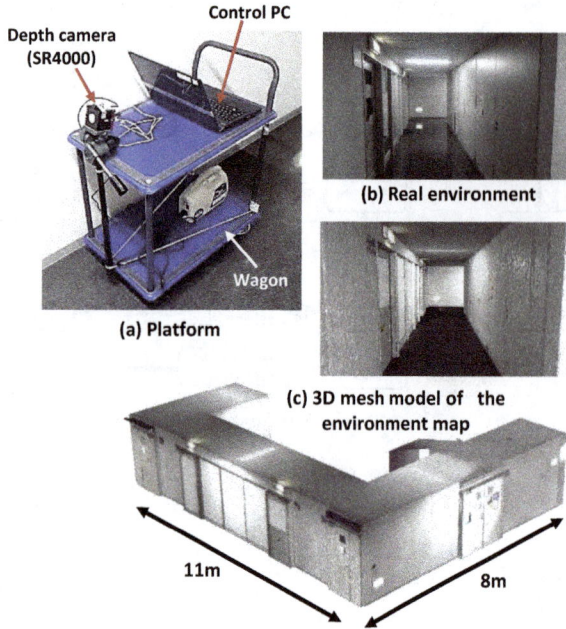

Figure 4. The platform and the environment map

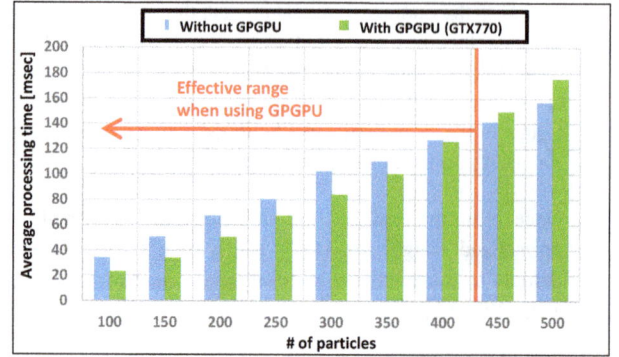

Figure 5. Processing time with and without GPGPU

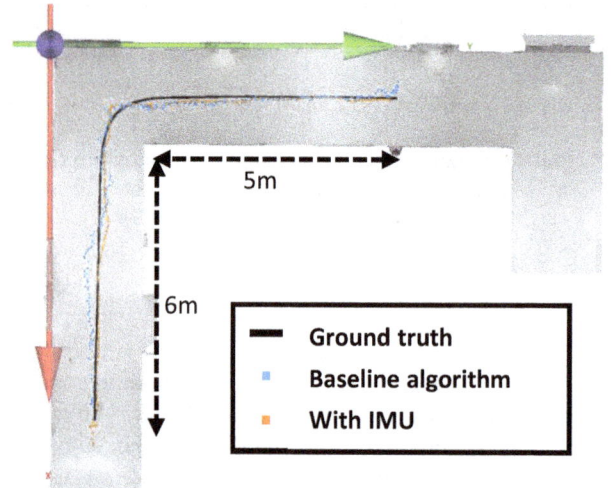

Figure 6. The traces of estimated camera positions
with and without IMU

anisotropic particle propagation based on an actual observation from the IMU can be introduced in the propagation step (Step1)) in MCL.

Different from Gaussian-based isotropic particle propagation in the baseline MCL, more particles are allocated adaptively along the direction of the measured acceleration and angular acceleration given from the IMU. In the propagation step in the MCL, this anisotropic particle propagation is easily implemented by replacing the original propagation equation Eq(4) with Eq(10).

$$x_{t+1}^{(i)} = x_t^{(i)} + \aleph(d_t, \hat{\sigma}^2) \qquad (10)$$

where, $d_t = [d_{xt}, d_{yt}, d_{zt}, d_{\varphi t}, d_{\theta t}, d_{\psi t}]$ is a 6D estimated positional and rotational displacements derived from the numerical integration of the measured acceleration and angular acceleration of IMU at time step $t$. We also assume from the observation that a standard deviation of each displacement in the system model is comparable to the estimated displacement $d_t$, and therefore set as $\hat{\sigma}^2 = [d_{xt}^2, d_{yt}^2, d_{zt}^2, d_{\varphi t}^2, d_{\theta t}^2, d_{\psi t}^2]$.

# 5. EXPERIMENTS

## 5.1 Setup

The improvement in efficiency and accuracy is verified in localization experiments in an indoor environment. Figure 4 shows our experimental setup. As shown in Figure 4-(a), a professional-level depth camera (SR4000, TOF camera, Pixel resolution: 176×144) and an IMU (ZMP IMU-Z Tiny Cube) were attached on a top surface of a small wagon. A laptop PC (Windows-7, Corei7-3.7GHz, and GeForce GT-650M) on the wagon recorded a time sequence both of raw depth image from the depth camera and acceleration data from the IMU on site. On the other hand, the localization calculation was done on the other desktop PC (Windows-7, Corei7-2.93GHz, and GeForce GTX-770).

As a preparation, 3D point clouds of corridor space (9× 11 × 2.4m) in a university building shown in Figure 4-(b) was first measured by a terrestrial laser scanner (FARO-Focus 3D), and a precise 3D mesh model with 141,899 triangles shown in Figure 4-(c) was created as an precise environmental map of the space using a commercial point cloud processing software. GeForce GTX-770 was used when investigating the effect of GPGPU programming on an efficiency of the localization.

## 5.2 Results

**Efficiency improvement by GPGPU programming:** Figure 5 compares the averaged processing time for single time step of MCL. 6-DOF depth camera pose is estimated in the experiment. From this figure, it is clearly shown that the proposed GPGPU-based algorithm is effective in reducing the time for localization when the number of particles remains below a few hundreds which is a general setting of this MCL.

However, the improvement in the processing speed was not so significant even when the GPGPU implementation is applied as shown in Figure 5. The reason for this behaviour is that GPGPU coding generally requires a sophisticated knowledge of parallel processing, and there is still a room for more efficient parallelization coding in likelihood estimation and weighting processing in GPU.

**Accuracy improvement by IMU:** Figure 6 compared the estimated camera positions using the baseline and proposed IMU-based algorithm in case of 200 particles. In this

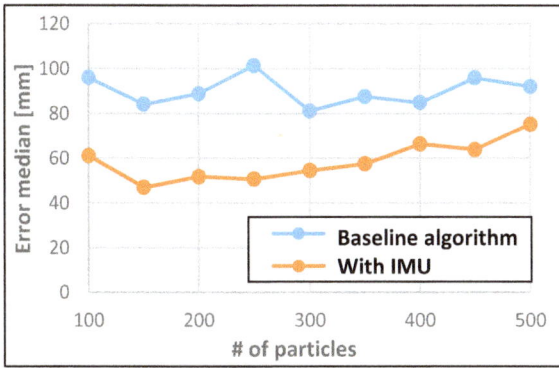

(a)   Median of error

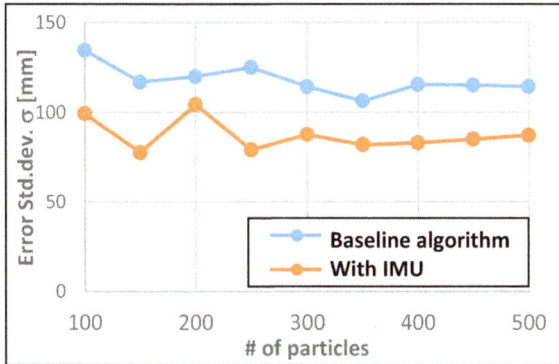

(b)   Standard deviation of error

Figure 7.  Localization error in different particle settings

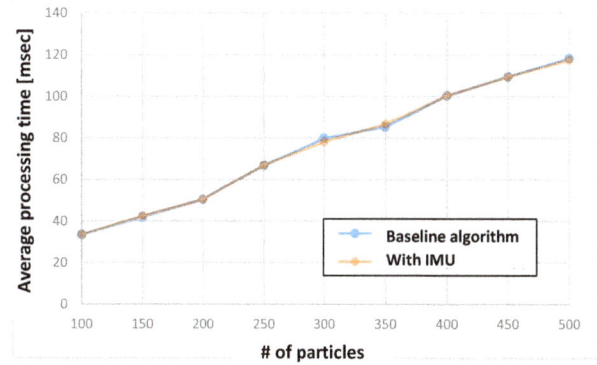

Figure 8.  Processing time per frame
in different particle settings

| | # of particles | Speed [FPS] | Localization Error | | |
|---|---|---|---|---|---|
| | | | Median[mm] | σ [mm] | 3σ % |
| Proposed method with IMU | 150 | 23.5 | 47 | 78 | 100 |
| [Fallon et al., 2012] | 350 | 7-8 | 300 | N/A | 90 |

Table 1. Comparison of the performances between
the proposed method and [Fallon et al., 2012]

experiment, 6-DOF depth camera pose was estimated, but the accuracy verification is verified only in a planer position $[x, y]$ of 2-DOF. The ground truth trace of the wagon was precisely collected from physical marker recording attached under the wagon using a terrestrial laser scanner.  The estimated positions with the aid of the IMU locate much closer to the ground truth than those without IMU.

Figure 7 shows the median and standard deviation of localization errors in different settings of the number of particles. In all settings, the IMU-based algorithm outperformed the baseline one, and the averaged median of the localization errors from the ground truth was reduced by 34%. And as shown in Figure 8, no significant decrease in the performance was observed when introducing the IMU.

Table 1 summarizes the performances between the proposed MCL method with IMU and the previous 3D MCL study (Fallon et al., 2012).  The table shows that our MCL method realized much smaller localization error (16%) even with 57% less particles.

## 6.   SUMMARY

Several methods were proposed to improve the accuracy and efficiency of 3D Monte Carlo Localization (MCL) for indoor localization. For the baseline algorithm, a terrestrial laser scanner was used for creating a precise 3D mesh model as an environment map, and a professional-level depth camera was installed as an outer sensor. Moreover, two original approaches were proposed for the improvement. As a result, it was confirmed that GPGPU-based algorithm was effective in increasing the computational efficiency to 10-50 FPS when the

number of particles remain below a few hundreds. On the other hand, inertia sensor-based algorithm reduced the localization error to a median of 47mm even with less number of particles. The results showed that our proposed 3D MCL method outperforms the previous one in accuracy and efficiency aspects.

## References

Borenstein, J., Everett, H.R., Feng, L., and Wehe, D., 1997, Mobile robot positioning: Sensors and techniques. *Journal of Robotic Systems*, 14(4), pp.231–249.

Dellaert, F., Fox, D., Burgard, W., and Thrun, S., 1999. Monte Carlo Localization for mobile robots. *IEEE International Conference on Robotics and Automation (ICRA)*, pp.1322–1328.

Doucet, A, Freitas, N., and Gordon, N., 2001. *Statistics Sequential Monte Carlo Methods in Practice*, Springer, New York, pp.437-439.

Eberly, D.H., 2014. *GPGPU Programming for Games and Science*, CRC Press, Boca Raton.

Fallon, M.F., Johannsson, H., and Leonard, J.J., 2012. Efficient scene simulation for robust monte carlo localization using an RGB-D camera. *IEEE International Conference on Robotics and Automation (ICRA)*, pp.1663–1670.

Hornung, A., Oswald S., Maier D., and Bennewitz, M., 2014. Monte carlo localization for humanoid robot navigation in complex indoor environments. *International Journal of Humanoid Robotics*, 11(2), pp. 1441002-1–1441002-27.

Jeong, Y., Kurazume, R., Iwashita Y., and Hasegawa T., 2013, Global Localization for Mobile Ro bot using Large-scale 3D

Environmental Map and RGB-D Camera. *Journal of the Robotics Society of Japan*, 31(9), pp.896–906.

Sanders, J., and Kandrot, E., 2010. *CUDA by Example: An Introduction to General-Purpose GPU Programming*, Addison-Wesley Professional, Boston.

Thrun, S., Fox, D., Burgard, W., and Dellaert, F., 2001. Robust Monte Carlo localization for mobile robots. *Artificial Intelligence*, 128(1), pp.99–141.

Thrun, S., Burgard, W., and Fox, D., 2005. *Probabilistic Robotics*, MIT Press, Cambridge, pp.189–276.

# 3

# AN AUTOMATIC 3D RECONSTRUCTION METHOD BASED ON MULTI-VIEW STEREO VISION FOR THE MOGAO GROTTOES

Jie Xiong [a,*], Sidong Zhong [a], Lin Zheng [a]

[a] School of Electronic Information, Wuhan University, Wuhan, Hubei 430072, China – xiongjiewhu1989@163.com, sdzhong@whu.edu.cn, zl@whu.edu.cn

**KEY WORDS:** 3D Reconstruction, Multi-view vision, Epipolar Constraint, Correlation, Texture mapping, Mogao Grottoes

**ABSTRACT:**

This paper presents an automatic three-dimensional reconstruction method based on multi-view stereo vision for the Mogao Grottoes. 3D digitization technique has been used in cultural heritage conservation and replication over the past decade, especially the methods based on binocular stereo vision. However, mismatched points are inevitable in traditional binocular stereo matching due to repeatable or similar features of binocular images. In order to reduce the probability of mismatching greatly and improve the measure precision, a portable four-camera photographic measurement system is used for 3D modelling of a scene. Four cameras of the measurement system form six binocular systems with baselines of different lengths to add extra matching constraints and offer multiple measurements. Matching error based on epipolar constraint is introduced to remove the mismatched points. Finally, an accurate point cloud can be generated by multi-images matching and sub-pixel interpolation. Delaunay triangulation and texture mapping are performed to obtain the 3D model of a scene. The method has been tested on 3D reconstruction several scenes of the Mogao Grottoes and good results verify the effectiveness of the method.

## 1. INTRODUCTION

With the rapid development of computer technology and sensing technology, three-dimensional (3D) digitization of objects has attracted more and more attention over the past decades. 3D modelling technology has been widely applied in various digitization fields, especially cultural heritage conservation. 3D digitization of cultural heritage is mainly used for digital recording and replication of cultural heritage. Considering the precious value of cultural heritage objects, non-contact and non-destructive measure approaches are generally taken to acquire 3D models. For realistic application, automatic, fast and low-cost 3D reconstruction methods with high precision are required.

A number of active and passive technologies (Pavlidis et al., 2007) are developed for 3D digitization of cultural heritage. Laser scanning methods (Huang et al., 2013) and structured light methods (Zhang et al., 2011) are typical active methods. The most significant advantage of laser scanning is high accuracy in geometry measurements. Nevertheless, the models reconstructed by laser scanning usually lack good texture and such devices have high cost. As passive methods, vision-based methods have the ability to capture both geometry information and texture information, requiring less expensive devices. According to the amount of cameras used, vision-based methods are divided into monocular vision, binocular vision and multi-view vision. Monocular vision methods can obtain depth information from two-dimensional characteristics of a single image or multiple images from a single view (Massot and Hérault, 2008; Haro and Pardàs, 2010). Such methods are usually not very robust to the environment. Moreover, monocular vision methods can gain 3D information from a sequence of images from different views (shape from motion,

SFM) (Chen et al., 2012). The SFM method has a high time cost and space cost. Binocular vision method can acquire 3D geometry information from a pair of images captured from two known position and angles. This method has high automation and stability in reconstruction. But this method easily leads to mismatched points due to repeatable or similar features of binocular images (Scharstein and Szeliski, 2002). In order to reduce the possibility of mismatching, 3D measurement systems based on multi-view vision have been developed (Setti et al., 2012). Generally, the systems have complex structure.

This paper presents an automatic 3D reconstruction method based on multi-view stereo vision. This method has reconstructed 3D models of several scenes of No.172 cave in the Mogao Grottoes using a portable four-camera photographic measurement system (PFPMS) (Zhong and Liu, 2012). The PFPMS is composed of four cameras to add extra matching constraints and offer redundant measurement, resulting in reducing the possibility of mismatching and improving measure accuracy relative to traditional binocular systems.

## 2. 3D RECONSTRUCTION METHODOLOGY

The authors take reconstruction of a scene of a wall for example to illustrate the whole process of 3D reconstruction, including multi-view images acquisition, multi-view image processing, triangulation and texture mapping.

### 2.1 Multi-view images acquisition

As the main hardware system, the PFPMS consists of four cameras with the same configuration parameters, which observes the target object at a distance of 2.0-5.0 m. Four

---

\* Corresponding author

images with a high image resolution of $6016 \times 4000$ can be captured synchronously by a button controller connected to the switch of shutters of the four cameras. The overall structure of the PFPMS is similar to a common binocular vision system and the difference is that two cameras with upper-lower distribution are substitute for each camera of a binocular system respectively, as shown in Figure 1. The four cameras have rectangular distribution and their optical axes are parallel to each other to minimize the impact of perspective distortion on feature matching. The distance between the left or right cameras is about 15 cm and the distance between the upper or lower cameras is about 75 cm. On the one hand, the baseline between the left two cameras or the right two cameras is short. As a result, the very small difference between the two images captured by them can help improve accuracy of matching. Furthermore, the left cameras and the right cameras can form four binocular systems with long baseline to calculate space position of feature points. Thus every point can be measured four times to improve precision. The corresponding parameters of the four cameras and the parameters of relative position of any two cameras need to be obtained before the measurement. The cameras can be calibrated with a tradition pinhole model (Tsai, 1987), and then 3D space coordinates of any point can be calculated with its coordinates in the four images.

Figure 1. The PFPMS

## 2.2 Multi-view images processing

The multi-view images processing is divided into extraction of feature points and matching of feature points. For convenience, let UL, LL, UR, LR image represent the upper-left, lower-left, upper-right, lower-right image respectively.

**2.2.1   Extraction of feature points:** With high detecting speed and high position accuracy, Harris corners (Harris and Stephens, 1988) are chosen as feature points for matching. We adopt corner extraction of image partition to ensure corner points' uniform distribution. Figure 2(a)(b) show the original UL image and its distribution of Harris corners respectively.

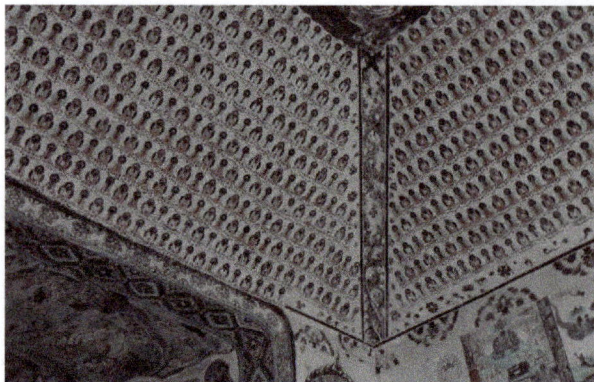

(a) The Original UL image

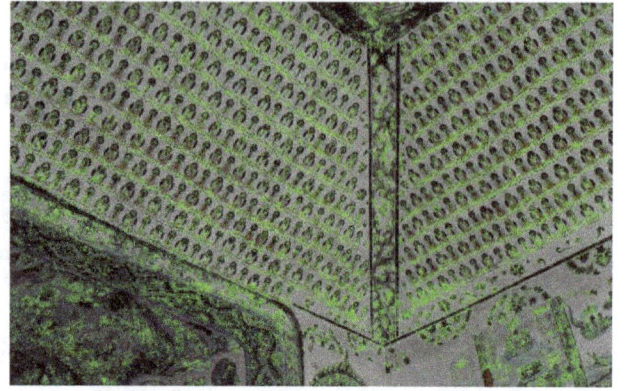

(b) Distribution of Harris Corners in UL image

Figure 2. Extraction of Harris Corners

In order to reduce the search range, the Harris corners detected are stored in a sub-regional way. The feature points in the four images can be extracted in this way.

**2.2.2   Matching of feature points:** Template matching methods are used to search the homologous image points in most stereo matching algorithms. Traditional template matching methods mainly include sum of squared differences (SSD), sum of absolute differences (SAD), normalized cross correlation (NCC) and zero mean normalized cross correlation (ZNCC) (Lazaros et al., 2008). These methods weigh the degree of similarity between two points by calculating the difference between the pixels inside the rectangle window around one point and the pixels inside the rectangle window around the other point. ZNCC is chosen for matching due to its stronger anti-noise ability. Let $I_1(x,y)$, $I_2(x,y)$ represent the intensity of the pixel at $(x,y)$ in image 1 and image 2 respectively and ZNCC can be given by the following expression.

$$ZNCC(x_1,y_1,x_2,y_2) =$$
$$\frac{\sum_{i=-N}^{N}\sum_{j=-N}^{N}\left[I_1(x_1+i,y_1+j)-\bar{I}_1\right]\left[I_2(x_2+i,y_2+j)-\bar{I}_2\right]}{\sqrt{\sum_{i=-N}^{N}\sum_{j=-N}^{N}\left[I_1(x_1+i,y_1+j)-\bar{I}_1\right]^2}\sqrt{\sum_{i=-N}^{N}\sum_{j=-N}^{N}\left[I_2(x_2+i,y_2+j)-\bar{I}_2\right]^2}} \quad (1)$$

$$\bar{I}_1 = \frac{1}{2N+1}\sum_{i=-N}^{N}\sum_{j=-N}^{N}I_1(x_1+i,y_1+j) \quad (2)$$

$$\bar{I}_2 = \frac{1}{2N+1}\sum_{i=-N}^{N}\sum_{j=-N}^{N}I_2(x_2+i,y_2+j) \quad (3)$$

where      $N$ = the half of the size of template window. ($N$ is set to 10 pixels in actually matching)

$x_1, y_1$ = coordinates of the matched point in image 1.

$x_2, y_2$ = coordinates of the matched point in image 2.

$\bar{I}_1$ = average intensity of the pixels inside the window around $(x_1, y_1)$.

$\bar{I}_2$ = average intensity of the pixels inside the window around $(x_2, y_2)$.

Figure 3 shows the main matching scheme based on epipolar constraint. Let $l_{UL-LL}$, $l_{UL-UR}$, $l_{UL-LR}$, $l_{LL-UR}$, $l_{LL-LR}$, $l_{UR-LR}$

represent the epipolar lines which can be obtained from the known parameters of the four cameras (Xu et al., 2012). The matching process is described as the following steps:

a)  For any point $P_{UL}$ in UL image, search its corresponding point along the epipolar line $l_{UL-LL}$ in LL image and find some possible points which have a ZNCC value above 0.9. Rank these points by ZNCC value from high to low and the top five points are chosen as the candidate matched points.

b)  Let $P_{LL}$ represent the first candidate point. $l_{UL-UR}$, $l_{UL-LR}$, $l_{LL-UR}$, $l_{LL-LR}$ can be obtained from the position of $P_{LL}$ and $P_{UL}$ respectively.

c)  Find the matched point $P_{UR}$ in the rectangle region (40 pixel × 40 pixel) around the intersection of $l_{LL-UR}$ and $l_{UL-UR}$ based on the maximum ZNCC value with $P_{UL}$. If the maximum value is less than 0.7, remove $P_{LL}$ from the candidate points and return to step (b).

d)  Find the matched point $P_{LR}$ in the rectangle region (40 pixel × 40 pixel) around the intersection of $l_{UL-LR}$ and $l_{LL-LR}$ based on the maximum ZNCC value with $P_{UL}$. If the maximum value is less than 0.7, remove $P_{LL}$ from the candidate points and return to step (b).

e)  Calculate the ZNCC value between $P_{UR}$ and $P_{LR}$. If the value is less than 0.9, remove $P_{LL}$ from the candidate points and return to step (b).

f)  Obtain $l_{UR-LR}$ from the position of $P_{UR}$ and calculate the distance from $P_{LR}$ to $l_{UR-LR}$. If the distance is less than 5 pixels, remove $P_{LL}$ from the candidate points and return to step (b).

g)  Calculate the matching error defined as the sum of the distance between each matched point and the intersection of the epipolar lines of two other matched points with long baseline relative to it. If the matching error is less than 20 pixels, $P_{UL}$, $P_{LL}$, $P_{UR}$, $P_{LR}$ can be regard as the homologous points and return step (a) for the matching of the next point. Otherwise, remove $P_{LL}$ from the candidate points and return to step (b).

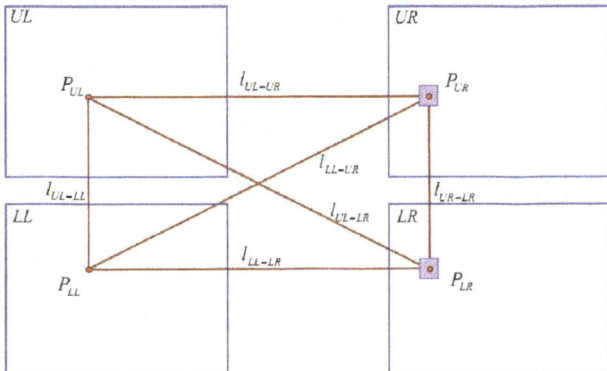

Figure 3. Matching scheme

Figure 4 shows the whole processing flow of the above-mentioned matching method.

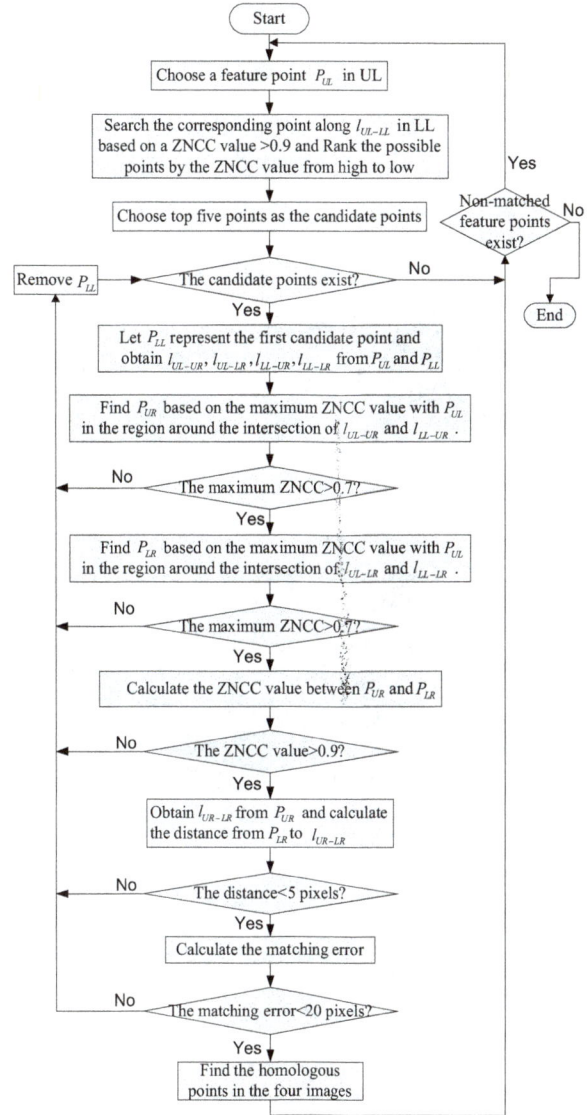

Figure 4. The processing flow of the matching method

After matching, sub-pixel interpolation operation can be performed to improve measurement precision. Bicubic interpolation is chosen to gain sub-pixel position of the homologous points because of its smooth interpolation effect. As the four cameras of the PFPMS form four binocular systems with a long baseline (UL-UR, UL-LR, LL-UR, LL-LR), for every matched point, take the average of four space coordinates respectively calculated from it and its homologous points as the final coordinates of its corresponding space point. Figure 5 shows the 3D point cloud.

## 2.3  Triangulation and texture mapping

Generally, the surface of the object can be expressed with a triangulated irregular net. Delaunay Triangulation (Tsai, 1993) is performed to process the point cloud obtained from stereo matching. In order to avoid appearance of some long and narrow triangles with long sides during triangulation, the length of every triangle's sides should be limited. Figure 6 shows Delaunay triangulation of the point cloud.

In order to reconstruct a model with texture, every point's colour information extracted from one of the captured images can be used for texture mapping. The UL image is selected as the texture image. To every triangle, each vertex's texture coordinates can be obtained from the image coordinates of its matched point in the texture image, and the texture coordinates of internal points can be calculated by linear interpolation of the vertexes' texture coordinates. The texture image is mapped automatically to a model in this way. Finally, the 3d model of the scene is generated, as shown in Figure 7.

Figure 5. The 3D point cloud (7484 points)

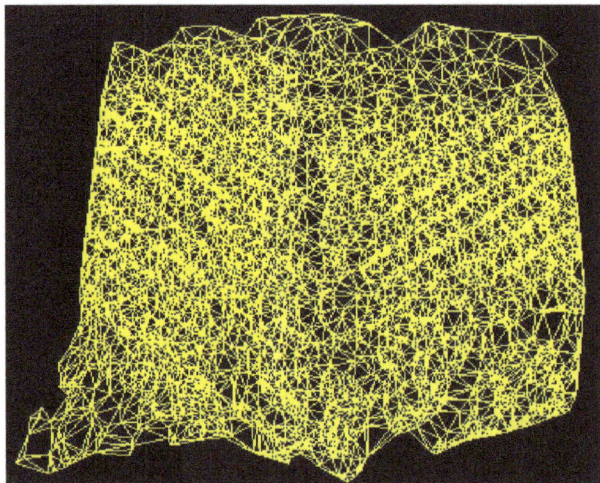

Figure 6. Delaunay Triangulation of the point cloud

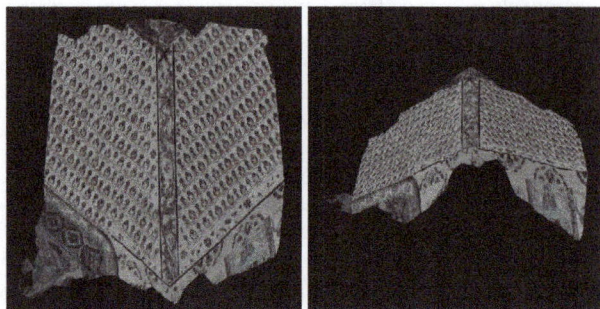

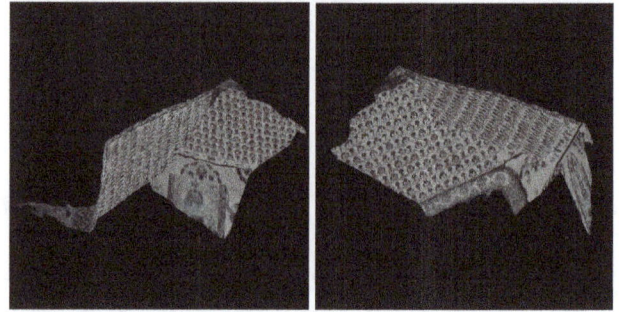

Figure 7. The 3D model of the scene from different views

## 3. EXPERIMENT RESULTS

The 3D model of the scene of a wall has been obtained with the matching method above and a good result is given. In order to test the stability and adaptability of the method, the 3D models of Scene A and Scene B are reconstructed respectively. Figure 8 and Figure 9 show the results. The results show these models describe geometric structure and texture of the scenes realistically as a whole. Not dense enough points are extracted from the region with poor texture. It leads to the loss of local model. For instance, incomplete models of the Buddhas indicate this point.

(a) The original UL image of Scene A

(b) The 3D model of Scene A

Figure 8. 3D Reconstruction of Scene A

(a) The original UL image of Scene B

(b) The 3D model of Scene B

Figure 9. 3D Reconstruction of Scene B

## 4. CONCLUTION

This paper proposes an automatic 3D reconstruction method based on multi-view stereo vision. 3D models of several scenes of No.172 cave in the Mogao Grottoes have been reconstructed using a portable four-camera photographic measurement system. The cameras of the measurement form two binocular systems with a short baseline and four binocular systems with a long baseline. The binocular system with a short baseline is used for rapidly matching with the small difference between the two images. The binocular systems with a long baseline are used for multiple measurements. Compared with a traditional binocular system, the PFPMS have the advantage of reducing the possibility of mismatching and improving measurement accuracy. The experiment results show the effective of this matching method.

The limitation of the method is that the point cloud is not enough dense in a region with poor texture. Besides, only several models of local scenes can be reconstructed but are not complete. The future work will be focused on obtaining a dense point cloud by introducing structured light and stitching of the models reconstructed from different perspectives.

## ACKNOWLEDGEMENTS

This work is supported by the 973 Program (2012CB725301) and National surveying and mapping geographic information public welfare industry special funding scientific research projects (210600001).

## REFERENCES

Chen, S., Wang, Y. and Cattani, C., 2012. Key issues in modeling of complex 3D structures from video sequences. http://www.hindawi.com/journals/mpe/2012/856523/.

Haro, G. and Pardàs, M., 2010. Shape from incomplete silhouettes based on the reprojection error. Image and Vision Computing, 28(9), pp. 1354-1368.

Harris, C. and Stephens, M., 1988. A combined corner and edge detector. In Proceedings of Alvey Vision Conference, The Plessey Company, ed., pp. 189–192.

Huang, H., Brenner, C. and Sester, M., 2013. A generative statistical approach to automatic 3D building roof reconstruction from laser scanning data. In: *ISPRS Journal of Photogrammetry & Remote Sensing*, 79, pp. 29-43.

Lazaros, N., Sirakoulis, G. C. and Gasteratos, A., 2008. Review of stereo vision algorithms: from software to hardware. International Journal of Optomechatronics, 2(4), pp. 435-462.

Massot, C. and Hérault, J., 2008. Model of frequency analysis in the visual cortex and the shape from texture problem. International Journal of Computer Vision, 76(2), pp. 165-182.

Pavlidis, G., Koutsoudis, A., Arnaoutoglou, F., Tsioukas, V. and Chamzas, C., 2007. Methods for 3D digitization of cultural heritage. Journal of cultural heritage, 8(1), pp. 93-98.

Scharstein, D. and Szeliski, R., 2002. A taxonomy and evaluation of dense two-frame stereo correspondence algorithms. International journal of computer vision, 47(1-3), pp. 7-42.

Setti, F., Bini, R., Lunardelli, M., Bosetti, P., Bruschi, S. and De Cecco, M., 2012. Shape measurement system for single point incremental forming (SPIF) manufacts by using trinocular vision and random pattern. Measurement science and technology, 23(11), 115402.

Tsai, R. Y., 1987. A versatile camera calibration technique for high-accuracy 3D machine vision metrology using off-the-shelf TV cameras and lenses. Robotics and Automation, IEEE Journal of, 3(4), pp. 323-344.

Tsai, V. J., 1993. Delaunay triangulations in TIN creation: an overview and a linear-time algorithm. International Journal of Geographical Information Science, 7(6), pp. 501-524.

Xu, S. B., Xu, D. S. and Fang, H., 2012. Stereo Matching Algorithm Based on Detecting Feature Points. In Advanced Materials Research, Vol. 433, pp. 6190-6194.

Zhang, K., Hu, Q. and Wang, S., 2011. A fast 3D construction of heritage based on rotating structured light. In International Symposium on Lidar and Radar Mapping Technologies.

International Society for Optics and Photonics, pp. 82861Z-82861Z.

Zhong, S. D. and Liu, Y., 2012. Portable four-camera three-dimensional photographic measurement system and method. http://worldwide.espacenet.com/publicationDetails/biblio?CC=CN&NR=102679961B&KC=B&FT=D.

**4**

# VIDEO-BASED POINT CLOUD GENERATION USING MULTIPLE ACTION CAMERAS

Tee-Ann Teo *

Dept. of Civil Engineering, National Chiao Tung University, Hsinchu, Taiwan 30010. – tateo@mail.nctu.edu.tw

**WG IV/7, WG V/4**

**KEY WORDS:** Action cameras, Image, Video, Point clouds

**ABSTRACT:**

Due to the development of action cameras, the use of video technology for collecting geo-spatial data becomes an important trend. The objective of this study is to compare the image-mode and video-mode of multiple action cameras for 3D point clouds generation. Frame images are acquired from discrete camera stations while videos are taken from continuous trajectories. The proposed method includes five major parts: (1) camera calibration, (2) video conversion and alignment, (3) orientation modelling, (4) dense matching, and (5) evaluation. As the action cameras usually have large FOV in wide viewing mode, camera calibration plays an important role to calibrate the effect of lens distortion before image matching. Once the camera has been calibrated, the author use these action cameras to take video in an indoor environment. The videos are further converted into multiple frame images based on the frame rates. In order to overcome the time synchronous issues in between videos from different viewpoints, an additional timer APP is used to determine the time shift factor between cameras in time alignment. A structure form motion (SfM) technique is utilized to obtain the image orientations. Then, semi-global matching (SGM) algorithm is adopted to obtain dense 3D point clouds. The preliminary results indicated that the 3D points from 4K video are similar to 12MP images, but the data acquisition performance of 4K video is more efficient than 12MP digital images.

## 1. INTRODUCTION

### 1.1 Motivation

Three-dimensional geospatial information of indoor environment can be generated from cameras and laser scanners. Laser scanners obtain 3D points directly while camera indirectly obtains 3D points via stereo image matching. Digital still cameras and digital videos are two possible ways to collect digital images for image matching. Nowadays, a lightweight action camera such as GoPro Hero 4 Black Edition is able to collect digital still images up to 12Mp (4000 x 3000) resolution and video up to 8.3MP (3840 x 2160) resolution at 30 frames per second. Although the spatial resolution of a digital still camera is higher than a digital video, the sampling rate of a digital video is better than a digital still camera. As the video data can be converted to frame images like digital still camera, these highly overlapped frame images from video provide high similarity and high redundancy for image matching. In addition, action camera is able to acquire both video and image (5 seconds per frame) simultaneously. Therefore, there is a need to compare these two strategies for indoor point clouds generation.

### 1.2 Action Cameras

With the development of camera technology, most action cameras provide both image and video functions. To compare the traditional consumer digital camera and action camera, the action camera, such as GoPro (GoPro, 2015), emphasizes on: light weight, small dimensions, waterproof, large field-of-view (FOV), 4K video recording and high burst frame rate. The comparison of up-to-date action cameras can be found at (Crisp, 2014; Staub, 2015). The action cameras are originally developed for sports and underwater usage. The user uses the action camera to record their activities during extreme sports or special events. Due to the light weight, low cost and high spatial resolution of video mode, the usage of action cameras are extended to unmanned aerial vehicle (UAV), mobile mapping system (MMS), and other photogrammetric purposes.

### 1.3 Related Works

The digital video devices record sequence images and these dynamic sampling images can be used for different applications. The traditional photogrammetry is mostly relied on high spatial resolution images. Due to the improvement of video's resolution and frame rate, the use of video technology for collecting geo-spatial data becomes an important trend. Many video-related applications are presented in different geoinformation-related domains. For example, the space borne Skybox[TM] constellation is capable of acquiring sub-meter satellite imagery and high-definition panchromatic video for earth monitoring; the video collected by UAV can be used to produce geospatial data via Full Motion Video (FMV) in ArcGIS[TM] software or other commercial software; the video of car cam recorder can be used for crowdsourced street level mapping via Mapillary.com or other online-mapping services.

Several photogrammetry studies used GoPro action cameras for 3D measurement purposes. Balletti et al. (2014) discussed different camera calibration methods using GoPro for 3D measurement purposes. Kim et al., (2014) construct the 3D point clouds of building façade using GoPro 1080P super-view stereo video. As the needs of stereo vision, the GoPro Company provide accessories (i.e. dual cameras stereo housing, synchronization cable, software) to capture and produce 3D movie. Because of water proof housing, this technology has also applied in underwater stereo vision. For example, Schmidt and Rzhanov (2012) used dual GoPro cameras to measure seafloor micro-bathymetry. The 4K stereo videos are able to generate 3mm resolution grid of seafloor at 70cm distance. Nelson et al., (2014) combined the sonar scanner and dual GoPro cameras in a remotely operated vehicle for underwater 3D reconstruction. The results showed the potential of combining 3D sonar data and 3D surface from image matching for underwater archaeological application. The previous studies indicated that GoPro stereo videos are suitable for close-range photogrammetry purposes.

## 1.4  Research Purposes

The objective of this study is to compare the image-mode and video-mode of multiple action cameras for 3D point clouds generation. Frame images are acquired from discrete camera stations while videos are taken from continuous trajectories. The proposed method includes five major parts: (1) camera calibration, (2) video conversion and alignment, (3) orientation modelling, (4) dense matching, and (5) evaluation. As the action cameras usually have large FOV in wide viewing mode, camera calibration plays an important role to calibrate the effect of lens distortion before image matching. A black and white chess box pattern and Brown equation are adopted in camera calibration. Once the camera has been calibrated, the author use these action cameras to take video in an indoor environment. The videos are further converted into multiple frame images based on the frame rates. In order to overcome the time synchronous issues between videos from different viewpoints, the author manually identify image scene to calculate the time shift factor between cameras in time alignment. A structure form motion (SfM) technique is utilized to obtain the image orientations. Then, semi-global matching (SGM) algorithm is adopted to obtain dense 3D point clouds (Remondino et al., 2014).

## 2.   EXPERIMENTS AND RESULTS

### 2.1  System Specifications

This study uses five GoPro Hero4 Black cameras for point clouds generation. These five cameras are integrated in a Freedom360$^{TM}$ mount to obtain data 360 degrees panorama image and controlled by a GoPro Remote Controller. The size of this multi-view camera is about 10cm x 10cm x 10cm cube (see Figure 1). The camera provides both camera and video modes. The highest spatial image resolution for a digital still image is 12MP (4000 x 3000) while the finer spatial image resolution for a digital video is 4K (3840 x 2160) at 30 frames per second (fps). As the shutter of 4K video (1/30 sec) might produce blur images, this study also consider 1080P (1920 x 1080) at 120fps to avoid image blur. Table 1 shows the related camera parameters.

The spatial resolution of action camera is usually lower than digital single-lens reflex (DSLR) cameras. In order to understand the suitability of using action camera in close-range photogrammetry, this study analyse the spatial resolution of action camera at different distances and different modes. Figure 2 summaries the spatial resolution of image and video at nadir and diagonal points. The action camera usually has large FOV and consequently the point near to image boundaries has larger spatial resolution. This issue should be taken into consideration in 3D measurement. To obtain at least 5cm resolution, the maximum distance for 12MP image and 4K video should be less than 20m. The action camera might not suitable for long-range photogrammetry, but it is suitable for indoor environment at near range distance (<20m). Therefore, the scope of this study is to use the multiple action cameras in an indoor environment.

Table 1. Related parameters for GoPro Hero Black

| Item | Description |
| --- | --- |
| Size | 41mm x 59mm x 30mm |
| Weight | 89g |
| CCDsize | 1/2.3" |
| Nominal focal length | 3mm |
| Image size (digital still image) | 4000 x 3000 |
| Image size (4K video) | 3840 x 2160 (max 30fps) |
| Image size (1080P video) | 1920 x 1080 (max 120fps) |

Figure 1. Multiview GoPro System.

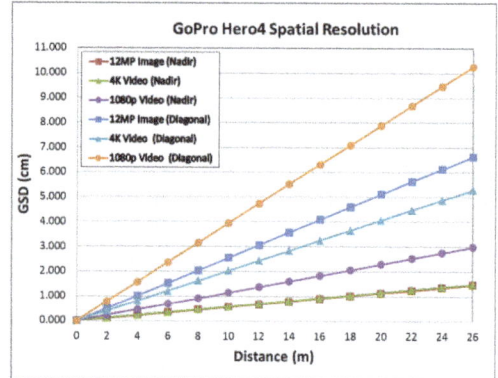

Figure 2. Ground sampling distances in different modes.

### 2.2  Camera Calibration

As the action cameras usually have large FOV in wide viewing mode, camera calibration plays an important role to calibrate the effect of lens distortion for image matching. This study uses Brown distortion model (equations (1) to (4)) (Brown, 1971) to determine the lens distortion. PhotoScan (Agisoft, 2015) and PhotoModeler (EOS System, 2015) are used to evaluate the results. PhotoScan uses regular chessboard pattern to obtain a large number of conjugate points in camera calibration. PhotoModeler uses circular signalized targets and self-calibration to determine the lens distortion parameters. Notice that, the radial distortion parameters K3 is needed for a large FOV camera.

$$\Delta x = \Delta x_r + \Delta x_d$$
$$\Delta y = \Delta y_r + \Delta y_d \tag{1}$$

$$\Delta x_r = \bar{x} \times (K_1 r^2 + K_2 r^4 + K_3 r^6)$$
$$\Delta y_r = \bar{y} \times (K_1 r^2 + K_2 r^4 + K_3 r^6) \tag{2}$$

$$\Delta x_d = P_1(r^2 + 2\overline{x}) + 2P_2\overline{xy}$$

$$\Delta y_d = P_2(r^2 + 2\overline{y}) + 2P_1\overline{xy} \tag{3}$$

$$r = \sqrt{(\overline{x})^2 + (\overline{y})^2} = \sqrt{(x - x_0)^2 + (y - y_0)^2} \tag{4}$$

Where, $(\Delta x, \Delta y)$ are total lens distortion; $(\Delta x_r, \Delta y_r)$ are radial distortion; $(\Delta x_d, \Delta y_d)$ are tangential distortion; $(K_1 \sim K_3)$ are coefficients of radial distortion; $(P_1 \sim P_2)$ are coefficients of tangential distortion; r is radial distance; $(x,y)$ are photo coordinate ; and $(x_0, y_0)$ are principal points.

This study performs the camera calibration for a 12MP image, a 4K video and a 1080P video separately. In video calibration, this study uses video mode to shoot the target code at different view angles and positions. Then, these video frames are converted into images at 1 image per second. Besides, the initial focal length and frame size (Kolor, 2015) are also written at EXIF for calibration purpose. The total errors of PhotoModeler are smaller than 2 pixels in all modes. However, the PhotoScan does not provide accuracy index in lens distortion correction. Table 2 shows the results of camera calibration for camera id 2 using Photomodeler. Figure 3 show the distortion curves of radial and tangential distortions. The impact of radial distortion is significantly larger than the tangential distortion. To compare the digital still image and video, the results of PhotoModeler show high consistence in radial distortion except the tangential distortion for 1080P.

To compare the results of PhotoModeler and PhotoScan, the radial distortion of PhotoModeler is larger than PhotoScan. This study also generates two undistorted images using these two methods (See Figure 4). The behavior of these two methods is similar at the center area. But for straight lines near to the corner area, the result of PhotoModeler is better than PhotoScan. Therefore, this study uses the lens distortion parameters from PhotoModeler.

Table 2. Results of camera calibration using Photomodeler

| Type | Image | Video | Video |
|---|---|---|---|
| Resolution | 12MP | 4K | 1080P |
| Width (pixel) | 4000 | 3840 | 1920 |
| Height (pixel) | 3000 | 2160 | 1080 |
| F (mm) | 2.752130 | 2.654308 | 2.821934 |
| xp (mm) | 3.171405 | 2.945599 | 3.139957 |
| yp (mm) | 2.412459 | 1.620624 | 1.717661 |
| Fw (mm) | 6.246702 | 5.881419 | 6.243885 |
| Fh (mm) | 4.686000 | 3.308000 | 3.514000 |
| k1 | 3.887E-02 | 3.659E-02 | 3.568E-02 |
| k2 | 6.895E-04 | 2.429E-03 | 8.153E-04 |
| k3 | 1.491E-04 | 0.000E+00 | 1.098E-04 |
| p1 | -1.193E-04 | 4.979E-04 | 3.126E-04 |
| p2 | 1.622E-04 | -4.025E-04 | 0.000E+00 |
| PixelSize | 0.001562 | 0.001532 | 0.003252 |

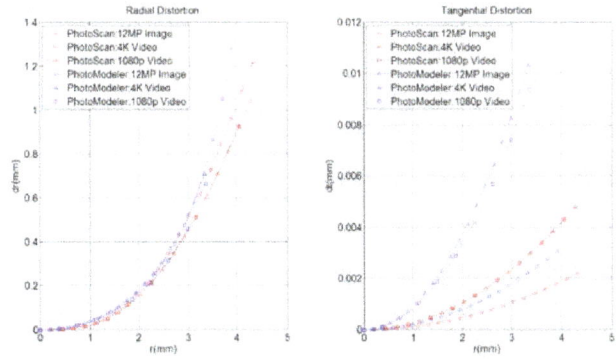

Figure 3. Results of lens distortions using different modes and different methods.

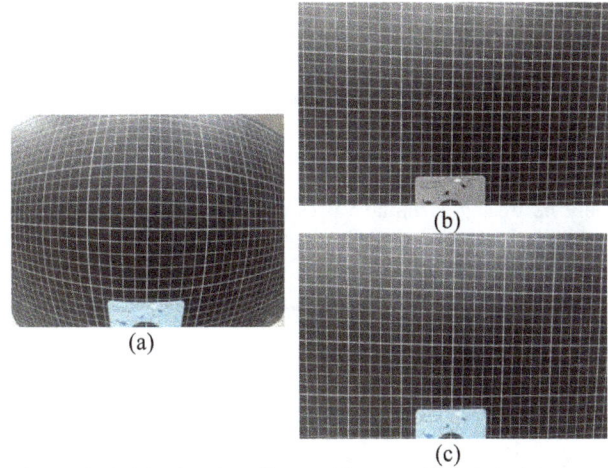

Figure 4. Original and undistorted images: (a) original image; (b) undistorted image using Photomodeler parameters; (c) undistorted image using Photoscan parameters.

## 2.3 Cameras Alignment

These five cameras are fixed together in a mount and a camera alignment is needed to determine the geometrical relationship between cameras. This study uses camera 1 as the master camera while the other 4 are the slave cameras. The transformation between master and slave cameras is descripted by two sets of parameters, i.e. lever-arms (dx, dy, dz) and boresight-angles (dω, dφ, dκ). In this study, 120 signalized targets (markers) are distributed on a 90cm x 90cm x 65cm box (see Figure 5a). Then, 80 images are taken from 16 stations by 5 cameras. These 80 images are used for bundle adjustment and determine their exterior orientations in mapping frame (see Figure 5b). The lever-arms and boresight-angles are calculated by equations (5) and (6) using exterior orientations (Rau et al., 2011).

$$R_{Slave}^{Master} = R_{Mapping}^{Master} \times R_{Slave}^{Mapping} \tag{5}$$

$$d_{Slave}^{Master} = R_{Master}^{Mapping}(d_{Slave}^{Mapping} - d_{Master}^{Mapping}) \tag{6}$$

Where, $R_{Slave}^{Master}$ is the rotation matrix between master and slave camera frames; $R_{Slave}^{Mapping}$ is the rotation matrix between slave camera frame and mapping frame from bundle adjustment; $R_{Master}^{Mapping}$ is the rotation matrix between master camera frame and mapping frame from bundle adjustment; $d_{Slave}^{Master}$ is the lever-arms between master and slave camera; ($d_{Slave}^{Mapping}, d_{Master}^{Mapping}$) are the vectors from slave/master to mapping frame.

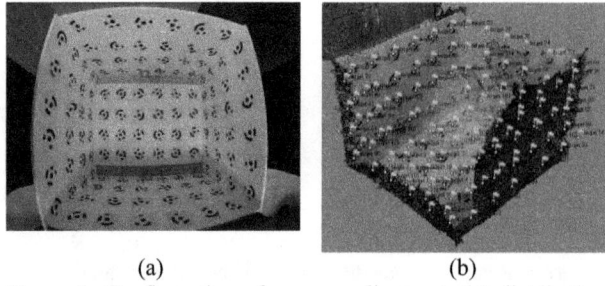

(a)            (b)

Figure 5. Configuration of cameras alignment: (a) distribution of markers; (b) result of bundle adjustment.

Table 3 summaries the results of cameras alignment. The standard deviations of boresight-angles are better than 0.6 degrees except for camera 5 on the top. The standard deviations of lever-arms are less than 1.9cm in all cases. It is about 19% of the size of this camera system (see Figure 1). In other words, the variation of lever-arms is around 1.9cm. These parameters can only be treated as initial values in orientation modelling and further investigation is needed.

Table 3. Estimated lever-arms and boresight-angles

|  | Lever-arms | | | Boresight-angles | | |
|---|---|---|---|---|---|---|
|  | dx (m) | dy (m) | dz (m) | dω (deg) | dφ (deg) | dκ (deg) |
| Cam1 & 2 | -0.024 ±0.019 | 0.081 ±0.006 | -0.001 ±0.006 | -0.917 ±0.247 | -0.121 ±0.499 | 179.518 ±0.510 |
| Cam1 &3 | 0.042 ±0.007 | 0.041 ±0.006 | -0.003 ±0.007 | 89.123 ±0.597 | -0.906 ±0.178 | 90.501 ±0.510 |
| Cam1 &4 | -0.052 ±0.015 | 0.041 ±0.015 | -0.009 ±0.011 | -93.035 ±0.420 | 1.599 ±0.449 | -90.166 ±0.501 |
| Cam1 &5 | -0.006 ±0.009 | 0.031 ±0.008 | -0.024 ±0.014 | 16.646 ±3.750 | 86.004 ±0.256 | 96.406 ±3.863 |

## 2.4 Data Synchronous

These five cameras are controlled by a remote control and no cables are connected between cameras. The author found a slightly time lag when triggering the camera to take image or video. This time lag does not affect the digital still image on a fixed tripod, but it might cause the data asynchronous in the video mode. As there is no cable to connect these cameras for synchronous purpose, the only way is using an additional timer to align the videos. Figure 6 shows the same timer taken from different cameras using video mode. A timer APP which has 1/100 sec precision in time alignment is used. All videos are shot to a same timer separately and the videos recorded times are shifted to the reference time of time. Although the timer may provide 1/100 sec precision, the time alignment precision is restricted by frame rate. For example, the time interval of 4K video frame is 1/30 sec. This method can only ensure 1/30 sec time synchronous for a 4K video.

## 2.5 Point clouds generation

After camera calibration, cameras and time alignment, video are converted into image frames at different sampling interval for 3D point clouds generation. The procedure includes: (1) structure from motion (SfM) technique for image orientations; (2) absolute orientation using control points; (3) semi-global matching (SGM) algorithm for dense point matching. This study utilizes a commercial Agisoft PhotoScan in 3D point clouds generation.

## 3. EVALUATION

The evaluation includes two cases, one is a stair and the other is a lobby.

### 3.1 Case 1. Stair

The 3D stair modelling is a challenging task in indoor modelling. A discrete digital still image usually cannot provide favourite intersection geometry due to the limited camera station. In the contrary, a digital video is able to take multi-view images effectively. The aim of this session is to compare the performance of a 12MP image, a 4K video and a 1080P video. Only one action camera is adopted in this section. For a digital still image, the author take the images for every steps of the stair. The duration of images is about 150 seconds. However, the duration of video is only 25 seconds for the same area. The data acquisition of video mode is much effective than image mode. Besides, the standard deviations of camera baseline are 15.1cm for 12MP, 7.9cm for 4K and 3.8cm for 1080P. The digital video may provide more uniform camera station than digital camera. To compare the 4K and 1080P videos, the resolution of 4K is higher than 1080P while the sampling rate of 4K (i.e. 30fps) is lower than 1080P (i.e. 120fps). Hence, the image quality (e.g. effect of motion blur) for 1080P is better than 4K visually.

The author use the image and video in relative orientation modelling and 4 control points are manually selected in absolute orientation. The residual of control points are less than 5cm in the three cases. Then, high density image matching is used to obtain point clouds of a stair. Table 4 summaries the results of these three modes. The point density of 12MP is the highest one, but the result of 4K video is similar to the results of 12MP. Figure 7 is a section of stair for comparison. The section includes 18 steps and the size of the stair is about 1.5m width, 4m length and 2.4m height. The shape of these three results shows high consistency. In other words, the 4K video is possible to produce similar results like 12MP images.

Table 4. Comparison of images and videos for a stair

|  | 12MP | 4K video | 1080P video |
|---|---|---|---|
| Number of pixel | 12MP | 8.29MP | 2.07MP |
| Duration of data acquisition | 150sec | 25sec | 25sec |
| Sampling rate | 3sec (manual) | 0.5sec | 0.25sec |
| Number of image | 46 | 50 | 100 |
| Estimated processing time | 1hr | 1hr | 2hr |
| Camera distance (m) | 38.2±15.1 | 27.5±7.9 | 15.0±3.8 |
| Point density (pts/cm²) | 65 | 60 | 44 |

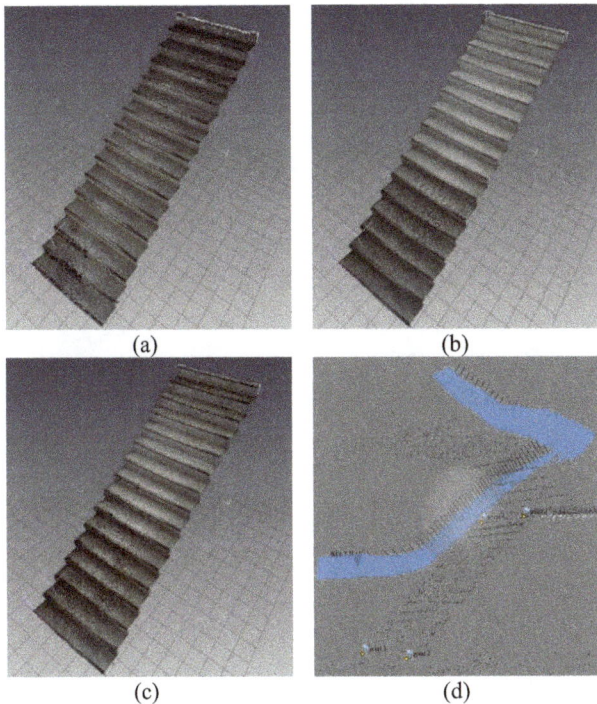

(a)　　　　　　　　　　(b)

(c)　　　　　　　　　　(d)

Figure 7. Results of a stair: (a) results of 12MP images; (b) results of 4K video; (c) results of 1080P video; (d) perspective centre of cameras for 1080P video.

## 3.2 Case 2. Lobby

In Case 2, the author uses the multi action cameras system to reconstruct the point clouds of a lobby. The test area is about 20m width, 15m length and 3m height. In order to have multi-view images for image matching, a tripod is used to take digital still images at five different heights (i.e. 1.0m, 1.25m, 1.50m, 1.75m, and 2.00m). The distance between cameras for the same station is about 0.25m while the distance between different stations is about 3m. The duration of image acquisition for these five stations is about 10 minutes. The duration of 4K video is just 22 seconds for the same area. Table 5 summaries the results of these two modes. The video mode obtains continuous image frames. The average camera centre of video mode is 32.2cm. Therefore, the number of frame from video is larger than traditional digital image (i.e. 376 images > 125 images). However, the video mode needs more computational time to produce point clouds (i.e. 4hrs > 2hrs).

Table 5. Comparison of images and videos for a lobby

|  | 12MP | 4K video |
|---|---|---|
| Number of pixel | 12MP | 8.29MP |
| Duration of data acquisition | 600sec | 47sec |
| Sampling rate | - | 0.5sec |
| Number of image | 125 | 376 |
| Estimated processing time | 2hrs | 4hrs |
| Camera distance (m) | Same station: 25cm Between station: 300cm | 32.2cm±11.6cm |
| Point density (pts/cm$^2$) | 78.6 | 72.8 |

Figures 8a and 8b show the distributions of 12MP images and 4K video frames. The numbers of camera station for 12MP and 4K video are 5 and 94. The video mode provides high overlapped multiview images for space intersection. Figures 8c and 8d compare the generated points from the same view point. The point clouds of 4K video are more complete than the 12MP image. Due to the limitation of image matching, the area without texture does not have 3D points after image matching.

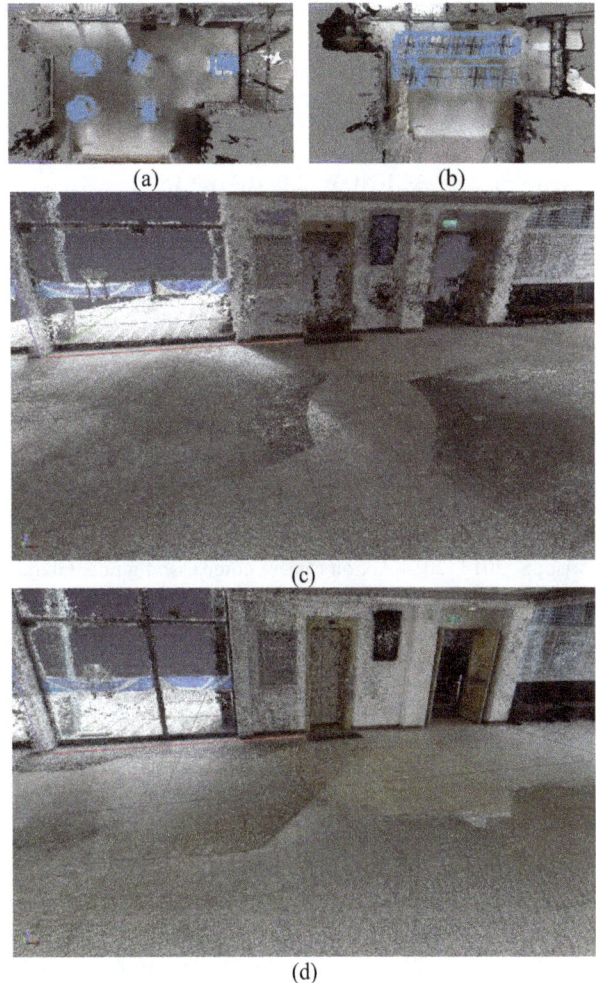

(a)　　　　　　　　　　(b)

(c)

(d)

Figure 8. Results of a lobby: (a) perspective centres of 12MP images; (b) perspective centres of 4K video frames; (c) points from 12MP image; (d) points from 4K video.

## 4. CONCLUTIONS AND FUTURE WORKS

This research proposed a multiple action cameras system for indoor mapping. The characteristic of this system is 360 degrees panorama imaging and 4K high resolution video. It is beneficial for data acquisition in an indoor environment as well as 3D point clouds generation. This study also demonstrated the results of camera calibration for image and video modes. The maximum radial distortion of a4K video reached 500 pixels at image boundary. The lens distortion should be pre-calibrated as the impact of lens distortion was significant in related to image frame. These five cameras were mounted together and the lever-arms and boresight-angles were calculated by cameras alignment. The results of cameras alignment can be used as the initial orientations in orientation modelling. The time synchronous was implemented by an additional timer in video mode. It can adjust the time tag issue of this system. Finally, the

3D point clouds were generated by orientation modelling and dense matching.

The preliminary result indicated that the 3D points from a4K video were similar to 12MP images. Besides the data acquisition performance of a4K video was faster than 12MP digital images, the limitation of this video-based point clouds generation is the huge computational time for large data set and low image quality caused by video compression and motion blur. Future works will evaluate the system in different scenarios and different parameters. As the radiometric performance of action camera will influence the geometrical performance, future works will focus on the radiometric performance for action cameras in image and video modes.

## ACKNOWLEDGEMENTS

This investigation was partially supported by the National Science Council of Taiwan under project number NSC 101-2628-E-009-019-MY3.

## REFERENCES

Agisoft, 2015. PhotoScan, URL: http://www.agisoft.com

Balletti, C., Guerra, F., Tsioukas, V. and Vernier, P., 2015. Calibration of action cameras for photogrammetric purposes, Sensors, 14: 17471-17490.

Brown, D.C., 1971, Close-range camera calibration, Photogrammetry Engineering, 37:855-866.

Crisp, S. 2014. 2014 Action camera comparison guide, Gizmag, URL: http://www.gizmag.com/compare-best-action-cameras-2014/34974/

EOS System, 2015, PhotoModeler Motion, URL: http://www.photomodeler.com

GoPro, 2015, GoPro Hero4 Black, URL: http://gopro.com

Kim, J.H., Pyeon, M.W., E.O, Y.D., and Jang, I.W., 2014. An experiment of three-dimensional point clouds using GoPro, International Journal of Civil, Architectural, Stuctural and Construction Engineering, 8(1): 82-85.

Kolor, 2015, About GoPro focal length and FOV, URL: http://www.kolor.com/wiki-en/action/view/Autopano_Video_-_Focal_length_and_field_of_view

Nelson, E.A. Dunn, I.T., Forrester, J., Gambin, T., Clark, C.M. and Wood, Z.J. 2014. Surface reconstruction of ancient water storage systems: an approach for sparse 3d sonar scans and fused stereo images. GRAPP, 161-168.

Rau, J.Y., Habib, A.F., Kersting, A.P. Chiang, K.W., Bang, K.I., Tseng, Y.H. and Li, Y.H. 2011. Direct sensor orientation of a land-based mobile mapping system, Sensors, 11: 7243-7261.

Remondino, F., Spera, M.G., Nocerino, E., Menna, F. and Nex, F. 2014. State of the art in high density image matching, Photogrammetric Record, 29 (6): 144-166.

Schmidt, V.E. and Rzhanov, Y., 2012 Measurement of micro-bathymetry with a GoPro underwater stereo camera pair, IEEE Ocean 2012, 1-6.

Staub, D., 2015. 2015 Best action camcorders review, Top Ten Reviews, URL: http://action-camcorders-review.toptenreviews.com/

# 5

# SCALABLE PHOTOGRAMMETRIC MOTION CAPTURE SYSTEM "MOSCA": DEVELOPMENT AND APPLICATION

V. A. Knyaz [a, *],

[a]St. Res. Institute of Aviation Systems (GosNIIAS), 125319, 7, Victorenko str., Moscow, Russia - knyaz@gosniias.ru

**Commission V, WG V/5**

**KEY WORDS:** Photogrammetry, accuracy, calibration, motion capture, tracking

**ABSTRACT:**

Wide variety of applications (from industrial to entertainment) has a need for reliable and accurate 3D information about motion of an object and its parts. Very often the process of movement is rather fast as in cases of vehicle movement, sport biomechanics, animation of cartoon characters. Motion capture systems based on different physical principles are used for these purposes. The great potential for obtaining high accuracy and high degree of automation has vision-based system due to progress in image processing and analysis. Scalable inexpensive motion capture system is developed as a convenient and flexible tool for solving various tasks requiring 3D motion analysis. It is based on photogrammetric techniques of 3D measurements and provides high speed image acquisition, high accuracy of 3D measurements and highly automated processing of captured data. Depending on the application the system can be easily modified for different working areas from 100 mm to 10 m. The developed motion capture system uses from 2 to 4 technical vision cameras for video sequences of object motion acquisition. All cameras work in synchronization mode at frame rate up to 100 frames per second under the control of personal computer providing the possibility for accurate calculation of 3D coordinates of interest points. The system was used for a set of different applications fields and demonstrated high accuracy and high level of automation.

## 1. INTRODUCTION

Nowadays motion capture as a process of acquiring real 3D movement of an object (or a set of points representing an object) for a further processing is of high demand by many applications. The most known fields of motion capture usage are movie and video game production where accurate registration of 3D motion provides a high impression of reality to virtual creatures.

### 1.1 Types of motion capture system

A few types of motion capture systems are now in use. Among them there are mechanical, acoustical, magnetic, optical systems.

Mechanical systems use potentiometers and sliders located in the required positions on an actor and provide registration of their spatial positions. They have some advantage such as an interface that is similar to stop-motion systems widely used in the film industry. Other advantages are independence from magnetic fields or reflections and short setting up time. Their main disadvantage is restriction caused by wires which connect sensors to registration system.

In acoustical system a set of acoustic receptors capture sounds from sound transmitters located on the object (actor). The specific sounds from emitters then picked up by receivers and 3D positions of emitters are calculated using registered times between emitting and receiving signal. To determine the 3D position of each transmitter, a triangulation of the distances between the emitter and each of the receptors is computed.

Acoustical motion capture systems have some problems which make them inconvenient in a number of cases. These problems are: the restrictions to the freedom of movement caused by the cables put on the actor, the limited number of transmitters that can be used and susceptibility to sound reflections or external noise.

Being not comparatively expensive magnetic systems are rather accurate and fast (about 100 fps) for simple movement capture. They use a set of magnets as markers of given points and a set of receivers for measuring the position and orientation of the markers relative to an antenna (Yabukami, 2000). The disadvantages of magnetic systems are also limitations caused by cables and possible interference in the magnetic field caused by various metallic objects and structures.

Typical vision-based motion capture systems usually include a set of cameras capturing video sequences of an actor/object on which special targets are placed. Then video sequences are processed for target detecting, identifying and tracing through the sequence. The level of accuracy of 3D point coordinates calculation is provided by calibration procedure and depends on application needs.

These systems are the most expensive ones in the market due to their cutting-end technological nature, such as the high-

---

* Corresponding author.

resolution cameras and sophisticated proprietary software. The cost reaches hundreds of thousands USD.

The doubtless advantages of such systems are possibility of capturing at very high speed, no limitation for actor moving in the working space, great potential for automation of the process.

## 1.2  Applications

The initial impulse for creating motion capture systems was done by entertainment industry which had a need for a mean of fast and accurate actor movements transfer into the movie or animation. And now movie and video game production industries are the main users of motion capture systems.

But the field of application for motion capture systems grows very dynamically. Among major areas of application there are medicine, sport, various branches of industry, scientific researches (Moeslund, 2006).

Motion capture systems in medical applications are used for accurate analysis of human motion which cannot be registered by other means. 3D study of human motion allows to find abnormalities and propose the way of rehabilitation.

In sport of high results motion capture systems are of great demand because they provide valuable information about high speed motion of a sportsman during competition. This information is the basis for improving the sport technic and achieving better results. Golf is one of the major users of motion capture systems for analysis and correction of sport technics.

Motion capture is often the single tool for scientific research of specific tasks where information about object 3D movement could not be obtained by other means. The typical examples are very fast dynamical process analysis, analysis of vehicle vibration, object 3D trajectory estimation and analysis and similar projects.

## 2.  SYSTEM OUTLINE

For a vision-based photogrammetric motion capture system a reliability and a convenience for a user are the key features defining the quality of the system. In this aspect a detection and tracking of given object points required for an application tasks play the essential role. Also if the area of application and scale of the captured process can be variable it is required that the motion capture system can work with different working area size and can be easily reconfigured and recalibrated. These are the main requirement which were in consideration in developing the system.

### 2.1  Hardware

The developed scalable 3D motion capture system "Mosca" is based on photogrammetric techniques for 3D measurements and provides high speed image acquisition, high accuracy of 3D measurements and high level of automation of captured data. Depending on the application the system can be easily modified for different working areas from 100 mm to 10 m. The developed motion capture system uses from 2 to 4 technical vision cameras for video sequences of object motion acquisition. All cameras work in synchronic mode at frame rate up to 100 frames per second under the control of personal

computer (PC) providing the possibility for accurate calculation of 3D coordinates of interest points. The system could be extended to more cameras by including an additional PC station in the system.

The original camera calibration and external orientation procedure is used to reach high accuracy of 3D measurements. The calibration procedure is highly automated due to applying original coded targets (Knyaz, 1998) for identifying and measuring image coordinates of reference points. The system calibration provides accuracy of 0.01% of working space (WS) of the motion capture system.

Technical characteristics of the motion capture system are presented in table 1.

| Camera resolution | 656 x 491 pixels |
|---|---|
| Acquisition speed | up to 100 fps |
| Number of tracking points | up to 200 |
| Working space (WS) | Scalable: |
|  | from 0.1x0.1x0.1m |
|  | to 10x10x10 m |
| 3D point coordinate accuracy | 0.01% of WS |

Table 1. Technical characteristics of the Mosca motion capture system

The possibility for varying the scale of imaging is provided by fast and highly automated procedures for calibration and exterior orientation.

### 2.2  Calibration

Calibration is performed using original technique and original software. Classical central projection model is used for camera imaging process. With given centre of projection $\mathbf{X}_O = (X_0, Y_0, Z_0)$ for object point a with spatial coordinates $\mathbf{X} = (X, Y, Z)$ its image coordinates $\mathbf{x} = (x, y)$ can be found from the co-linearity equation:

$$\mathbf{X} = \mathbf{X}_O - \mu \mathbf{A}^T (\mathbf{x} - \mathbf{x}_p)$$

Where $\mathbf{A}$ is transition matrix, $\mathbf{x}_p$ – image coordinates of principal point, $\mu$ – scale factor.

The additional parameters describing CCD camera model in co-linearity conditions are taken in form:

$$\Delta x = a\overline{y} + \overline{x}r^2 K_1 + \overline{x}r^4 K_2 + \overline{x}r^6 K_3 + (r^2 + 2\overline{x}^2)P_1 + 2\overline{xy}P_2$$

$$\Delta y = a\overline{x} + \overline{y}r^2 K_1 + \overline{y}r^4 K_2 + \overline{y}r^6 K_3 + 2\overline{xy}P_1 + (r^2 + 2\overline{y}^2)P_2$$

$$\overline{x} = m_x(x - x_p); \ \overline{y} = -m_y(y - y_p); \ r = \sqrt{\overline{x}^2 + \overline{y}^2}$$

where  $x_p, y_p,$-the coordinates of principal point,
  $m_x, m_y$ - scales in $x$ and $y$ directions,
  $a$ – affinity factor,
  $K_1, K_2, K_3$ – the coefficients of radial symmetric distortion
  $P_1, P_2$ - the coefficients of decentring distortion

The common procedure for determining unknown parameters of camera model is bundle adjustment procedure using observations of test field reference points with known spatial coordinates (Knyaz, 2002).

Image interior orientation and image exterior orientation ($X_i$, $Y_i$, $Z_i$ – location and $\alpha_i, \omega_i, \kappa_i$ and angle position in given coordinate system) are determined as a result of calibration. The residuals of co-linearity conditions for the reference points after least mean square estimation $\sigma_x$, $\sigma_y$ are concerned as precision criterion for calibration.

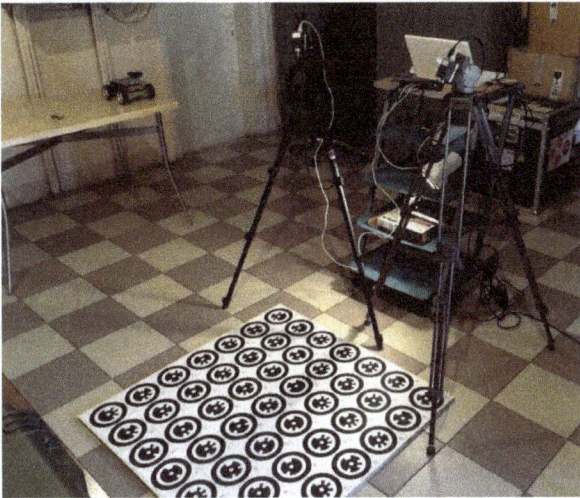

Figure 1. Exterior orientation

The results of the cameras interior orientation parameters estimation by described technique are presented in Table 2.

|  | Camera 1 | Camera 2 | Camera 2 |
|---|---|---|---|
| $m_x$ | 0.00943414 | 0. 00935769 | 0. 00929585 |
| $m_y$ | 0. 00944242 | 0. 00935618 | 0.00928185 |
| $b_x$ | 291.4501 | 298.077039 | 289.09868 |
| $b_y$ | 255.45588 | 255.501334 | 253.84673 |
| $a$ | -0.00668 | -0.00617081 | -0.005961 |
| $K_1$ | 0.0008712 | 0.001612895 | 0.0007685 |
| $K_2$ | -0.0000182 | 0.000091961 | 0.00001594 |
| $K_3$ | 0.00000032 | -0.000007424 | -0.00000041 |
| $P_1$ | 0.0002827 | -0.000030034 | -0.000110274 |
| $P_2$ | 0.00004459 | 0.00014336 | 0.000081377 |

| $\sigma_x$ | 0.30 mm |
|---|---|
| $\sigma_y$ | 0.31 mm |

Table 2. Sample of calibration results for system configuration for 3x3x3 m working area

The external orientation of the motion capture system is performed after choosing a working space and camera configuration for motion capture. For external orientation the same test field is used. It defines the global coordinate system in which 3D coordinates are calculated.

## 3. ALGORITHMS FOR AUTOMATION

The problem of given object point detection and tracing is of great importance for the motion capture system. For most part of applications it is needed to measure 3D coordinates of given points so these points have to be marked on the objects by special targets which should be detected and identified in the image. Coded targets could not be applied in this case because of their rather large size required for reliable identification. So for automation of the target detection and identification some techniques were developed.

### 3.1 Algorithms for target detection

**3.1.1 Algorithm assumption:** Algorithm works in assumption that the target is a connected region in the image which meets to three conditions:

1. There is a single maximum of intensity for any section through the center of the probable region.
2. The value of this maximum is greater than given threshold B.
3. Dimensions of the region belong to given range between $D_{min}$ and $D_{max}$.

**3.1.2 Algorithm description.** Algorithm is based on image $I(x,y)$ binarization by the sequence of thresholds $h_{max}$, $h_{max}$-s, $h_{max}$-2s,..., $h_{max}$-ns; finding all connected regions in every binary image and selecting only such regions which meet to conditions 1-3.

Algorithm's parameters:
$h_{max}$ – maximum value of intensity for binary image. The recommended value of $h_{max} = I_{max}$ - B/2.
$h_{min}$ - minimum value of intensity for current binary image.
s – a step in intensity increasing.
B – minimal intensity of target to search.
$D_{min}$ and $D_{max}$ - minimum and maximum values of a target dimension.

**3.1.3 Algorithm steps:** Algorithm includes the following steps:

1. Building the intensity histogram HIST[0.. $I_{max}$] for initial image $I(x,y)$.

2. Choosing the initial intensity level:

$$h = h_{max}$$

3. Choosing the new intensity level of binary image:

$$until\ HIST[h]=0\ h:=h-1$$

4. Image $I(x,y)$ binarization at threshold h:

$$if\ I(x,y) > h\ then\ B(x,y):=1$$
$$else\ B(x,y):=0$$

5. Searching for all connected regions in image $B(x,y)$ and creating the array of descriptors for every detected region R:

- coordinates of upper left $(x_1,y_1)$ and lower right $(x_2,y_2)$ corners of the minimal rectangle containing detected region R
- maximal $B_{max}$ and minimal $B_{min}$ values of intensity in detected region R;
- the number N of regions detected at previous binary image, belonging to region R;
- unique number $M(x,y)$ of region R which allow to check if any pixel $(x,y)$ belongs to the region R

6. Output image $R(x,y)$ includes only that regions for which the condition 1-3 are carried out:

$$if\ (x_2 - x_1 > A_{min})\ and\ (y_2 - y_1 > A_{min})$$
$$and$$
$$(x_2 - x_1 < A_{max})\ and\ (y_2 - y_1 < A_{max})$$
$$and$$
$$(B_{max} - B_{min} > B)\ and\ (N < 2)$$
$$than\ R(x,y):=1$$
$$else\ R(x,y):=0$$

7. $h := h - s$

8. If $h > h_{min}$ then go to 3.

9. Finding all connected regions in the image $R(x,y)$.

10. For every region R in the binary image $R(x,y)$ coordinates of its center of mass are calculated. The resulting sub-pixel coordinates $(x^*, y^*)$ are the coordinates of the target:

$$x^* = \frac{\sum_{x,y \in R} xI(x,y)}{\sum_{x,y \in R} I(x,y)}$$

$$y^* = \frac{\sum_{x,y \in R} yI(x,y)}{\sum_{x,y \in R} I(x,y)}$$

## 3.2 Algorithm for searching corresponding points in the captured images

After detection all $n_t$ targets by described above algorithm their coordinates $(x_i^j, y_i^j)$ for every target $t_i$, i=1, …, $n_t$ are known for every image $I_j$, j=1,…, $n_l$. Also parameter of exterior orientation for every image $I_j$, j=1,…, $n_l$ are known due to preliminary exterior orientation procedure.

For determining 3D coordinates of the target it is required to find the correspondence between detected targets in different images. Because all target have the same shape it is impossible to apply correlation or descriptor-based methods. If the number of cameras more than two epipolar geometry could be used for identification of similar targets in different images.

For point $p_i^1$ in the frame from the first camera its image in the frame from second camera should lay on the epipolar line $r_i^1$ which is an intersection of the plane defined by three points (center of projection of the first camera, center of projection of the second camera and the image of the point $p_i^1$ in the frame from the first camera) and projection plane of the second camera (figure 2).

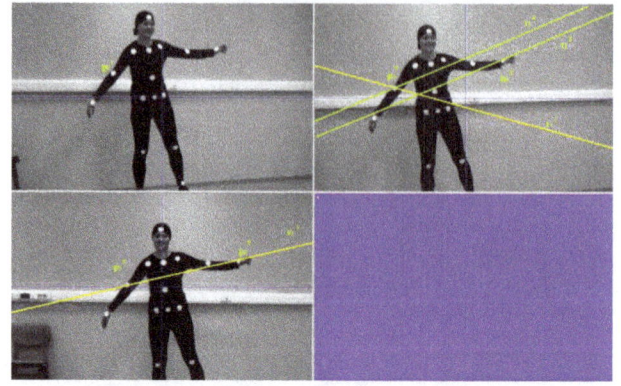

Figure 2. Epipolar points searching

So the targets $p_i^2$ and $p_j^2$ detected in the frame from the second camera and lying on the epipolar line $r_i^1$ are potential images of point $p_i$. In a similar way point $p_i^1$ will be imaged in the frame from the third (fourth, etc) camera as epipolar line and the points regarded as possible images of $p_i^1$ will be presented in the frame from the third camera as a set of epipolar lines $r_i^2$ $r_j^2$ from the second camera. So the corresponding point $p_i^3$ to point $p_i^1$ in the third image will be the point of target location in which epipolars are intersected. Figure 2 illustrates the algorithm for points correspondence searching.

This algorithm allows to establish the correspondence between targets images from more than two cameras and then provide the possibility to resolve collision when tracing target along the recorded video sequence.

### 3.3 Software

The original software for synchronic video sequences capture and their automated processing was developed. The software supports a set of procedures for motion capture and processing such as:

- automated system orientation
- video sequences capture in synchronic mode
- automated target detection and correspondence problem solution
- automated target tracing
- 3D trajectory calculation and visualisation

## 4. APPLICATIONS

The developed photogrammetric motion capture system is applicable for wide variety of applications where accurate, fast and reliable data on object (object points) is needed. Among application in which Mosca was used are biometry and biomechanics, robot dynamical model identification, unmanned aerial vehicle (UAV) self-orientation accuracy evaluation, virtual objects control.

### 4.1 Human motion capture

Human biomechanics is one of the important applications which need a mean for fast and accurate human motion in different modes. In this case usually it is required to obtain accurate 3D trajectories of given points of a human body. Depending on the task to solve the Mosca could capture full human body motion or movements of body parts (e.g., facial expressions) with higher accuracy.

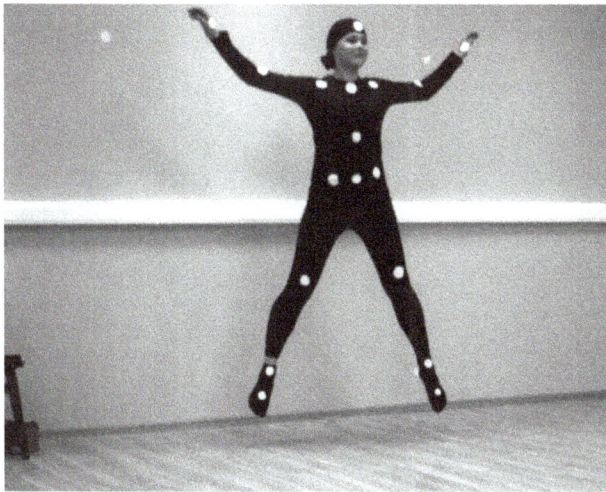

Figure 3. An actor with targets placed according BVH model

For human motion capture the Mosca is configured for working space of about 2.5x2.5x2.5 m so that the required movement of an actor could be captured. Figure 3 presents an actor with targets placed according BVH model during the acquisition process.

In figure 4 software interface is shown with detected and identified targets and biped 3D model generated using captured data.

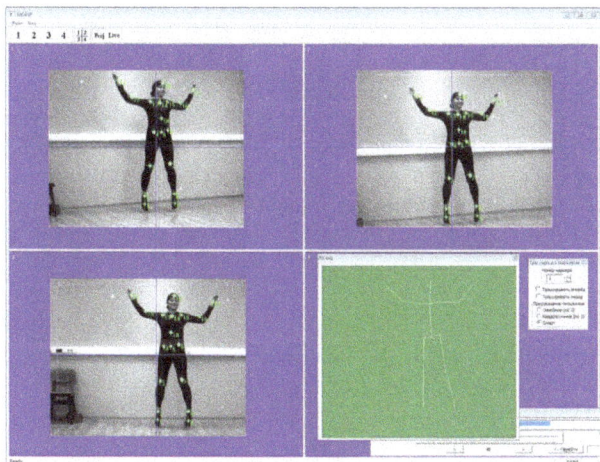

Figure 4. Software interface with detected and identified targets and biped 3D model

## 4.2 UAV accuracy estimation

The Mosca system was used for estimation the accuracy of self-orientation of unmanned aerial vehicle (UAV). UAV Parrot AR.Drone2.0 was used for experiments.

Its basic technical characteristics are given in Table 3.

| Parameter | Value |
|---|---|
| Velocity | 0.5 m/s |
| Weight | 366 g |
| Dimensions | 515x515 mm |

| Frontal camera | |
|---|---|
| - Frame rate | 30 fps |
| - Resolution: | 1280x720 pix |
| Vertical camera | |
| - Frame rate | 60 fps |
| - Resolution: | 320x240 pix |

Table 3. UAV Parrot AR.Drone2.0 technical characteristics

Figure 5 presents Parrot AR.Drone2.0 bottom view with indication of targets' placement (1, 2, 3, 4, 5) and the position of the frontal camera (*a*).

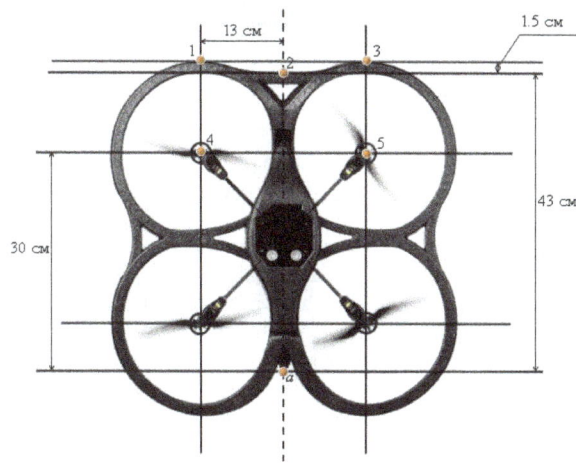

Figure 5. Parrot AR.Drone2.0 bottom view

The problem to be solved was to determine with what accuracy the UAV could determine its own position and orientation basing on processing video information from the frontal camera. Preliminary calibration of UAV frontal camera was carried out using a set of images of the test field captured by UAV frontal camera. Figure 6 shows the test field image acquired by the frontal camera. The accuracy of the UAV camera calibration was at the level of 0.5 mm.

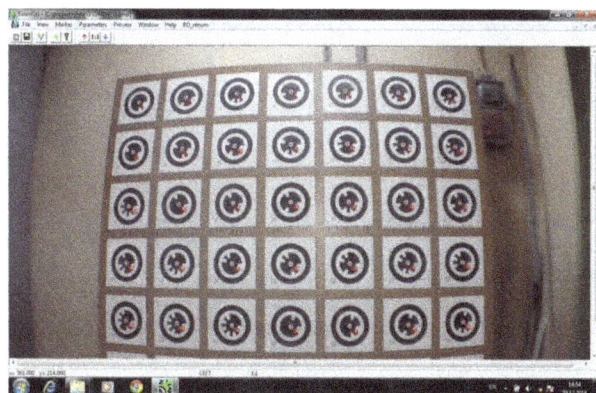

Figure 6. The test field image acquired by frontal camera.

Then the flight trajectory of the set of targets placed on the UAV was captured by the Mosca along with acquiring video sequences of this flight by UAV frontal camera. Both data sets were processed resulting in two 3D trajectories: captured by the motion capture system and self-estimated by UAV. Both trajectories were registered in common coordinate system defined by the test field. Coded targets were used to provide high accuracy and automation of the process. The synchronization between these two trajectories was performed using special light marker in the acquired video sequences.

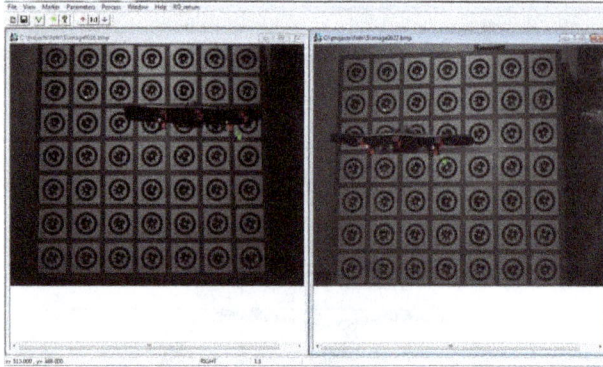

Figure 7. The images of UAV during the flight acquired by the motion capture system.

The position of the UAV frontal camera was estimated by using coordinates of the targets in coordinate system connected with the UAV and coordinates of the targets in the motion capture coordinate system. The mean errors in UAV position and UAV angular orientation is given in table 3.

| α | ω | κ |
|---|---|---|
| 0.0232 | -0.08392 | 0.03281 |
| **X**, mm | **Y**, mm | **Z**, mm |
| -5.2618 | 8.0955 | 27.2022596 |

Table 4. The mean errors in UAV position and UAV angular orientation

The results of verifying the UAV self-orientation by video from frontal camera demonstrated a high potential for vision-based techniques for UAV navigation.

### 4.3 Skied-steered robot dynamic model identification

In system identification problem it is important to have accurate data about system output on given input. In case of skied-wheel robot it is needed to register output velocity and angular orientation of the vehicle and this data has to be synchronized with input commands. For obtaining the required information accelerometers could be used but the accuracy of state vector components is not enough and synchronized problem requires some additional hardware.

The developed Mosca system was used for synchronized output registering during model identification of Hercules skied-wheel robot. The accuracy and sample rate of the motion capture system are adequate to the task of dynamic model identification. Special program block for synchronization of the robot input commands and capturing frames was developed and implemented.

For registering the robot moving during the experiment a set of circular targets located on the robot upper deck. The central target (#8) defines the centre of the robot coordinate system.

Targets #6 and #10 defines X axis of the robot coordinate system, targets #3 and #13 defines Y axis. These points were used for calculation of output parameters $v_x$ and θ.

Figure 8. Hercules skied-steered robot with targets

To estimate the full dynamics model of Hercules robot a white noise command sequence was generated. Resulting motion was recorded using motion capture system. The input signal used for the experiment is given by:

$$U(t) = \begin{bmatrix} a_m \cdot rand(t) \\ d_m \cdot rand(t) \end{bmatrix}$$

Here $a_m$ – is maximum average PWM value, $d_m$ – is maximum command difference, rand(t) – random number uniformly distributed over an interval [-1, 1]. The resulting command sequence, longitudinal speed $v_x$ and rotational speed θ captured using the motion capture system for an experiment with $a_m = 30$ PWM and $d_m = 60$ are shown on figure 9.

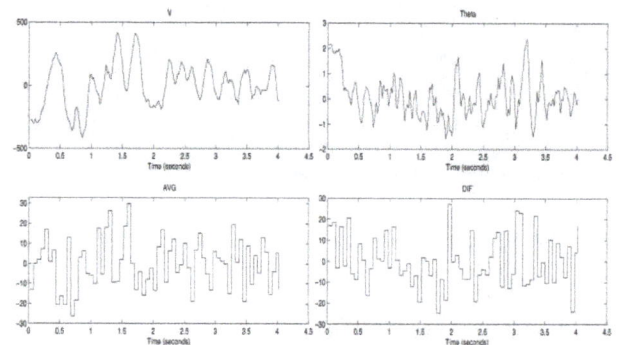

Figure 9. Output $v_x$, θ and input commands used for white noise signal experiment

## 5. CONCLUSION

The photogrammetric scalable motion capture system Mosca is developed. The Mosca system uses from 2 to 4 technical vision cameras for video sequences of object motion acquisition. It can be easily modified for different working areas from 100 mm to 10 m. The original algorithms for object point detecting, identifying and tracking is developed which provides high level of automation for video sequence processing.

Some results of using the Mosca photogrammetric system in various fields of applications such as biomechanics, dynamical model identification, self-orientation accuracy estimation are presented and discussed. The application results show high accuracy and high reliability of the developed photogrammetric system for 3D motion capture.

## ACKNOWLEDGEMENTS

The work was performed with the support by Grant of RF President for Leading Science schools of Russia (НШ-3477.2014.8) and by Grant 15-08-99580a of Russian Foundation for Basic Research (RFBR).

## REFERENCES

Knyaz V.A., Sibiryakov A.V., 1998. The Development of New Coded Targets for Automated Point Identification and Non-contact 3D Surface Measurements.   In: *The International Archives of the Photogrammetry, Remote Sensing and Spatial Information Sciences*, Hakodate, Japan, Vol. XXXII, part 5, pp. 80-85.

Knyaz V.A., 2002. Method for on-line calibration for automobile obstacle detection system. In: *International Archives of Photogrammetry and Remote Sensing*, Proceedings of ISPRS Commission V Symposium "CLOSE-RANGE IMAGING, LONG-RANGE VISION", Vol. XXXIV, part 5, Commission V, September 2-6, Corfu, Greece, pp. 48-53

Moeslund, T. B., Hilton, A., and Kruger, V., 2006. A survey of advances in vision-based human motion capture and analysis. In: *Computer Vision and Image Understanding* 104, pp. 90–126.

Yabukami S., Kikuchi H., Yamaguchi M., Arai K. I., Takahashi K., A. Itagaki, Wako N., 2000. Motion capture system of magnetic markers using three-axial magnetic field sensor. In: *IEEE Transactions on Magnetics*, Vol: 36, Issue: 5, pp: 3646-3648. 2000.

# 6

# CT ANATOMY OF BUCCAL FAT PAD AND ITS ROLE IN VOLUMETRIC ALTERATIONS OF FACE

R. A. Guryanov[a], A. S. Guryanov[b]

[a] Sechenov's First Moscow State Medical University, Moscow, Russia – robert.gurianov@gmail.com
[b] MEDLAZ Clinic, Moscow, Russia – a_gurianov@mail.ru

**Commission V, WG V/5**

**KEY WORDS:** Buccal Fat Pad Anatomy, Photogrammetry, Computed Tomography, Medical Visualisation

**ABSTRACT:**

The aim of our study is the revision of the anatomy of buccal fat pad and its role in a volumetric pattern of face. Bichat fat pad is a fatty anatomical structure with body and numerous process enclosed between the bony and muscular structures in temporal, pterygopalatine fossae and extents to the cheek area. Nevertheless, the opinion about its structure and role in forming of volume pattern of face sometimes could be controversial. The Bichat fat pad consists on predominately hormone insensitive fat tissue with underdeveloped stroma, this leads to the stability of the fat pad volume and lesser radiodensity in contrast to the subcutaneous fat. Moreover, the buccal fat pad is delimited from the subcutaneous fat of cheek area by the strong capsule. This feature allows us to use CT to divide the Bichat fat pad from the surrounding tissues. The thorough embryological data provide the distinction of Bichat fat pad from the subcutaneous fat of cheek area even at the stage of development. On the other hand, the border between the masticatory muscles and the processes of the fat pad is not evident and resembles cellular spaces in the other anatomical areas. To elicit the role of the buccal fat pad in volume pattern of face and its function we have performed the several experiments, analyzed the postoperative results after Bichat fat pad resection using surface scanner and CT data. At first, we have performed the gravity test: the patient's face photogrammetry scanning in horizontal and vertical position of head and it revealed the excess of volume in temporal area in horizontal position. To exclude mechanism of overflowing of the skin and subcutaneous fat over the zygomatic arch we have placed the markers on the skin surface at the different areas of face including the projection of ligaments and found out that the migration of soft tissue over the zygomatic arch is about 3-5 mm and almost the same in temporal area. However, the acquired result was unsatisfying because cannot exclude completely the migration of superficial tissues. In following experiments it was shown that the intensive pressure on the cheek area in vertical position produce the volume excess in the temporal area similar and more exaggerate than in gravity test. To correlate the excess of tissue with underlying anatomical structures we had acquired the CT's of some probationers, performed 3D reconstruction of bony structures, Bichat fat pad, and aligned with the previous surface scans. The projection of this excess in both experiments corresponds with the temporal process of Bichat fat pad. That means that the leading mechanism of these changes is protrusion of temporal process of Bichat fat pad through the leaves of temporal fascia due to pressure on the buccal extension: in these conditions, the buccal fat pad works as a communicating vessel between the cheek area and temporal fossa. This fact has suggested us that the phenomenon of the deepening of temporal area during the ageing could be produced as by the atrophy of buccal fat pad as by the migration of the fat pad to the cheek area due to ptosis.

## 1. INTRODUCTION

The volumetric appearance of the human face is initially defined by the bony structures that serve as the rigid frame for the soft tissues. The soft tissues of the face is a complex system of subcutaneous fat, SMAS, mimic muscles ligaments also includes the deep fat compartment of face is known as the buccal [Bichat] fat pad. The main feature of the soft tissues is their higher inconstancy, mobility in comparison with bones so the former are vastly subjected to adaptive and age related changes. Bachat fat pad (BFP) consists of a main part – the body, and several extensions arising from it: masseteric extension, preigomandibular extension, orbital extension, pterygopalatine extension, temporal extension (Kahn, J.L., Wolfram-Gabel, R. & Bourjat, P., 2000, Yousuf, S. et al., 2010). The extension`s anatomy is spatially complex, because they fill the space between the masticatory muscles (m. masseter, m. pterygoideus medialis et lateralis, m. temporalis). Nowadays, the distinct physiological role of Bichat fat pad is not clearly defined, surgeons consider the fat pad as the source of plastic material and as a body-to-resect in cases when patient unsatisfied by the cheek plumpness, also the buccal fat pad regarded by some surgeons as the morphological substance for jowls formation (Zhang, H.-M. et al., 2002, Yousuf, S. et al., 2010). Nevertheless, the volumetric role of Bichat fat pad is significant and takes sufficient part in forming the midface appearance. The aim of our work is to revise the anatomy of Bichat fat pad and to clarify its role in the face volumetric appearance using such noninvasive visualization technics as the CT reconstruction and photogrammetry.

## 2. MATERIALS AND METHODS

### 2.1 Photogrammetry study

Photogrammetry allows us to analyze the changes in volumetric pattern of face under the variety of factors as the gravitation or premediated mechanical impact. This factors open up the possibility to understand the general regularities of the soft tissue migration and correlate our findings with the anatomical and clinical observations.

### 2.1.2 Gravity test

The main force that effects constantly on the soft tissues of face is gravity. We have developed the test that could cast the light on the mechanisms of aging and ptosis, and reveals the dynamic of volumes migration, the zones of adhesion of soft tissues to the bony framework. We have called that test the gravity test. Using photogrammetry the scanning of probationers face is performed in vertical and horizontal positions. The acquired scans are aligned regarding the reference points – the points or surfaces in which volume changing and migration of tissues is predictably insignificant due to reliable fixation to the bones.

Then on the aligned models, we compute the geodesic map (distance map) of difference between shapes of face in vertical in horizontal position, so the migration of volume reveals. We have analyzed the pattern of volumetric changes in 20 of different age groups, including 7 congener pairs (mother-daughter). In addition, we perform the modification of test (5 cases) in which the marks are placed on the surface of skin, thus the migration of skin surface can be observed. This modification of gravity test allows us to estimate the mobility of skin surface regarding the bony structures, hence in immovable areas correlate with ligaments between the derma and bones and the migration of soft tissues under the force of gravitation in that area is confined. In the point of maximal protrusion, we calculate the distance between the aligned models what permits us to access the grade of protrusion. In view of that the elasticity of skin is able to effect on the grade of protrusion we have used the indirect parameter – relative extension coefficient to estimate he elasticity of the skin.

### 2.1.2 Cheek-pressing test

The temporal fossa and the cheek area communicates with the cheek area through the masticatory space enclosed between the masseter laterally. Cheek pressing test is performed in vertical position, the probationer should push the cheek in projection of buccal extension of Bichat fat pad by his own finger, and the photogrammetry scans are taking at this moment. The scan is aligned with the scan in vertical position and the geodesic map is computing. The acquired geodesic map is compared with that of gravity test. This test imitates the mechanism of soft tissues' pressure on the buccal fat pad.

### 2.1.3 Clench test

When we clench our teeth, the masticatory muscles increase their volume due to contraction and gain of blood inflow. The similar effect occurs during the act of mastication. Thus, the clench test shows how the muscles effect on the volumetric pattern of human face, and how the mastication effect on the Bichat fat pad volume migration.

### 2.2 CT reconstruction

The method of 3D reconstruction has been used to access the projection of the buccal fat pad on the surface of face and to elicit the topography of fat pad. The separation of Bichat fat pad in can arouse some complexities: first, we have to mind that in a cheek area the buccal fat pad has a strong capsule that delimit the fat pad from the subcutaneous fat; second, in the temporal area the fat pad should be distinguished from the interfascial fat between the two leaves of temporal fascia (Cho, K.H. et al., 2013). The fascia's and BFC (-69 to -49 HU) is much dense than a fatty tissue consisting predominately of adipocytes (-179 to – 124 HU) in radiodensity scale. This fact allows us to perform the virtual dissection of the buccal fat pad. We had analysed the series of 10 patient's CTs of head, including two patients after BFP resection. The examination of BFP has being performed on the multiplane and 3D reconstructions: the

capsule of the Bichat fat pad, topographic anatomy features, position of processes were accessed.

## 3. RESULTS

The gravity test has shown the significant increase of volume in the temporal area. The mean value of increase was 2.35 mm, and can be observed visually. The specific pattern of volumetric changes in the temporal area has been observed (fig.1). The area with volume excess has the evident limit below over the zygomatic arch. The footprint of the protrusion varies in individuals, but never fill the whole temporal fossa and never appears at the areas where we can confidently palpate the temporalis muscle. This phenomenon has stimulate us to find the anatomical base for it. We have performed the CTs of our patients and have revised the anatomy of buccal fat pad on multiplane reconstruction and 3D reconstruction. In the cheek area the buccal fat pad has a considerable capsule, which allows us to divide it from the subcutaneous fat. On the other hand, in temporal area there is no distinct border between the masticatory muscles and BFP it and the latter is directly adjacent to the former with no distinguishable capsule, but laterally it is strongly delineated by the deep leaf of temporal fascia from interfascial fat compartment enclosed between the superficial and profound leaves of temporal fascia.

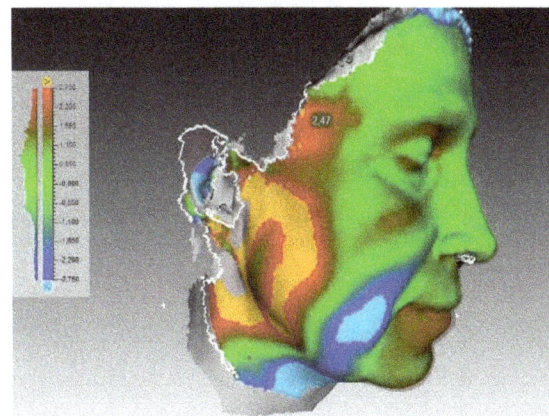

Figure 1. The gravity test. Geodesic measurements. Green – no volume changes. Blue – diminishing of volume. Red – augmentation of volume. The isolated excess of volume in temporal area marked with a surface distance value.

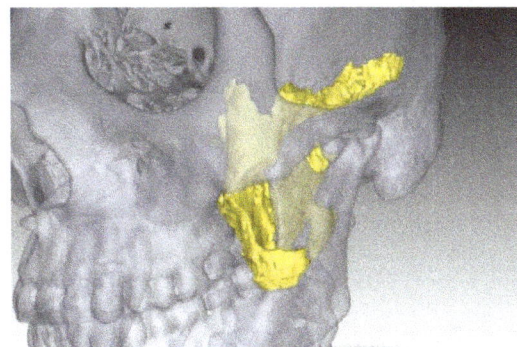

Figure 2. Reconstruction of the buccal fat pad. The connection between the buccal process and temporalis process is shown.

3D reconstruction of BFP allowed us to estimate the spatial orientation of the processes and understand how the BFP projects on the skin surface. The buccal process is connected with the temporal process via the body of buccal fat pad. (fig. 2)The area of protrusion in the temporal area corresponds with the projection of temporal process of BFP on the skin surface (fig.3). This permits us to conclude that the protrusion in the temporal area at gravity test appears due to the migration of Bichat fat pad and the pressure of overlying midface tissues on it. Nevertheless, we presumed that the increase of volume in temporal area can be conditioned by two mechanisms the migration of the tissues over or under the zygomatic arch.

To exclude mechanism of overflowing of the skin and subcutaneous fat over the zygomatic arch we have placed the markers on the skin surface at the different areas of face including the projection of ligaments and found out that the migration of soft tissue over the zygomatic arch an in projection of malar bone is about 3-5 mm and almost the same in temporal area (fig. 4). However, the acquired result was unsatisfying because cannot exclude completely the migration of superficial tissues (skin, subcutaneous fat), but it strongly demonstrate us that the migration over that structures is limited as the volumetric changes.

At the gravity test in the congener pairs, it has become obvious that the area of protrusion in temporal area definitely expands in age. This can occur due to the diminishing of elasticity of the soft tissue during the ageing.

The cheek-pressing test revealed that the pressing in a projection of body of the Bichat fat pad can produce the protrusion in the temporal area (fig. 5), less prominent though (approximately 58% of temporal excess in gravity test). This fact confirm our supposition that the pressure applied to the buccal part of the BFP's body is able to cause the protrusion in the temporal area.

At the clench test the protrusion also appears in the temporal area, but also the increase of volume is observed at the beginning of temporalis muscle, where the buccal fat pad is absent, and at the contour of masseter muscle. We suppose that the part of the protrusion just above the zygomatic arch concerns to the buccal fat pad.

## 4. DISCUSSION

This study is trying to explain the volumetric alteration of face under the action of different forces. It is known that the atrophy of buccal fat pads develops during the aging, but being hormone insensitive the Bichat fat pad volume do not depends of grade of obesity, thus even on the cachectic patients it saves its volume constant, in contrast to subcutaneous fat (Loukas, M. et al., 2006). In embryogenesis, the BFP also develops separately from the other fatty structures: the capsule of BFP forms first and only after that, the differentiation of the fatty tissue begins (Cho, K.H. et al., 2013). This suggests that the BFP has its own individual function (because the function is always connected with morphology) which is not clearly defined nowadays (Yousuf, S. et al., 2010).

It is noticeable that the excess of volume in temporal area at the accomplished experiments has a mediated mechanism. At first the force of gravity acts on the soft tissues, which effects on the buccal process and the body of Bichat fat pad. Bichat fat pad conducts the pressure to the deep leaf of temporal fascia, interfascial fat, superior leaf of temporal fascia and, finally, the skin. The fatty tissue of BFP, having lack of stroma, consisting of lipids and being incompressible, playing the role of volumetric buffer of temporal, infratemporal areas, thus the buccal fat pad works as a communicating vessel between the cheek area and temporal fossa. That means that at the act of mastication the muscles would not be restrained in their compartments during the chewing. The thick fascial leaves of temporalis fascia limits the temporal fossa, so the temporalis muscle is enclosed in a kind of hard compartment, limited laterally by the temporalis fascia, medially by the synergic pterygoid muscles. With no presence of BFP the masticatory muscles due to ingrowing inflow and function would be squeezed in their receptacles arousing pain and subsequent inability to chew.

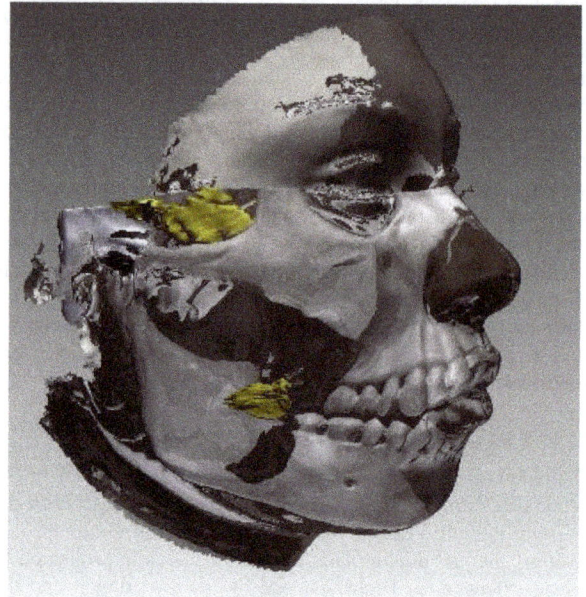

Figure 3. The temporal process of buccal fat pad: temporal and buccal processes.

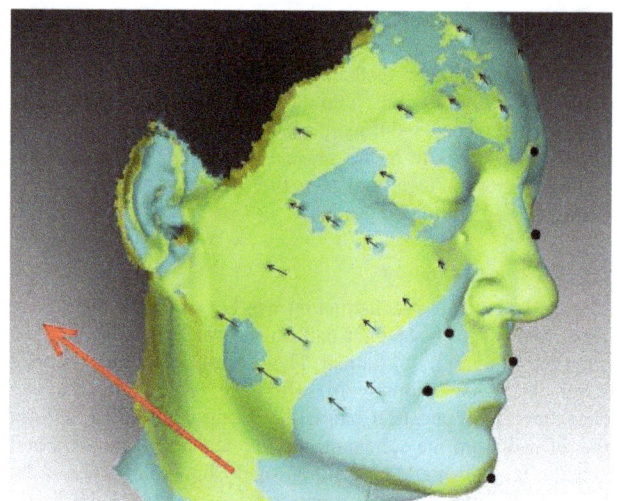

Figure 4. The variation of gravity test with marks. The little migration over the zygomatic arch is present.

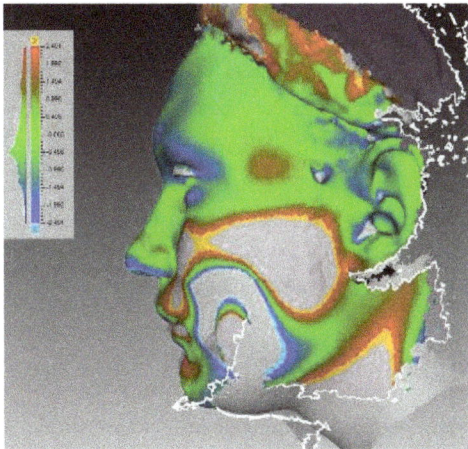

Figure 5. Pressing test. The protrusion over the zygomatic arch.

Computed tomography reconstructions demonstrates the connection between the temporal and cheek areas by the BFP. The temporal, infraorbital, pterygopalatine processes of buccal fat pad connect the deep areas of face with the superficially lying cheek area, so the septic process can spread along this fatty tract and involve the structures as the nerves, venous plexuses, arteries and can spread on the other adjacent structures.

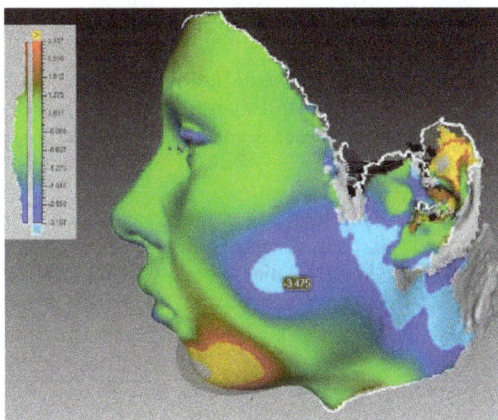

Figure 6. Geodesic map after the buccal fat pad resection and augmentation mentoplasty. The decrease of volume in the check area in front of anterior margin of masseter muscle.

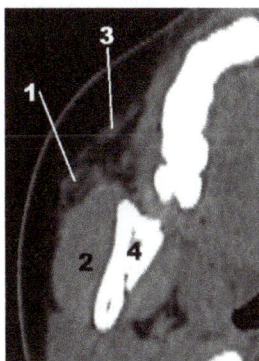

Figure 7. BFP after resection. 1 - fibrosis in the projection of BFP; 2 – masseter muscle; 3 – parotid duct; 4 – ramus mandibuli

From the view of plastic surgery, the Bichat fat pad and its size plays the significant role in the appearance of face. The resection of fat pad decrease the volume in the specific area at the front margin of the masseter muscle, but surprisingly not affecting the temporal area (fig. 6). The only fact that can explain this is that on the CT's the strong fibrosis (fig.7) of the resected buccal part was observed, so it hinders the temporal and the latter parts to prolapse into the freed space. On the other hand, we know that the Bichat fat pad has a developed capsule only in the cheek area, the part that is enclosed between the muscles has loose septs connected directly to the epimiseum of masticatory muscles (Kahn, J.L., Wolfram-Gabel, R. & Bourjat, P., 2000), and they could prevent the alteration in the volumetric appearance of the temporal area.

The ptosis of the soft tissues appears due to continuous action of gravitation during the aging. Decreasing elasticity of the connective tissue cannot prevent the migration of the soft tissues downwards. Many reports that the volume of the buccal fat pad and another fatty tissue of face decreases in age (Gierloff, M. et al., 2012, Loukas, M. et al., 2006, Xiao, H., M. Bayramiçli, I.T. Jackson, 1999), but not takes into account that the enhancement migration ability may be caused by the decreasing elasticity of the soft tissue compartment, including the compartment of the whole body – the skin. The expanding of protrusion area in age may be connected with decreased elasticity of connective tissue strictures, thus the deepening in the temporal area has the mechanism inverse to that in gravity test.

## REFERENCES

Cho, K.H. et al., 2013. Deep fat of the face revisited. *Clinical Anatomy*, 26(3), pp.347–356.

Gierloff, M. et al., 2012. Aging Changes of the Midfacial Fat Compartments. *Plastic and Reconstructive Surgery*, 129(1), pp.263–273.

Kahn, J.L., Wolfram-Gabel, R. & Bourjat, P., 2000. Anatomy and imaging of the deep fat of the face. *Clinical Anatomy*, 13(5), pp.373–382.

Loukas, M. et al., 2006. Gross anatomical, CT and MRI analyses of the buccal fat pad with special emphasis on volumetric variations. *Surgical and Radiologic Anatomy*, 28(3), pp.254–260.

Xiao, H., M. Bayramiçli, I.T. Jackson, 1999. Volumetric analysis of the buccal fat pad Eur J Plas Surg 22:177-181

Yousuf, S. et al., 2010. A review of the gross anatomy, functions, pathology, and clinical uses of the buccal fat pad. *Surgical and Radiologic Anatomy*, 32(5), pp.427–436.

Zhang, H.-M. et al., 2002. Anatomical structure of the buccal fat pad and its clinical adaptations. *Plastic and reconstructive surgery*, 109(7), pp.2509–2518; discussion 2519–2520.

# 7

# APPLICATIONS OF PANORAMIC IMAGES: FROM 720° PANORAMA TO INTERIOR 3D MODELS OF AUGMENTED REALITY

I-C. Lee, and F. Tsai

Center for Space and Remote Sensing Research, National Central University, Taiwan - (iclee, ftsai)@ csrsr.ncu.edu.tw

**KEY WORDS:** Panoramic Images, Close-range Photogrammetry, 3D Point Cloud, 3D Indoor Modeling

**ABSTRACT:**

A series of panoramic images are usually used to generate a 720° panorama image. Although panoramic images are typically used for establishing tour guiding systems, in this research, we demonstrate the potential of using panoramic images acquired from multiple sites to create not only 720° panorama, but also three-dimensional (3D) point clouds and 3D indoor models. Since 3D modeling is one of the goals of this research, the location of the panoramic sites needed to be carefully planned in order to maintain a robust result for close-range photogrammetry. After the images are acquired, panoramic images are processed into 720° panoramas, and these panoramas which can be used directly as panorama guiding systems or other applications.

In addition to these straightforward applications, interior orientation parameters can also be estimated while generating 720° panorama. These parameters are focal length, principle point, and lens radial distortion. The panoramic images can then be processed with close-range photogrammetry procedures to extract the exterior orientation parameters and generate 3D point clouds. In this research, VisaulSFM, a structure from motion software is used to estimate the exterior orientation, and CMVS toolkit is used to generate 3D point clouds. Next, the 3D point clouds are used as references to create building interior models. In this research, Trimble Sketchup was used to build the model, and the 3D point cloud was added to the determining of locations of building objects using plane finding procedure. In the texturing process, the panorama images are used as the data source for creating model textures. This 3D indoor model was used as an Augmented Reality model replacing a guide map or a floor plan commonly used in an on-line touring guide system.

The 3D indoor model generating procedure has been utilized in two research projects: a cultural heritage site at Kinmen, and Taipei Main Station pedestrian zone guidance and navigation system. The results presented in this paper demonstrate the potential of using panoramic images to generate 3D point clouds and 3D models. However, it is currently a manual and labor-intensive process. A research is being carried out to Increase the degree of automation of these procedures.

## 1. INTRODUCTION

### 1.1 Background

Point cloud is one of the intermediate data for mapping and three dimensional (3D) modeling. The increase of the popularity using point cloud comes after the Light Detection and Ranging (LiDAR) technology. In the field of digital archiving of architectures, LiDAR technology is widely used due to it's simple to operate machines and highly automated processing software. However, the price of the LiDAR system is pretty high compare with photogrammetry/computer vision solutions. As a result, our research team is on the quest of developing a streamline procedure for close-range photogrammetry generating 3D building models. This paper demonstrate our intermediate result, prove-of-concept, of creating point cloud from panoramic images and forming 3D model with this point cloud.

The technique used to create 3D models of architectures in the past decade is mainly LiDAR systems. LiDAR system scans the surface of the object and create point cloud. Point clouds are then processed with semi-automatic procedures to generate parametric 3D models. The procedure of acquiring point clouds is relatively streamlined, however, creating model from point cloud is not. After the 3D model is created, photos acquired on site are needed for the texturing process. Since the photos are acquired randomly without a standard operation procedure (SOP), the orientation of the photos are unknown, and spots of the building might be missing. These issues resulted in a painstaking manual operated texturing process. Even with systems having camera mounted directly on top of LiDAR scanner, the photo

acquired by the camera is not a good candidate to use for texturing due to the long distance between LiDAR system and the target result in low resolution of the texturing photo. Hence, a set of photos acquired individually for model texturing is needed.

Photogrammetry technique is another way of extracting 3D information to create 3D models. The reason photogrammetry technology loses its popularity in close-range mapping to LiDAR system is because the high cost of human labor and low degree of automatic process. Efforts made by researchers around the world regarding close-range photogrammetry have significantly improve the degree of automation recently. However, there are still processes cannot be automated due to the in-situ conditions. One possible solution to minimize these unknown conditions is the use of panorama. In this research, we investigate the possibility to use panoramic images as data source of close-range photogrammetry, develop a procedure to process these panoramic images, create point cloud from this data source and finally create indoor 3D building models.

Point cloud is not the only way of creating 3D models. It is because the popularity of LiDAR systems, which creates point clouds, encourage researchers to develop point cloud processing and modeling software. From the photogrammetry point of view, line or plane features can be extracted directly from the images. Hence point cloud is not a data source in necessity for 3D modeling. Since the goal of this research is prove-of-concept for the use of panoramic images, we will stick with the mainstream procedure for generating 3D models. Which is creating point cloud from the panoramic images, and generating 3D models from this point cloud.

## 1.2 Terrestrial LiDAR vs. Close-range Photogrammetry

Although both LiDAR and photogrammetry systems can create point clouds, the mechanism is entirely different. LiDAR system measure distance accurately and robustly within its measuring range. However, the measurement of scan angles are relatively inaccurate. On the other hand, photogrammetry uses the angles measured from images of the same point to intersect the rays in the object space to calculate the coordinate. As a result, they are having totally different error models. Generally speaking, the accuracy of LiDAR point cloud only related to the LiDAR system itself and the distance between the laser source and the target. On the contrary, point cloud created by photogrammetry method is highly dependent on the intersection condition of the photos. The distribution of LiDAR points is a result of scanning process. The points are uniformly spaced along a sphere from the perspective of the system center. Point cloud generated from photogrammetry uses dense matching technique, it creates matched corresponding points between images pixel by pixel. In texture rich areas, the matching result is accurate and robust, however, in areas where textures are lacking, it is less accurate and the results are noisy.

Generally speaking, the point cloud acquired by close-range photogrammetry technology is noisy and less robust than the point cloud extracted by LiDAR system. Under this circumstances, most of the software creating 3D models from point clouds does not work well with photogrammetry point cloud. For instance, Pointools plug-in for Sketup is one of the software we have been using by or laboratory for reconstruct 3D models from LiDAR point cloud in the past, because the noisiness of the point cloud created by photogrammetry technique, it is difficult to lock on to the correct corner point. This example shows that these LiDAR point cloud processing software need to be improved specifically for photogrammetry point cloud in order to streamline the process flow and achieve the best accuracy of 3D models.

## 1.3 Panoramic Images

A set of panoramic images are the data source to form a panorama. Within a panorama, the panoramic images cover the zenith and nadir is usually called a 720° panorama. While acquiring panoramic images, in order to maintain a gap-free panorama, projection center of all images have to be stationary in theory. However, under this circumstances, the depth of a feature point in a set of panoramic images cannot be acquired. Multiple panorama sites are needed to acquire the depth information. In this research, we used the panoramic images acquired for tour guiding system. There are multiple sites within an indoor environment, however, the determining of the location of panorama sites is purely based on the aspect of exhibition. Without considering the image network or intersection condition.

## 2. METHODOLOGY

## 2.1 Procedure

First step of the procedure is the acquisition of panoramic images. Second, stitching panoramic images into panoramas, to retrieve the interior orientation parameters. Next, panoramic images been processed with aerial triangulation procedure. Then, dense matching is performed to create point cloud. Finally, process for 3D building model creating and texturing is performed.

## 2.2 Data Acquisition

Multiple hardware devices can be used to acquire panoramic image. There are motorized and manual panorama heads from varies of manufactures. In this research, Gigapan, a motorized panorama head is used to capture the images. Depend on different camera lens used, adjustment is needed for Gigapan in order to keep the projection center stationary while the panorama head is rotating. The number of images for a panorama site is depend on the focal length of the lens used. In this research, a 45mm lens is used with a full frame Digital Single-Lens Reflex (DSLR) camera. Under this circumstances, 96 photos are needed to establish a panorama (without nadir). Since the images will be use as textures of 3D models, High Dynamic Range (HDR) mode is used for texture rich representation. In HDR mode, three images are taken instead of one, result in 288 images acquired for each panorama site.

## 2.3 Stitching of Panoramic Images

There are two purposes for this process, first, estimate the interior orientation of the camera, and second, create panorama photo for latter 3D model texturing process. In this research, image stitching is done using Kolor Autopano software. After processing, the software can provide focal length, coordinates of principal point, and parameters of lens distortion. Since no control points are needed in this process, accurate interior parameters is a must in order to create a 3D model with correct scale.

## 2.4 Aerial Triangulation

Aerial Triangulation is a process of estimating the position and attitude, namely exterior orientation parameters, of each photo for latter point cloud generation process. A structure from motion (SFM) software VisualSFM (Wu, 2011, and Wu et al., 2011) is used in this research.

## 2.5 Dense Matching

In this research, CMVS (Furukawa and Ponce, 2010) software is used to create point cloud. CMVS is integrated seamlessly with VisualSFM, and the point cloud can also be display in VisualSFM. There are multiple algorithm and software for dense matching. CMVS is capable of handling large amount of input photos and create billions of points within a project with ordinary personal computers.

## 2.6 3D Model Creating and Texturing

Pointools plugin for SketchUp is been tested to establish 3D model manually. However, after several attempt identifying a corner point with Pointools and failed, we realize that the noisiness of the point cloud created by photogrammetry technique made the digitizing process almost impossible. In order to overcome the issue of noisy point cloud, RANdom SAmple Consensus (RANSAC, Fischler and Bolles, 1981) algorithm is used to find planes. After the plane with majority of points is been selected, all the points belong to this plan is removed and reprocess the point cloud with RANSAC, until all major planes are been identified. Next, the intersecting line of connected planes, for example ground surface and wall surface, is calculated. Then, these lines are drawn manually in Sketchup. With manual editing of the lines into surfaces, the 3D building model is established.

Texturing process uses images from panorama as data source. When texturing ground surface, panorama is projected with planer transformation and select the projection axial as nadir (Figure 1). Other surfaces are processed in a similar fashion, by projecting the panorama with a planer transformation, with the projection axial parallel to the normal of the texturing surface. Then, these projected panorama images are use as texture images in Sketchup.

Figure 1. Example of ground surface used as textures.

## 3. EXPERIMENT

This 3D building modeling and texturing process is been tested at a culture heritage site at Kinmen called Tsai family ancestral shrine. It is a traditional southern Fukien styled architecture. All the original photographs are selected to perform the panorama stitching process, and the interior orientation parameters are calculated. Then, the original photograph with Exposure Value (EV) of zero is selected to perform the aerial triangulation process by using the previously calculated parameters as initial value. The dense matching process is performed afterwards. Since traditional southern Fukien architecture is more complicated than modern concrete building, the 3D building modeling process is trickier than we have discussed previously.

There are three levels of ground surfaces connected with stairs, and two gable roof main structure connected with two gable roofed corridor. After 20 surfaces been identified by RANSAC algorithm, we have determine the seven major planes to construct the building. Three ground level surfaces, two walls and two surfaces belong to the gable roof of the main court room. Small surfaces such as a step within stairs is not identified. To accurately determine the location of the break line of a step, points belong to a ground level is compressed into two dimension along the normal vector of the surface and rasterized to create an image (Figure 2). Then this image is textured on to the 3D model (Figure 3) and the edges of the point cloud image is the break line of the stairs. This technique can be used to determine the location of the object that can be seen from point cloud but cannot retrieve the plane equation by RANSAC. Similar process is performed and pillars, furniture, and decoration items can be located.

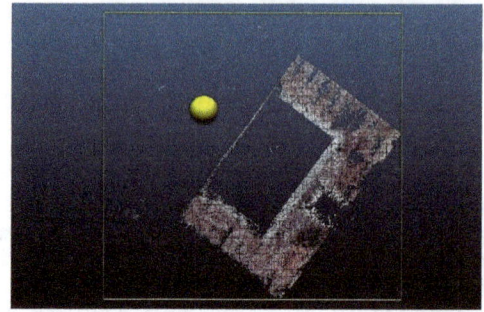

Figure 2. Point cloud rasterized image of one of the ground levels.

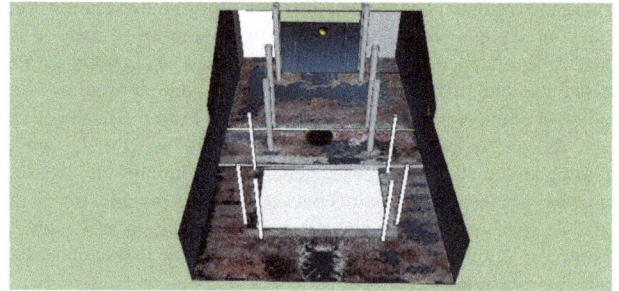

Figure 3. Point cloud rasterized image used as texture to determine the break line of stairs and edge of stage.

Finally, the panorama images are used as model textures. However, pillars may create a deformed column on the texturing image, and the void of the nadir of the panorama needed to be fixed. These areas needed to be patched with another panorama image from different site. The 3D building model of Tsai family ancestral shrine constructed is shown in Figure 4.

Figure 4. 3D building model constructed by proposed procedure.

## 4. CONCLUSION

Please keep in mind that this research is a prove-of-concept for using panoramic images to create 3D building model. All of the processes are done semi-automatically or manually, and the details are not attended. The preliminary result shows that the 3D building model can be created using this procedure, and panorama image can be used as texture as well. In the future, we will be investigating the accuracy of the created model and the possibility of automate the procedure of 3D building modeling from panoramic images.

### References

Fischler, M. and Bolles, R., 1981. Random sample consensus: a paradigm for model fitting with applications to image analysis and automated cartography. *Commun. ACM* 24(6), pp. 381-395.

Furukawa, Y. and Ponce, J., 2010. Accurate, Dense, and Robust Multi-View Stereopsis. *IEEE Transactions on Pattern Analysis and Machine Intelligence*, Vol. 32, Issue 8, pp. 1362-1376.

Wu, C., 2011. "VisualSFM: A Visual Structure from Motion System", http://ccwu.me/vsfm/ (1 March 2015)

Wu, C., Agarwal, S., Curless, B., and Seitz S. M., 2011. Multicore Bundle Adjustment, *IEEE Conference on Computer Vision and Pattern Recognition (CVPR)*, Providence, RI, USA, pp. 3057- 3064.

# 8

# DEVELOPMENT OF INTEGRATION AND ADJUSTMENT METHOD FOR SEQUENTIAL RANGE IMAGES

K. Nagara [a], T. Fuse [a]

[a] Dept. of Civil Engineering, University of Tokyo, 7-3-1 Hongo Bunkyo Tokyo 113-8656, Japan
nagara@trip.t.u-tokyo.ac.jp

**Commission V, WG V/4**

**KEY WORDS:** 3D data, Range image, Point cloud, ICP algorithm, Self-calibration bundle adjustment

**ABSTRACT:**

With increasing widespread use of three-dimensional data, the demand for simplified data acquisition is also increasing. The range camera, which is a simplified sensor, can acquire a dense-range image in a single shot; however, its measuring coverage is narrow and its measuring accuracy is limited. The former drawback had be overcome by registering sequential range images. This method, however, assumes that the point cloud is error-free. In this paper, we develop an integration method for sequential range images with error adjustment of the point cloud. The proposed method consists of ICP (Iterative Closest Point) algorithm and self-calibration bundle adjustment. The ICP algorithm is considered an initial specification for the bundle adjustment. By applying the bundle adjustment, coordinates of the point cloud are modified and the camera poses are updated. Through experimentation on real data, the efficiency of the proposed method has been confirmed.

## 1. INTRODUCTION

With increasing widespread use of three-dimensional data, the demand for simplified data acquisition is also increasing. Even so far, three-dimensional data has been acquired by image processing (Photogrammetry) or laser range scanners. Image processing is an easy way for end users because users only have to take photos of the object. However, it is difficult to acquire a dense-range image in general. On the other hand, laser range scanners can acquire dense and high precision range data. However, only a few people use it because it is expensive and time-consuming. Therefore, a data acquisition method that is easy to use and can acquire dense and high precision range data is desired. Recently, the range camera, which is a simplified sensor, is widely noticed (Remondino and Stoppa, 2013). It is the camera that can obtain range images, and is much more inexpensive than laser range scanners. Its inner structure is the almost same as original cameras, and it is equipped with some infrared LEDs and the special image sensor. It measures distance by a time-of-flight method (Mutto *et al.*, 2012). Since each element of the image sensor measures distance independently, the range camera can acquire a dense-range image in a single shot. Although it is usually used for object recognition and estimation of human pose, it can be used for three-dimensional data acquisition. However, the range camera has two drawbacks: its measuring coverage is narrow and its measuring accuracy is limited. The former drawback can be overcome by integrating sequential range images, and the latter drawback can be probably overcome by adjusting the error of point clouds. Our goal is to develop a method for integrating and adjusting sequential range images.

## 2. METHOD

### 2.1 Overview

This method integrates and adjusts range images taken by a TOF (Time-of-flight) range camera. Figure 1 shows the overview of the proposed method. The number in the parenthesis is corresponding section in this paper. Firstly, point clouds are generated from range images, and registered initially to be integrated. ICP algorithm (Besl and McKay, 1991) is employed to register. ICP algorithm assumes that the point cloud is error-free. However, the measurement accuracy of the range camera is not high. Therefore, we adjust the coordinate values of the point clouds. The coordinate values are assumed to be unknown parameters, and adjusted by bundle adjustment (Luhmann *et al.*, 2014; Triggs *et al.*, 2000). It re-estimates the coordinate values and camera poses. Self-calibration bundle adjustment is employed because the lens distortion of range cameras cannot be ignored (Lichti et al., 2010).

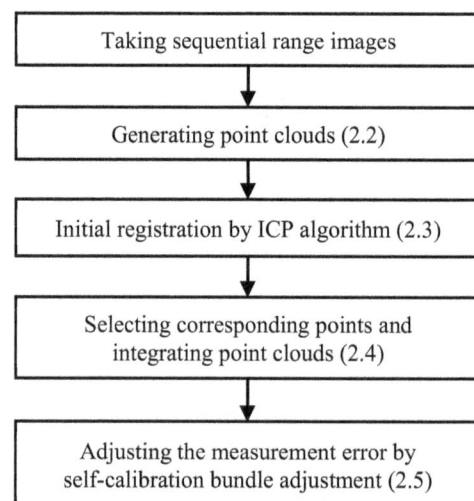

Figure 1. The overview of the proposed method

### 2.2 Generating point clouds

Range image is an image that each element has distance data. 3-d coordinate values of each element can be calculated by using this distance data (Eq. 1). Point clouds are generated as the set of these 3-d coordinate values.

$$X = \frac{d \cdot x}{\sqrt{x^2 + y^2 + f^2}}$$

$$Y = \frac{d \cdot y}{\sqrt{x^2 + y^2 + f^2}} \qquad (1)$$

$$Z = \frac{d \cdot f}{\sqrt{x^2 + y^2 + f^2}}$$

where   $(X, Y, Z)$ = 3-d coordinate in the camera coordinate
$(x, y)$ = 2-d image coordinate
$d$ = distance
$f$ = focal length

$f$ can be acquired by camera calibration. We employed Zhang's method (Zhang, 2000) for camera calibration by taking shots of the checkerboard and using their intensity images. However, since regular range cameras don't have much picture elements and the measurement accuracy is not high, the focal length acquired here is used as default value, and will be re-estimated by self-calibration bundle adjustment.

## 2.3   Initial registration by ICP algorithm

Each point cloud is registered to a common coordinate by ICP algorithm. This algorithm is used to register two point clouds and estimate orientation and position with respect to one of the cameras. Firstly, each pair is registered and its transformation is estimated. Finally, all point clouds are registered to the first point cloud. This means they are registered to the common coordinate (Figure 2).

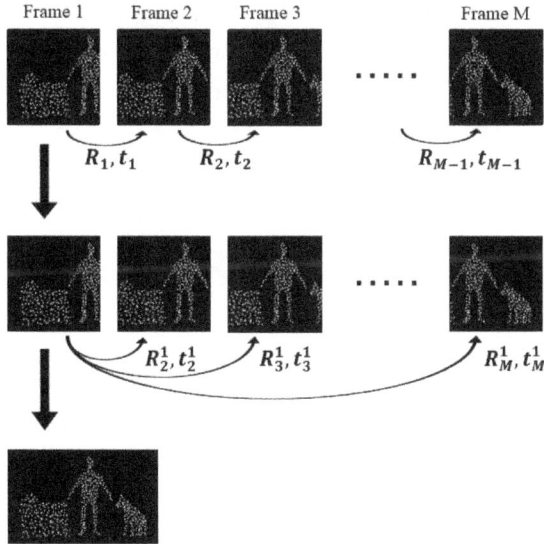

Figure 2. Registration to the first frame

Let $\vec{p_k}$ be the point cloud of the frame $k$. $\vec{p_k}$ is the list of (x,y,z) of the points in the $k$-th camera's coordinate. If the total number of frames is $M$, transformations to be estimated are $(\boldsymbol{R_1}, \boldsymbol{t_1}), ..., (\boldsymbol{R_{M-1}}, \boldsymbol{t_{M-1}})$. All points clouds are registered to the first point cloud by using these transformations. For example, in the case of $\vec{p_k}$, firstly it is registered to the frame $k$-1 by using $(\boldsymbol{R_{k-1}}, \boldsymbol{t_{k-1}})$. Let $\vec{p_k}^{k-1}$ be the registered point cloud, it is shown as Eq. 2.

$$\vec{p_k}^{k-1} = \mathbf{R}_{k-1}\vec{p_k} + \mathbf{t}_{k-1} \qquad (2)$$

Secondly, it is registered to the frame $k$-2 by using $(\boldsymbol{R_{k-1}}, \boldsymbol{t_{k-1}})$. Let $\vec{p_k}^{k-2}$ be the registered point cloud, it is shown as Eq. 3.

$$\vec{p_k}^{k-2} = \mathbf{R}_{k-2}(\mathbf{R}_{k-1}\vec{p_k} + \mathbf{t}_{k-1}) + \mathbf{t}_{k-2} \qquad (3)$$

By repeating the same process, $\vec{p_k}$ is registered to the point cloud of the frame $1$ finally. All point clouds are registered to the common coordinate by applying these processes to the all point clouds.

In addition, each transformation from the frame $1$ to the frame $k$ ($k$=2,...,$M$) is needed to apply bundle adjustment. Let $\boldsymbol{R_k^1}$ and $\boldsymbol{t_k^1}$ be the transformation from the frame $1$ to the frame $k$, they are shown as Eq.4.

$$\boldsymbol{R_k^1} = \boldsymbol{R_{k-1}^1}\mathbf{R}_k$$
$$\boldsymbol{t_k^1} = \boldsymbol{R_{k-1}^1}\mathbf{t}_k + \mathbf{t}_{k-1}^1 \qquad (4)$$

Since $\boldsymbol{R_1^1} = \boldsymbol{I}$ and $\boldsymbol{t_1^1} = \boldsymbol{0}$, $\boldsymbol{R_2^1}$ and $\boldsymbol{t_2^1}$ are calculated firstly. $\boldsymbol{I}$ is 3*3 identity matrix. All $\boldsymbol{R_k^1}$ and $\boldsymbol{t_k^1}$ are also calculated recursively. These transformations are used as default values in bundle adjustment.

## 2.4   Selecting corresponding points and integrating point clouds

All point clouds are integrated by unifying corresponding points. Corresponding points between point clouds are already known by applying ICP algorithm. These points are not exactly the same point because their correspondence is based on nearest neighbor search. However, they are assumed to be the same point if the distance between the points is sufficiently small.

We made the histogram of the distance between corresponding points (Figure 3). As a result, approximately 3,000 corresponding points are selected, and the other points are discarded.

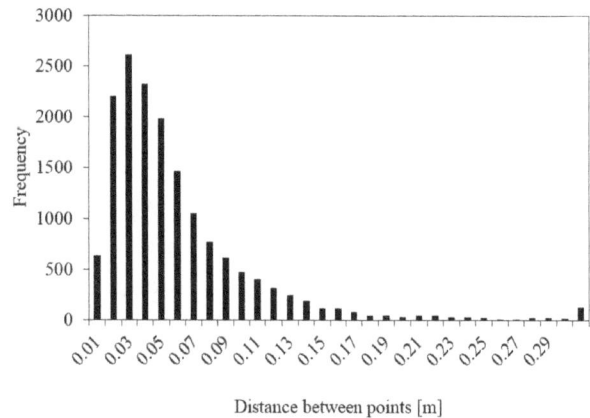

Figure 3. Histogram of the distance between corresponding points

After selecting corresponding points in each pair of frames, these points are corresponded among more than two frames. Finally, these points are unified. The coordinate values of unified points are set to the mean of values of the original points. After this section, the integrated point cloud means the set of these unified points.

## 2.5 Adjusting the measurement error by self-calibration bundle adjustment

The camera's position and orientation on each frame $(\mathbf{R}, \mathbf{t})$, the coordinate values of the integrated point cloud $(X, Y, Z)$, and the image coordinate values of the corresponding range image $(x, y)$ are already known in the sections so far. We apply self-calibration bundle adjustment by using these data.

The default parameters of inner orientation elements are defined as follows. The focal length is estimated by the camera calibration (Section 2.2). The principal point is set to the center of the image. We employ Brown's lens distortion model (Fryer and Brown, 1986), and all parameters $(K_1, K_2, K_3, P_1, P_2)$ are set to zero.

$$
\begin{aligned}
\delta_x &= (K_1 r^2 + K_2 r^4 + K_3 r^6)(x - c_x) + \\
&\quad P_1\{r^2 + 2(x - c_x)^2\} + 2P_2(x - c_x)(y - c_y) \\
\delta_y &= (K_1 r^2 + K_2 r^4 + K_3 r^6)(y - c_y) + \\
&\quad 2P_1(x - c_x)(y - c_y) + P_2\{r^2 + 2(y - c_y)^2\}
\end{aligned}
\tag{5}
$$

where    $r^2 = (x - c_x)^2 + (y - c_y)^2$
$\delta_x, \delta_y$ = lens distortion
$x, y$ = image coordinates
$c_x, c_y$ = principal point
$K_1, K_2, K_3$ = radial distortion coefficients
$P_1, P_2$ = tangential distortion coefficients

Let the total number of frames be M, and the total number of corresponding points be N. Collinearity equation of the frame $\kappa$ and the corresponding point $\alpha$ is shown as Eq. 6 ($\kappa=1,...,M$, $\alpha=1,...,N$).

$$
\begin{aligned}
x_{\alpha\kappa} &= \frac{P_\kappa^{11} X_\alpha + P_\kappa^{12} Y_\alpha + P_\kappa^{13} Z_\alpha + P_\kappa^{14}}{P_\kappa^{31} X_\alpha + P_\kappa^{32} Y_\alpha + P_\kappa^{33} Z_\alpha + P_\kappa^{34}} \\
y_{\alpha\kappa} &= \frac{P_\kappa^{21} X_\alpha + P_\kappa^{22} Y_\alpha + P_\kappa^{23} Z_\alpha + P_\kappa^{24}}{P_\kappa^{31} X_\alpha + P_\kappa^{32} Y_\alpha + P_\kappa^{33} Z_\alpha + P_\kappa^{34}}
\end{aligned}
\tag{6}
$$

where    $(x_{\alpha\kappa}, y_{\alpha\kappa})$ = $\alpha$-th corresponding point in the frame $\kappa$
$\boldsymbol{P}_\kappa = \boldsymbol{K}_\kappa[(\boldsymbol{R}_\kappa^1)^t - \boldsymbol{t}_\kappa^1]$
$\boldsymbol{K}_\kappa = \begin{bmatrix} f & 0 & c_x \\ 0 & f & c_y \\ 0 & 0 & 1 \end{bmatrix}$
$(X_\alpha, Y_\alpha, Z_\alpha)$ = 3-d coordinate values of the corresponding point $\alpha$
$\boldsymbol{P}_\kappa$ = projection matrix of the frame $\kappa$

The evaluation function to be minimized is defined as Eq. 7. This is the sum of square errors between $(x_{\alpha\kappa} + \delta_x, \; y_{\alpha\kappa} + \delta_y)$ and $(x_{\alpha\kappa}^{real}, y_{\alpha\kappa}^{real})$.

$$
E = \sum_{\kappa=1}^{M} \sum_{\alpha=1}^{N} I_{\alpha\kappa} \left\{ \left(x_{\alpha\kappa} + \delta_x - x_{\alpha\kappa}^{real}\right)^2 + \left(y_{\alpha\kappa} + \delta_y - y_{\alpha\kappa}^{real}\right)^2 \right\}
\tag{7}
$$

where    $(x_{\alpha\kappa}^{real}, y_{\alpha\kappa}^{real})$ = real image coordinate values
$I_{\alpha\kappa} = \begin{cases} 1 & \text{if point } \alpha \text{ is on the frame } \kappa \\ 0 & \text{if point } \alpha \text{ is not on the frame } \kappa \end{cases}$

This evaluation function $E$ is minimized where $(X_\alpha, Y_\alpha, Z_\alpha)$, $\boldsymbol{R}_\kappa^1$, $\boldsymbol{t}_\kappa^1$, $\boldsymbol{K}_\kappa$, and $(K_1, K_2, K_3, P_1, P_2)_\kappa$ are unknown parameters. Since $E$ is a non-linear function, it is solved with Levenberg–Marquardt algorithm, which is one of the non-linear least squares methods.

## 3. EXPERIMENTAL SETUP AND RESULTS

In this chapter, we show how we acquired data and evaluated the method.

### 3.1 Experimental setup

We took range images of a cube whose size was already measured (Figure 4). We acquired 9 sequential range images by moving the range camera around the cube. In addition, 8 patterns of data that were different in distance were acquired (Figure 5). We prepared a small cube additionally because some part of the big cube was out of the camera view when the camera was near to the cube. The spec of the range camera is shown in Table 1. The size of the cube is shown in Table 2.

Figure 4. Illustration of the cubes

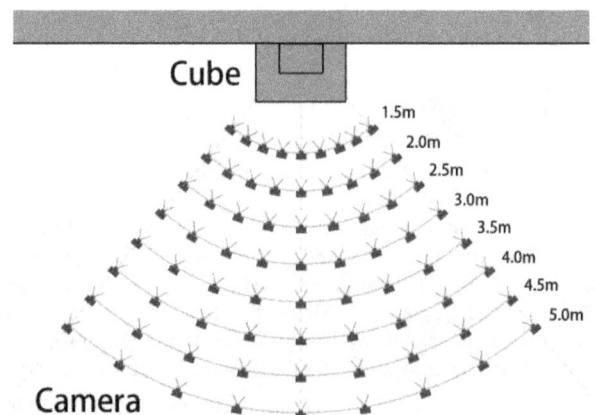

Figure 5. Eight patterns of data

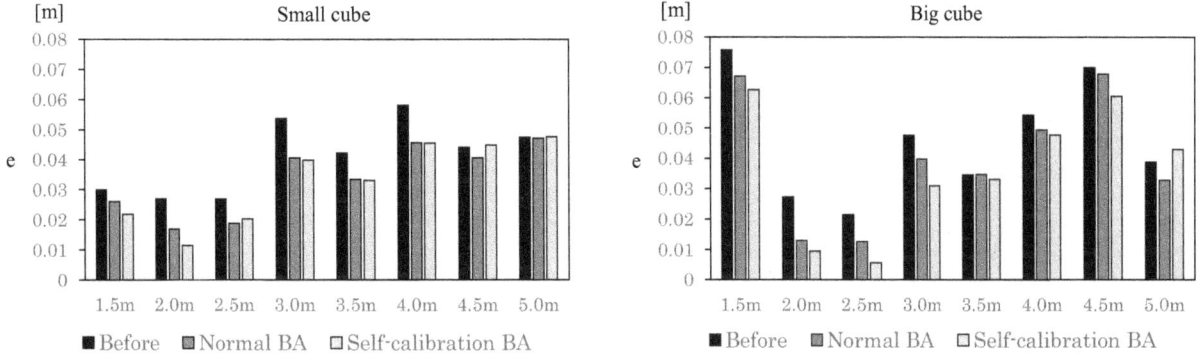

Figure 6. Error value $e$ (BA = Bundle Adjustment)

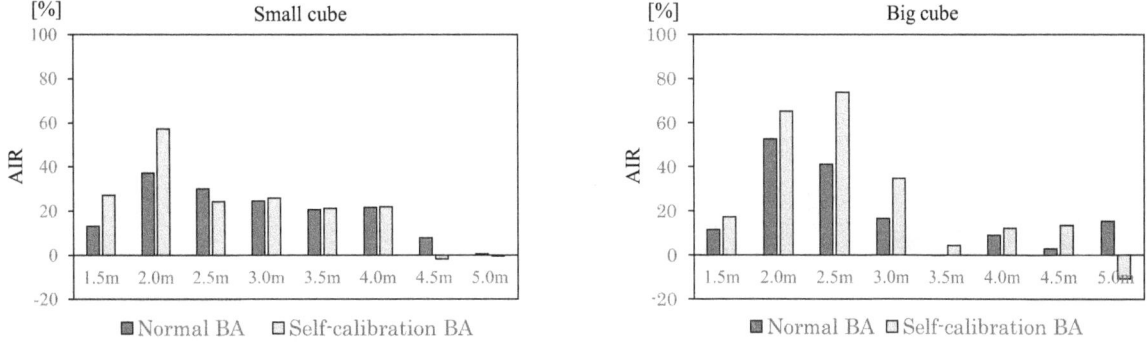

Figure 7. Accuracy improvement ratio (AIR)

| Name | DISTANZA P-401D |
|------|-----------------|
| Resolution | 128px * 126px |
| Angle of view | 27° |
| Measurable distance | 0.5m – 15m |
| Measurement error | 1% on the center of image and 3m distance |
| Size | Width: 160mm Height: 75mm Depth: 47mm |
| Weight | 650g |

Table 1. Spec of the range camera

|  | Width | Height | Depth |
|--|-------|--------|-------|
| Big cube | 62.0 | 56.0 | 40.5 |
| Small cube | 26.5 | 13.0 | 20.0 |

[cm]

Table 2. Size of the cube

## 3.2 Results

We applied our method to 8 patterns of data and evaluated it. In addition, we applied normal bundle adjustment and compared the result to the result of self-calibration bundle adjustment. We evaluated the error value $e$, which is the difference between the real size of the cube and the size measured on the integrated 3-d point cloud (Eq. 8), and the accuracy improvement ratio, which is defined as Eq. 9.

$$e = \frac{|w_p - w| + |h_p - h| + |d_p - d|}{3} \qquad (8)$$

where $w_p, h_p, d_p$ = the size measured on the point cloud
$w, h, d$ = the real size of the cube
($w$: width, $h$: height, $d$: depth)

Accuracy Improvement Ratio (AIR)
$$= \frac{e_b - e_a}{e_b} \times 100 \qquad (9)$$

where $e_b$ = the error before applying an adjustment method
$e_a$ = the error after applying an adjustment method

The results are shown on Figure 6 and Figure 7. Figure 6 shows the error value $e$ of Before (before applying an adjustment method), Normal BA (after applying the normal bundle adjustment), and Self-calibration BA (after applying the self-calibration bundle adjustment) with each pattern. Figure 7 shows the accuracy improvement ratio of Normal BA and Self-calibration BA with each pattern. Note that the reliability of the results where the distance is 4.5m – 5.0m is not high because the density of point cloud is not high. In addition, we measured only width and depth of the big cube where the distance is 1.5m – 2.5m because the some part of the cube is out of the camera view.

The results show that our proposed method is efficient, especially where the distance is 2.0m – 3.0m. The accuracy improvement ratio is more than 50%. Moreover, it shows that self-calibration bundle adjustment is better than normal bundle adjustment. Also, it is found that the accuracy improvement ratio becomes lower as the distance becomes further. However, the ratio is low where the distance is 1.5m. In this case, ICP algorithm might not work properly. It is suggested by Figure 8. It shows the initial re-projection error before applying bundle adjustment with each pattern. It is found that the initial re-projection error of the distance 1.5m is much higher than that of the other patterns. It is not effective to apply bundle adjustment if ICP algorithm fails and points properly match each other. The reason is described later.

The cause of the decline of the accuracy improvement ratio may be wrong corresponding points. Generally, the measurement error of a range camera becomes larger as the distance becomes further, and point matching tends to fail in ICP algorithm if the measurement error is large. Bundle adjustment cannot correct wrong correspondence and be strongly affected by it because bundle adjustment assumes that the correspondence is true. Since correspondence is based on nearest neighbor search, not on image features, it is difficult to distinct wrong correspondence. This problem is a future issue. At last, the visualization of the integrated point clouds of the cubes are shown in Figure 9. The point clouds which are applied self-calibration bundle adjustment are shown on 'After' column.

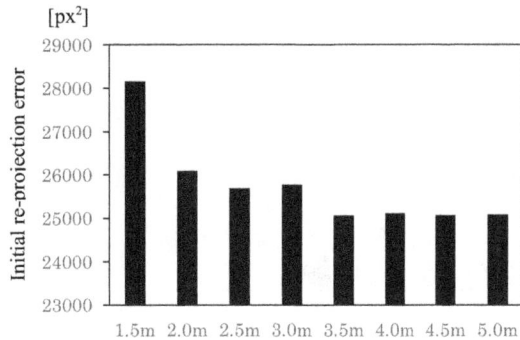

Figure 8. Initial re-projection error

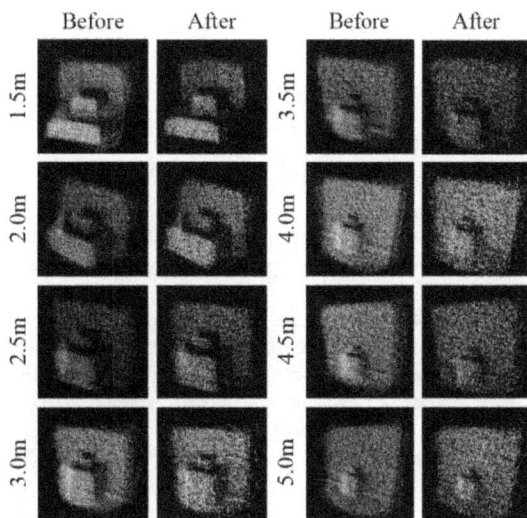

Figure 9. Visualization of the integrated point clouds

## 4. CONCLUSION

Our research developed an integration and adjustment method for sequential range images taken by a range camera. The proposed method consists of ICP algorithm and self-calibration bundle adjustment. Through experiments with the real data, the efficiency of the proposed method was confirmed. We also evaluated the influence of the distance between a camera and a target. It is found that the accuracy became lower as the distance becomes further. This is because the measurement error of the range camera becomes too large to be adjusted. In this case, outlier removal or smoothing of raw data may be effective. 3-d data acquisition by a range camera will be effective for additional measurement when data is not acquired at sufficient density or measurement is lacking. Development of a method for integrating point cloud into another point cloud with different density will enhance the value of our research.

## REFERENCES

Besl, P.J. and McKay, N.D., 1991. A method for registration of 3-D shapes, *IEEE transactions on Pattern Analysis and Machine Intelligence*, 14(2), pp.239-256.

Fryer, J.G. and Brown, D.C., 1986. Lens distortion for close-range photogrammetry, *Photogrammetric engineering and remote sensing*, 52(Jan), pp.51-58.

Lichti, D. D., Kim, C. and Jamtsho, S., 2010. An integrated bundle adjustment approach to range camera geometric self-calibration, *ISPRS Journal of Photogrammetry and Remote Sensing*, 65(4), pp.360-368.

Luhmann, T., Robson, S., Kyle, S. and Boehm, J., 2014. *Close Range Photogrammetry and 3D Imaging [2nd Edition]*, Walter de Gruyter GmbH, Berlin.

Mutto, C.D, Zanuttigh, P. and Cortelazzo G.M., 2012. *Time-of Flight Cameras and Microsoft Kinect*, Springer, New York.

Remondino, F. and Stoppa, D. eds, 2013. *TOF Range-Imaging Cameras*, Springer, New York.

Triggs, B., Mclauchlan, P. F., Hartley, R. I. and Fitzgibbon, A. W., 2000. *Bundle Adjustment - A Modern Synthesis*, In Vision Algorithms: Theory and Practice, pp.298-372.

Zhang, Z., 2000. A flexible new technique for camera calibration, *IEEE Transactions on Pattern Analysis and Machine Intelligence*, 22(11), pp.1330-1334.

# 9

# DIFFUSION BACKGROUND MODEL FOR MOVING OBJECTS DETECTION

Boris V. Vishnyakov, Sergey V. Sidyakin, Yuri V. Vizilter

State Research Institute of Aviation Systems (FGUP GosNIIAS), Moscow, Russia – vishnyakov@gosniias.ru

**Commission III, WG V/5 and WG III/3**

**KEY WORDS:** Background modelling, movement detection, diffusion maps, regression, LBP descriptor

**ABSTRACT:**

In this paper, we propose a new approach for moving objects detection in video surveillance systems. It is based on construction of the regression diffusion maps for the image sequence. This approach is completely different from the state of the art approaches. We show that the motion analysis method, based on diffusion maps, allows objects that move with different speed or even stop for a short while to be uniformly detected. We show that proposed model is comparable to the most popular modern background models. We also show several ways of speeding up diffusion maps algorithm itself.

## 1. INTRODUCTION

Recently a lot of background models intended for moving object detection and foreground segmentation were introduced. Moreover, changedetection.net benchmark was created (Goyette, 2012, Y. Wang, 2014) for testing and ranking existing and new algorithms for change and motion detection.

In the paper (Vishnyakov, 2012) we introduced the regression pseudospectrum background model, that was based on the fast way of accumulating the layers of the spectrum using the regression model. However, despite the very high computation speed, that approach lacked quality in complex conditions (swaying branches, changeable lighting conditions, slow objects).

The idea of our new approach is as follows: we improve regression background model using diffusion maps and diffusion regressive filtering.

Below in this paper we describe our approach and provide evaluation results on the changedetection.net database for the most suitable categories: «baseline», «thermal», «bad weather», «low framerate».

## 2. DIFFUSION BACKGROUND MODEL

### 2.1 Diffusion map

Let $I(k)$ be the input image on the frame $k$. Let us assume that $I(k)$ is a grayscale image (or it has been converted to grayscale), $I(k, q)$ is the brightness value of the image $I$ in pixel $q$.

The basis of our approach is the diffusion morphology (Vizilter, 2013) which allows comparing images by shape matching using the projection of image one on image two. In this morphology the projection is evaluated using diffusion maps, that were introduced in (Lafon, 2004, Coifman, 2006a, Coifman, 2006b):

$$P_{I_2}^I(k, p) = \sum_q A_{p,q}(k) I_2(q),$$

where $p, q$ – points,

$$A_{p,q}(k) = \frac{\exp\left(-\frac{\|v(p)-v(q)\|^2}{\varepsilon}\right)}{\sum_q \exp\left(-\frac{\|v(p)-v(q)\|^2}{\varepsilon}\right)}$$ – a diffusion operator (diffusion map),

$v(p)$ – a feature vector, calculated in the point $p$ of the image $I$ with the square kernel,

$\|\cdot\|$ – a distance between two feature vectors.

### 2.2 Complex LBP Descriptor

In the papers (Gorbatsevich, 2014, Vishnyakov, 2014) the complex LBP descriptor was introduced as a feature vector $v(\cdot)$ along with the hamming distance $\|\cdot\|$ as a distance between feature vectors for the diffusion maps. This allows very fast computation of the diffusion map.

Our implementation provides a possibility for real time image processing. The computation of diffusion filtering with heat kernel in its original form is an extremely time-consuming procedure even for reasonable neighbourhood of $p$. We propose to substitute such computationally unpleasant descriptors by the combination of intensity $I(p)$ and threshold LBP (Ahonen, 2004) for $v(p)$. In our experiments, the mean value of intensity in the p neighbour was used. Mean value is computed by a fast algorithm with sliding sum recalculation (but this is not presented in code below). The local binary pattern (LBP) is calculated as a 64-bit vector for each pixel $p$ based on a comparison of its value and values of its neighbours in sliding window. If the value of neighbour pixel is less than the value of central pixel and the difference between them is greater than threshold, then the corresponding bit is set to 1, otherwise – to 0. We substitute the original neighbourhood matching metrics by LBP matching metric - Hamming distance, the mean values of intensities are compared by threshold. As local binary patterns are stored as bit fields, the computation of Hamming distance is performed via bitwise XOR operation. The exponent is calculated using table values. Due to this, the usage of our complex LBP descriptor allows both increasing the computational speed and obtaining heat kernels very similar to original.

### 2.3 Background model

The idea of the paper is as follows: the approach for the image comparison in the diffusion morphology can be also used for the moving object detection. For robustness of the approach we propose regression accumulators (Vishnyakov, 2012) $m_n(k)$ and $m_n^*(k)$ for both the original image and the filtered image respectively:

$$m_n(k) = \alpha \cdot m_n(k-1) + (1-\alpha) \cdot I(k),$$
$$m_n^*(k) = \alpha \cdot m_n^*(k-1) + (1-\alpha) \cdot I^*(k),$$

where $I^*(k) = P_{m_n(k)}^{I(k)}(k)$ is a diffusion filter of the $I(k)$, which is the projection of $I(k)$ on the memory $m_n(k)$, $n$ is a parameter, related to the memory length, $\alpha = \alpha(n)$.

After computing $m_n(k)$ and $m_n^*(k)$, we compare the difference

$$D(k) = |m_n^*(k) - I^*(k)|$$

to the threshold in each pixel and get the binary moving object mask $M(k)$:

$$M(k) = \begin{cases} 1, D(k) \geq thr \\ 0, D(k) < thr \end{cases}$$

## 2.4 Illustrations

On figure 1 we show smoothed accumulator (memory) $m_n^*(k)$ using regression model for the projection image $I^*(k)$ – a diffusion filter of the $I(k)$. As you can see, no moving object is present on this accumulator. In addition, the accumulator is smoothed because of smoothing properties of the diffusion map. On figure 2 we show the projection image $I^*(k)$ itself. One moving object is present on this image. The projection image is also smoothed.

On figure 3 we show the difference $D(k)$ between projection image $I^*(k)$ and smoothed accumulator $m_n^*(k)$.

On figure 4 we show image $M(k)$ – binarized using threshold difference image $D(k)$.

Figure 1. Smoothed memory $m_n^*(k)$.

Figure 2. Image projection $I^*(k)$.

Figure 3. Background difference $D(k)$.

Figure 4. Binarized background difference $M(k)$.

## EXPERIMENTS

In order to demonstrate the quality characteristics of the proposed approach we use the common benchmark videos from changedetection.net database (changedetection.net, 2014). Changedetection.net provide a realistic, camera-captured, diverse set of 31 videos (~70,000 frames). They have been selected to cover a wide range of detection challenges and are representative of typical indoor and outdoor visual data captured today in surveillance, smart environment and video database scenarios. The 2014 dataset consist of different categories. We choose «baseline», «thermal», «bad weather», «low framerate» categories, because the proposed approach can be directly applicable to them in the presented straightforward form (regression model of background learning). Other «camera jitter», «shadow», «PTZ» (pan-tilt zoom) scenarios are not considered, because they require some special modification of the basic algorithm, we hope to explore them in the future. «Baseline» category represents a mixture of moderate challenges. Some videos have subtle background motion, others have isolated shadows, some have an abandoned object and others have pedestrians that stop for a short while and then move away. «Thermal» category includes videos that have been captured by far-infrared cameras. These videos contain typical thermal artifacts such as heat stamps (e.g., bright spots left on a seat after a person gets up and leaves), heat reflection on floors and windows, and camouflage effects, when a moving object has the same temperature as the surrounding regions. «Bad Weather» category includes outdoor videos captured in challenging winter weather conditions, i.e., snow storm, snow on the ground, fog. «Low Frame-Rate» category contains videos capture at varying frame-rates between 0.17 fps and 1 fps. For «baseline» and «thermal» categories the results of proposed approach was compared to 37 other existed methods results available from change detection 2012 benchmark. For «Bad Weather», «Low Frame-Rate» categories the results of proposed approach was compared to 22 other existed methods results available from change detection 2014 benchmark.

The following methods were used for comparative analysis: SUBS (St-Charles, 2014), PAWCS (St-Charles, 2015), SOBS1 (L. Maddalena, 2012), SOBS2 (L. Maddalena, 2010), SOBS3 (L. Maddalena, 2008), PSP (A. Schick, 2012), S-360_1 (M. Sedky, 2011), MLBS (Jian Yao, 2007), GPGMM (T. S. F. Haines, 2012), SGMM (R. Heras, 2011), PBAS2 (M. Hofmann, 2012), KDE1 (A. Elgammal, 2000), BB (F. Porikli, 2005), CDPS (Francisco J. Hernandez, 2013), HIST (J. Zheng, 2006), CWD (M. De Gregorio, 2013), KNN (Z. Zivkovic, 2006), GMM1 (P. KaewTraKulPong, 2001), MAHAL (Y. Benezeth, 2010), CHEB1 (A. Morde, 2012), RMOG (S. Varadarajan, 2013), LSS (J-P Jodoin, 2012), GMM2 (Z. Zivkovic, 2004), ED (Y. Benezeth, 2010), GMM3 (D. Riahi, 2012), SGMM (Heras Evangelio, 2012), UBA (D. Park, 2009), PROST (F. Seidel, 2014), GMM4 (C. Stauffer, 1999) TUBI (L. Dar-Shyang, 2005), KDE2 (Y. Nonaka, 2012), KDE3 (S. Yoshinaga, 2013), QCHMD (O. Strauss, 2012), CHEB2 (A. Morde, 2012), FTSG (R. Wang, 2014), SS (Zhenkun, 2015), BWA (B. Wang, 2014), EFIC (G. Allebosch, 2015), S-360_2 (M.Sedky,2014), CWDH (M. De Gregorio, 2014), UBSS (H. Sajid, 2015), MSTBGM (Xiqun Lu, 2014).

Quantitative results for «baseline» category are shown in Tables 1, 2, 3. The following metrics are used: Re – Recall, SP - (Specificity), FPR - False Positive Rate, FNR - False Negative Rate, PWC - Percentage of Wrong Classifications, F - F-Measure, PR – Precision, TP - True Positive, FP - False Positive, FN - False Negative, TN - True Negative, Rank – rank among all comparative methods in chosen category (38 or 22).

| DIFF | TP | FP | FN | TN |
|---|---|---|---|---|
| pedestrians | 619693 | 122498 | 51169 | 67446609 |
| PETS2006 | 2915070 | 295655 | 1913120 | 366417649 |
| highway | 4728297 | 1220961 | 729692 | 85433508 |
| office | 7932008 | 303147 | 710086 | 116235539 |
| *baseline* | *16195068* | *1942261* | *3404067* | *635533305* |

Table 1. Baseline category. Results for each video in this category.

| Method | RE | SP | FPR | FNR | PWC | F | PR |
|---|---|---|---|---|---|---|---|
| SUBS | 0,95 | 0,99 | 0,002 | 0,05 | 0,35 | 0,95 | 0,94 |
| PAWCS | 0,94 | 0,99 | 0,002 | 0,05 | 0,44 | 0,93 | 0,93 |
| SOBS1 | 0,93 | 0,99 | 0,002 | 0,06 | 0,37 | 0,93 | 0,93 |
| CDET | 0,97 | 0,99 | 0,002 | 0,02 | 0,35 | 0,94 | 0,92 |
| SOBS2 | 0,93 | 0,99 | 0,002 | 0,06 | 0,39 | 0,92 | 0,92 |
| SOBS3 | 0,91 | 0,99 | 0,002 | 0,08 | 0,43 | 0,92 | 0,93 |
| GPRMF | 0,9 | 0,99 | 0,002 | 0,09 | 0,46 | 0,92 | 0,95 |
| PSP | 0,93 | 0,99 | 0,002 | 0,06 | 0,41 | 0,92 | 0,96 |
| S-360_1 | 0,96 | 0,99 | 0,003 | 0,03 | 0,42 | 0,93 | 0,90 |
| MLBS | 0,84 | 0,99 | 0,001 | 0,15 | 0,89 | 0,90 | 0,96 |
| DPGMM | 0,96 | 0,99 | 0,003 | 0,03 | 0,49 | 0,92 | 0,89 |
| PBAS1 | 0,95 | 0,99 | 0,002 | 0,04 | 0,48 | 0,92 | 0,89 |
| SGMM | 0,93 | 0,99 | 0,002 | 0,06 | 0,54 | 0,92 | 0,91 |

| | | | | | | | |
|---|---|---|---|---|---|---|---|
| PBAS2 | 0,95 | 0,99 | 0,003 | 0,04 | 0,48 | 0,92 | 0,89 |
| KDE1 | 0,89 | 0,99 | 0,002 | 0,10 | 0,54 | 0,90 | 0,92 |
| BB | 0,73 | 0,99 | 0,001 | 0,26 | 0,90 | 0,82 | 0,96 |
| CDPS | 0,94 | 0,99 | 0,003 | 0,05 | 0,62 | 0,92 | 0,89 |
| HIST | 0,87 | 0,99 | 0,002 | 0,12 | 0,66 | 0,90 | 0,92 |
| CWD | 0,89 | 0,99 | 0,002 | 0,10 | 0,66 | 0,90 | 0,91 |
| KNN | 0,79 | 0,99 | 0,002 | 0,20 | 1,28 | 0,84 | 0,92 |
| GMM1 | 0,58 | 0,99 | 0,001 | 0,41 | 1,93 | 0,71 | 0,95 |
| MAHAL | 0,88 | 0,99 | 0,003 | 0,11 | 0,72 | 0,89 | 0,90 |
| CHEB1 | 0,82 | 0,99 | 0,003 | 0,17 | 0,83 | 0,86 | 0,91 |
| RMOG | 0,70 | 0,99 | 0,001 | 0,29 | 1,59 | 0,78 | 0,91 |
| LSS | 0,97 | 0,98 | 0,013 | 0,02 | 1,33 | 0,84 | 0,75 |
| GMM2 | 0,80 | 0,99 | 0,002 | 0,19 | 1,32 | 0,83 | 0,89 |
| ED | 0,83 | 0,99 | 0,004 | 0,16 | 1,02 | 0,87 | 0,91 |
| GMM3 | 0,66 | 0,99 | 0,002 | 0,33 | 1,53 | 0,75 | 0,91 |
| SGMM | 0,86 | 0,99 | 0,005 | 0,13 | 1,24 | 0,85 | 0,85 |
| **DIFF\*** | **0,82** | **0,99** | **0,004** | **0,17** | **0,94** | **0,84** | **0,87** |
| UBA | 0,90 | 0,99 | 0,008 | 0,09 | 1,01 | 0,81 | 0,74 |
| PROST | 0,84 | 0,99 | 0,006 | 0,15 | 1,15 | 0,82 | 0,81 |
| GMM4 | 0,81 | 0,99 | 0,005 | 0,18 | 1,53 | 0,82 | 0,84 |
| TUBI | 0,89 | 0,98 | 0,015 | 0,10 | 2,08 | 0,76 | 0,67 |
| KDE2 | 0,74 | 0,99 | 0,004 | 0,25 | 1,80 | 0,73 | 0,79 |
| KDE3 | 0,75 | 0,99 | 0,00 | 0,24 | 1,91 | 0,75 | 0,78 |
| QCHMD | 0,70 | 0,99 | 0,007 | 0,29 | 2,21 | 0,66 | 0,70 |

Table 2. Baseline category. DIFF* - proposed method.

| Ranks among 38 methods | | | | | | | |
|---|---|---|---|---|---|---|---|
| | RE | SP | FPR | FNR | PWC | F | PR | Overall |
| DIFF | 26 | 29 | 29 | 26 | 22 | 24 | 28 | 30/38 |

Table 3. Baseline category. Rank for each metric for proposed method presented.

Results for «Thermal» category is shown in Table 4, 5, 6.

| DIFF | TP | FP | FN | TN |
|---|---|---|---|---|
| corridor | 10667192 | 7675519 | 1672965 | 352311506 |
| dining | 16264143 | 4775408 | 3200154 | 202408859 |
| lakeSide | 5035690 | 26973588 | 2937391 | 381184826 |
| library | 58508839 | 3889517 | 3452426 | 255590998 |
| park | 384658 | 48641 | 330479 | 34446499 |
| *thermal* | *90860522* | *43362673* | *11593415* | *1225942688* |

Table 4. Thermal category. Results for each video in this category.

| Method | RE | SP | FPR | FNR | PWC | F | PR |
|---|---|---|---|---|---|---|---|
| DIFF | 0,76 | 0,98 | 0,025 | 0,23 | 3,31 | 0,67 | 0,67 |

Table 5. Thermal category.

| Ranks among 38 methods | | | | | | | |
|---|---|---|---|---|---|---|---|
| RE | SP | FPR | FNR | PWC | F | PR | Overall |

| DIFF | 7 | 37 | 37 | 7 | 24 | 24 | 37 | 32/38 |

Table 6. Thermal category. Rank for each metric for proposed method presented.

Results for «Bad Weather» category is shown in Table 7, 8, 9.

| DIFF | TP | FP | FN | TN |
|------|-----|-----|-----|-----|
| blizzard | 8105421 | 114550 | 4151098 | 1038823675 |
| skating | 11950348 | 2149566 | 2823328 | 281628841 |
| snowFall | 5100041 | 931585 | 2656794 | 975209889 |
| wetSnow | 3439753 | 694655 | 2575110 | 460144956 |
| *weather* | *28595563* | *3890356* | *12206330* | *2755807361* |

Table 7. Bad Weather category. Results for each video in this category.

| Methods | RE | SP | FPR | FNR | PWC | F | PR |
|---------|-----|-----|------|------|------|-----|-----|
| DIFF | 0,68 | 0,99 | 0,002 | 0,32 | 0,78 | 0,76 | 0,88 |

Table 8. Bad Weather category.

| | Ranks among 22 methods | | | | | | | |
|------|-----|-----|------|------|------|-----|-----|---------|
| | RE | SP | FPR | FNR | PWC | F | PR | Overall |
| DIFF | 12 | 16 | 16 | 12 | 14 | 9 | 11 | 15/22 |

Table 9. Bad Weather category. Rank for each metric for proposed method presented.

Results for «Low Frame-Rate» category is shown in Table 10, 11, 12.

| DIFF | TP | FP | FN | TN |
|------|-----|-----|-----|-----|
| port | 19425 | 83230 | 26366 | 153835893 |
| tram | 290806 | 96304 | 64765 | 12357342 |
| tunnel | 1283430 | 135444 | 3702953 | 176739284 |
| turn | 1355782 | 259399 | 450049 | 22203801 |
| low fr. r. | *2949443* | *574377* | *4244133* | *365136320* |

Table 10. Low Frame-Rate category. Results for each video in this category.

| Methods | RE | SP | FPR | FNR | PWC | F | PR |
|---------|-----|-----|------|------|------|-----|-----|
| DIFF | 0,56 | 0,99 | 0,005 | 0,43 | 1,59 | 0,55 | 0,67 |

Table 11. Low Frame-Rate category.

| | Ranks among 22 methods | | | | | | | |
|------|-----|-----|------|------|------|-----|-----|---------|
| | RE | SP | FPR | FNR | PWC | F | PR | Overall |
| DIFF | 18 | 10 | 10 | 18 | 13 | 8 | 8 | 13/22 |

Table 12. Low Frame-Rate category. Rank for each metric for proposed method presented.

From conducted experiments we conclude that the proposed approach can be used for background modelling, performs better than several well-known methods or is comparative to them in case of different shooting conditions.

## CONCLUSION

## REFERENCES

Ahonen, T., Hadid, A., Pietikainen, M., 2004. Face recognition with local binary patterns. ECCV, pp 469-481

G. Allebosch, D. Van Hamme, F. Deboeverie, P. Veelaert and W. Philips "Edge based foreground background segmentation with interior/exterior classification". To be published in proceedings of VISAPP 2015.

Y. Benezeth, P.-M. Jodoin, B. Emile, H. Laurent, and C. Rosenberger. Comparative study of background subtraction algorithms.J. of Elec. Imaging, 19(3):1–12, 2010.

St-Charles, P.-L., Bilodeau, G.-A., Bergevin, R., "SuBSENSE : A Universal Change Detection Method with Local Adaptive Sensitivity". Accepted for IEEE Transactions on Image Processing, Nov. 2014.

St-Charles, P.-L., Bilodeau, G.-A., Bergevin, R., "SuBSENSE: A Universal Change Detection Method with Local Adaptive Sensitivity". Accepted for IEEE Transactions on Image Processing, Nov. 2014.

P.-L. St-Charles, G.-A. Bilodeau, R. Bergevin, "A Self-Adjusting Approach to Change Detection Based on Background Word Consensus" in IEEE Winter Conference on Applications of Computer Vision (WACV). Big Island, Hawaii, USA, Jan. 6-9, 2015.

Coifman, R., Lafon, S., Maggioni, M., Keller, Y., D Szlam, A, Warner, F., Zucker, S. Geometries of sensor outputs, inference and information processing // Storage and Re-trieval for Image and Video Databases, edited by Intelligent Integrated Microsystems – 2006. – Vol. 6232 – P. 623209.

Coifman, R., Lafon, S. Diffusion maps // Applied and Computational Harmonic Analysis. – 2006. – Vol. 21(1). – P. 5-30.

Lee, Dar-Shyang. "Effective Gaussian mixture learning for video background subtraction." Pattern Analysis and Machine Intelligence, IEEE Transactions on 27.5 (2005): 827-832.

Dong Liang, Shun'ichi Kaneko, "Improvements and Experiments of a Compact Statistical Background Model", arXiv:1405.6275, 2014.

A. Elgammal, D. Harwood, and L. Davis, "Non-parametric model for background subtraction," in Proc. Eur. Conf. on Computer Vision, Lect. Notes Comput. Sci. 1843, 751-767 2000.

Francisco J. Hernandez-Lopez and Mariano Rivera. Change Detection by Probabilistic Segmentation from Monocular View. Machine Vision and Applications, pages 1-21, 2013.

M. De Gregorio and M. Giordano "A WiSARD-based approach to CDnet" 1-st BRICS Countries Congress (BRICS-CCI) – 2013.

M. De Gregorio and M. Giordano "Change Detection with Weightless Neural Networks", in proc of IEEE Workshop on Change Detection, 2014.

T. S. F. Haines and T. Xiang "Background Subtraction with Dirichlet Processes" ECCV 2012.

R. Heras and T. Sikora "Complementary Background Models for the Detection of Static and Moving Objects in Crowded Environments", in 8th Proceedings of the IEEE International Conference on Advanced Video and Signal-Based Surveillance (AVSS), 2011.

Heras Evangelio, R. and Pätzold, M. and Sikora, T. "Splitting Gaussians in Mixture Models", Proceedings of the 9th IEEE International Conference on Advanced Video and Signal-Based Surveillance, 2012.

V. Gorbatsevich, D. Pronin, Y. Vizilter Diffusion maps fast algorithm implementation // TCVS-2014 theses – 2014. In Russian.

N. Goyette, P.-M. Jodoin, F. Porikli, J. Konrad, and P. Ishwar, changedetection.net: A new change detection benchmark dataset, in Proc. IEEE Workshop on Change Detection (CDW-2012) at CVPR-2012, Providence, RI, 16-21 Jun., 2012.

M. Hofmann, P. Tiefenbacher, G. Rigoll "Background Segmentation with Feedback: The Pixel-Based Adaptive Segmenter", in proc. of IEEE Workshop on Change Detection, 2012.

J-P Jodoin, G-A Bilodeau, N Saunier "Background subtraction based on Local Shape", arXiv:1204.6326v1 , 27 Apr 2012.
Jian Yao and Jean-Marc Odobez, Multi-Layer Background Subtraction Based on Color and Texture. Visual Surveillance workshop (CVPR-VS), Minneapolis, June 2007.

P. KaewTraKulPong and R. Bowden, "An Improved Adaptive Background Mixture Model for Real-time Tracking with Shadow Detection", in proc. of Workshop on Advanced Video Based Surveillance Systems, 2001.

D. Kit, B. T. Sullivan, and D. H. Ballard. Novelty detection using growing neural gas for visuo-spatial memory. In IROS, pages 1194–1200. IEEE, 2011.

Lafon, S. Diffusion maps and geometric harmonics // PhD thesis. – Yale University, Dept of Mathematics & Applied Mathematics. – 2004.

L. Maddalena, A. Petrosino, A Self-Organizing Approach to Background Subtraction for Visual Surveillance Applications, IEEE Transactions on Image Processing, Vol. 17, no.7, 2008, p1168-1177.

L. Maddalena, A. Petrosino, A Fuzzy Spatial Coherence-based Approach to Background/ Foreground Separation for Moving Object Detection, Neural Computing and Applications, Springer London, Vol. 19, pp. 179–186, 2010.

L. Maddalena, A. Petrosino, "The SOBS algorithm: what are the limits", in proc of IEEE Workshop on Change Detection, CVPR 2012.

A. Morde, X. Ma, S. Guler [IntuVision] "Learning a background model for change detection", in proc. of IEEE Workshop on Change Detection, 2012.

Y. Nonaka, A. Shimada, H. Nagahara, R. Taniguchi "Evaluation Report of Integrated Background Modeling Based on Spatio-temporal Features", in proc. of IEEE Workshop on Change Detection, 2012

D. Park, H. Byun, Object-wise multilayer background ordering for pubic area surveillance", in IEEE International Conference on AVSS, 2009

F. Porikli and O. Tuzel. "Bayesian background modeling for foreground detection" in proc. of ACM Visual Surveillance and Sensor Network, 2005.

Riahi, D., St-Onge, P.L., Bilodeau, G.A. (2012). RECTGAUSS-Tex: Block-based Background Subtraction, Technical Report, École Polytechnique de Montréal, EPM-RT-2012-03.

Satoshi Yoshinaga, Atsushi Shimada, Hajime Nagahara, Rin-ichiro Taniguchi, Background Model Based on Intensity Change Similarity Among Pixels, the 19th Japan-Korea Joint Workshop on Frontiers of Computer Vision, pp.276-280, 2013.01.

A. Schick, M. Bäuml, R. Stiefelhagen "Improving Foreground Segmentations with Probabilistic Superpixel Markov Random Fields", in proc of IEEE Workshop on Change Detection, 2012.

SEDKY Mohamed Hamed Ismail, MONIRI Mansour and CHIBELUSHI Claude Chilufya "Object Segmentation Using Full-Spectrum Matching of Albedo Derived from Colour Images" UK patent application no. 0822953.6 16.12.2008 GB, 2008, PCT patent application international application no. PCT/GB2009/002829, EP2374109, 2009, US patent no. 2374109 12.10.2011 US, 2011.

M. Sedky, M. Moniri and C. C. Chibelushi "Spectral-360: A physical-based technique for change detection", in proc of IEEE Workshop on Change Detection, 2014.

F. Seidel, C. Hage, M. Kleinsteuber "pROST - A Smoothed Lp-norm Robust Online Subspace Tracking Method for Real time Background Subtraction in Video". In Machine Vision and Applications, 25(5):1227-1240, 2014.

C. Stauffer and W. E. L. Grimson, "Adaptive background mixture models for real-time tracking," in Proc. Int. Conf. on Computer Vision and Pattern Recognition, Vol. 2, IEEE, Piscataway, NJ (1999).

O. Strauss, D. Sidibé, W. Puech "Quasi-continuous histogram based motion detection",Technical Report, LE2I, 2012.

Varadarajan, S.; Miller, P.; Huiyu Zhou, "Spatial mixture of Gaussians for dynamic background modelling," Advanced Video and Signal Based Surveillance (AVSS), 2013 10th IEEE International Conference on , pp.63,68, 27-30 Aug. 2013.

Vishnyakov, B., Vizilter, Y., Knyaz V. Spectrum-Based Object Detection And Tracking Technique For Digital Video Surveillance // ISPRS Archives – XXXIX-B3 – 2012. – P. 579-583.

Vishnyakov, B., Gorbatsevich, V., Sidyakin, S., Malin, I., Egorov, A Fast moving objects detection using iLBP background model // Int. Arch. Photogramm. Remote Sens. Spatial Inf. Sci., XL-3 – 2014 – P. 347-350.

Vizilter Y., Gorbatsevich V., Rubis A., Vygolov O. Image comparison using shape matching diffusion correlation morthology // Computer Optics, 2013, Vol. 37, 1. In Russian.

R. Wang, F. Bunyak, G. Seetharaman and K. Palaniappan "Static and Moving Object Detection Using Flux Tensor with Split Gaussian Models", in proc of IEEE Workshop on Change Detection, 2014.

Y. Wang, P.-M. Jodoin, F. Porikli, J. Konrad, Y. Benezeth, and P. Ishwar, CDnet 2014: An Expanded Change Detection Benchmark Dataset, in Proc. IEEE Workshop on Change Detection (CDW-2014) at CVPR-2014, pp. 387-394. 2014.

B. Wang and P. Dudek "A Fast Self-tuning Background Subtraction Algorithm", in proc of IEEE Workshop on Change Detection, 2014.

Xiqun Lu "A multiscale spatio-temporal background model for motion detection", ICIP 2014.

Yin, Baocai; Zhang, Jing; Wang, Zengfu: 'Background segmentation of dynamic scenes based on dual model', IET Computer Vision, 2014, DOI: 10.1049/iet-cvi.2013.0319 IET Digital Library, http://digital-library.theiet.org/content/journals/10.1049/iet-cvi.2013.0319.

J. Zheng, Y. Wang, N.L. Nihan, and M.E. Hallenbeck. Extracting roadway background image : Mode-based approach. Transportation Research Record : Journal of the Transportation Research Board, 1944(-1) :82–88, 2006.

Zhenkun huang Ruimin Hu and Shihong Chen, paper has been submitted to CVPR2015.

Z. Zivkovic, "Improved adaptive Gaussian mixture model for back-ground subtraction," in Proc. Int. Conf. Pattern Recognition, pp. 28-31, IEEE, Piscataway, NJ 2004.

Z. Zivkovic, F. van der Heijden "Efficient adaptive density estimation per image pixel for the task of background subtraction", Pattern Recognition Letters, vol. 27, no. 7, pages 773-780, 2006.

## APPENDIX

### C++ - LIKE SOURCE CODE FOR OUR COMPLEX LBP DIFFISUION FILTERING

Pseudo-code of Diffusion filtering and processing that is used in Diffusion Background Model.

```cpp
class Diffusion
{
public:
unsigned char**  Image1;
unsigned char**  Image2;
int**  OutImage1;
int**  OutImage2;
int**  summs;
__int64**  LBP;
int ExpTableInt[65];
int lbpthr, ithr;
int BlockSize;
int step;
int WSize;
int imW, imH;
double E;

Diffusion(int W, int H)
{
// a small parameter comparable to the
smallest // distances between two feature
vectors v(p) // and v(q)
```

```cpp
E = 10;
// LBP threshold
lbpthr = 10;
// Intensity threshold
ithr = 50;
// Half of the sliding window (neighbourhood)
BlockSize = 5;
WSize = BlockSize; step = 1;
// Pre-calculating table values for exponent
for (int i = 0; i < 65; i++)
 ExpTableInt[i] = (int)(exp(-i / E) * 1024);
 imW = W; imH = H;
Image1 = new unsigned char*[H];
Image2 = new unsigned char*[H];
LBP = new __int64*[H];
LBP[0] = new __int64[W*H];
summs = new int*[H];
summs[0] = new int[H*W];
OutImage1 = new int*[H];
OutImage1[0] = new int[H*W];
OutImage2 = new int*[H];
OutImage2[0] = new int[H*W];
for (int i = 1; i < H; i++)
{
 LBP[i] = &LBP[0][i*W];
 OutImage1[i] = &OutImage1[0][i*imW];
 OutImage2[i] = &OutImage2[0][i*imW];
 summs[i] = &summs[0][i*imW];
}
}

~Diffusion()
{
 delete[] Image1; delete[] Image2;
 delete[] LBP[0]; delete[] LBP;
 delete[] summs[0]; delete[] summs;
 delete[] OutImage1[0]; delete[] OutImage1;
 delete[] OutImage2[0]; delete[] OutImage2;
}

//diffusion filtering of image1 by itself
void Process(unsigned char* image1)
{
 for (int i = 0; i < imH; i++)
  Image1[i] = &image1[i*imW];
  BuildLBPMap();
  Filter();
}

//diffusion projection of image2 on the form of
//image1
void Project(unsigned char* image1, unsigned
          char* image2)
{
 for (int i = 0; i < imH; i++)
 {
  Image1[i] = &image1[i*imW];
  Image2[i] = &image2[i*imW];
 }
 BuildLBPMap();
 MorphProject();
}

// LBP pre-calculation with lbpthr threshold
void BuildLBPMap()
{
```

```
if (WSize >= 4)
 WSize = 3; WSize *= step;
 for (int x = WSize; x < imW - WSize; x++)
 for (int y = WSize; y < imH - WSize; y++)
 {
  int ind = 0;
  __int64 Res = 0;
  int Pix = Image1[y][x];
  for (int x2 = -WSize; x2 < WSize; x2 +=
step)
  for (int y2 = -WSize; y2 < WSize; y2 +=
step)
  {
   if (Pix - Image1[y + y2][x + x2] > lbpthr)
   Res += ((__int64)2) << ind;
   ind++;
  }
  LBP[y][x] = Res;
 }
}

// bit counting
unsigned int POPCNT(__int64 Arg)
{ unsigned int* t = (unsigned int*)&Arg;
  return (__popcnt(t[0]) + __popcnt(t[1]));}

// heat kernel calculation
int CompareHammingint(__int64 A, __int64 B)
{ return ExpTableInt[POPCNT(A^B)]; }

void Filter(){
memset(&summs[0][0], 0, imW*imH * 4);
memset(&OutImage1[0][0], 0, imW*imH * 4);
int Wd2 = BlockSize; int x, y;
for (int y = 0; y < imH - Wd2; y++){
for (int x = Wd2; x < imW - Wd2; x++){
for (int y2 = 0; y2 < BlockSize; y2 += step)
for (int x2 = -BlockSize; x2 < BlockSize; x2
+= step){
int t = CompareHammingint(LBP[y][x], LBP[y +
y2][x + x2]);
if (abs(Image1[y][x] - Image1[y + y2][x + x2])
> ithr) t = 0;
summs[y][x] += t;
summs[y + y2][x + x2] += t;
OutImage1[y][x] += t * Image1[y + y2][x + x2];
OutImage1[y + y2][x + x2] += t * Image1[y][x];
} OutImage1[y][x] /= (summs[y][x]);
}}}

void MorphProject(){
memset(&summs[0][0], 0, imW*imH * 4);
memset(&OutImage1[0][0], 0, imW*imH * 4);
memset(&OutImage2[0][0], 0, imW*imH * 4);
int Wd2 = BlockSize; int x,y;
for (int y = 0; y < imH - Wd2; y++){
for (int x = Wd2; x < imW - Wd2; x++){
for (int y2 = 0; y2 < BlockSize; y2 += step)
for (int x2 = -BlockSize; x2 < BlockSize; x2
+= step){
int t = CompareHammingint(LBP[y][x], LBP[y +
y2][x + x2]);
if (abs(Image1[y][x] - Image1[y + y2][x + x2])
> ithr) t = 0;
summs[y][x] += t;
summs[y + y2][x + x2] += t;

OutImage2[y][x] += t * Image2[y + y2][x + x2];
OutImage2[y + y2][x + x2] += t * Image2[y][x];
OutImage1[y][x] += t * Image1[y + y2][x + x2];
OutImage1[y + y2][x + x2] += t * Image1[y][x];
}
OutImage1[y][x] /= (summs[y][x]);
OutImage2[y][x] /= (summs[y][x]);
}}}};
```

# 10

# MODELING OF BIOMETRIC IDENTIFICATION SYSTEM USING THE COLORED PETRI NETS

G. R. Petrosyan [a], L. A. Ter-Vardanyan [a], A.V. Gaboutchian [b,]

[a] Institute for Informatics and Automation Problems of NAS RA, International Scientific - Educational Center of NAS RA
Armenia, Yerevan, petrosyan_gohar@list.ru, lilit@sci.am,

Russian Federation, Moscow, [b] armengaboutchian@mail.ru

**KEY WORDS: Petri Net (PN), Colored Petri Net (CPN), position, transition, guard, identification, biometric system**

**ABSTRACT:**

In this paper we present a model of biometric identification system transformed into Petri Nets. Petri Nets, as a graphical and mathematical tool, provide a uniform environment for modelling, formal analysis, and design of discrete event systems. The main objective of this paper is to introduce the fundamental concepts of Petri Nets to the researchers and practitioners, both from identification systems, who are involved in the work in the areas of modelling and analysis of biometric identification types of systems, as well as those who may potentially be involved in these areas. In addition, the paper introduces high-level Petri Nets, as Colored Petri Nets (CPN). In this paper the model of Colored Petri Net describes the identification process much simpler.

## 1. Introduction

Petri Nets (PN) are a graphical tool for formal description of the flow of activities in complex systems. Compared to other more popular techniques of graphical system representation (for instance, block diagrams or logical trees), PN are particularly matched for representation of logical interactions among parts or activities in a system in a natural way. Typical situations that can be modelled by PN are: synchronization, concurrency and conflict [1,2].

**Definition.** Petri Net $M(C,\mu)$ pair, where $C=(P,T,I,O)$ is the network structure and $\mu$ is the network condition. In structure $C$ of a $P$-positions, $T$-transitions are finite sets. $I:T \rightarrow P^{\infty}, O:T \rightarrow P^{\infty}$ are the input and output functions, respectively, where $P^{\infty}$ are all possible collections (repetitive elements) of $P$. $\mu:P \rightarrow N_0$ is the function of condition, where $N_0 = \{0,1,...\}$ is the set of integers. We determine (in a known manner) the allowed transitions of Petri Nets and the transitions from one state to another, as well the set of reachable states.

Places, transitions, and arcs are the basic Petri Net components. A Petri Net can be thought of as a bipartite graph consisting of two types of nodes, places and transitions. Places are displayed pictorially as circles (or ovals) and transitions are displayed as rectangles. An example Petri Net consisting of two places P1 and P2 and one transition T2 is shown in Figure 1. Note that arcs connect a place to a transition or a transition to a place, but they do not connect two places or two transitions.

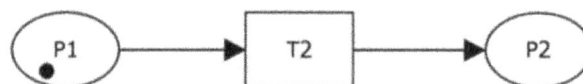

Fig. 1. Basic Petri Net configuration

The interpretation of places and transitions depends on the system being modeled. Places could represent resource status or operations. Arcs often represent the flow of data or resources. Transitions could represent the start/finish of processes. In terms of simulations, transitions can be used to model both *activities* and *events*. Activities can be thought of as the processes and logic of the system, while events occur at a single point in time and cause a change in the state of a system (White and Ingalls 2009). In fact, a transition may act as a super-process consisting of many sub-processes. This is where hierarchical nets come into play (which we will explain later). Often transitions can change the state of a net through the manipulation of *tokens* via the *firing rule* which is explained next.

Colored Petri Net (CPN) are considered as Classical Petri Net modern expansion which was created by K. Jensen [3]. Colored Petri Net (CPN) is a graphical oriented language for design, specification, simulation and verification of systems. It is in particular well-suited for systems that consist of a number of processes which communicate and synchronize.

Typical examples of application areas are communication protocols, distributed systems, automated production systems, work flow analysis. The CPN language allows the model to be represented as a set of modules, allowing complex nets (and systems) to be represented in a hierarchical manner. In the classical or traditional Petri Net tokens do not differ from each other, we can say that they are colorless. Unlike Classical Petri

Net in Colored Petri Net of a position can contain tokens of arbitrary complexity, example, lists, etc., that enables modeling more reliable models.

**Definition.** The mathematical definition of Colored Petri Net: CPN is a nine-tuple

$$CPN = (\Sigma, P, T, A, N, C, G, E, I), \text{ where:}$$

1. $\Sigma$ is a finite set of non-empty types, also called color sets. In the associated CPN Tool, these are described using the language CPN-ML [6]. A token is a value belonging to a type.

2. $P$ is a finite set of places. In the associated CPN Tool these are depicted as ovals/circles.

3. $T$ is a finite set of transitions. In the associated CPN Tool these are depicted as rectangles.

4. $A$ is a finite set of arcs. In the associated CPN Tool these are depicted as directed edges. The sets of places, transitions, and arcs are pairwise disjoint, that is

$$P \cap T = P \cap A = T \cap A = \varnothing.$$

$N$ is a node function. It is defined from $A$ into $P \times T \cup T \times P$. In the associated CPN Tool this depicts the source and sink of the directed edge.

5. $C$ is a color-function, $C : P \to \Sigma$.

6. $G$ is a guard function. It is defined from $T$ into expressions such that:

$$t \in T : [Type(G(t)) = B \& Type(Var(G(t))) \subseteq \Sigma]$$

7. $E$ is an arc expression function. It is defined from A into expressions such that:

$$\forall a \in A : [Type(E(a)) =$$
$$= C(p)_{MS} \& Type(Var(E(a))) \subseteq \Sigma],$$

where $p$ is the place of $N(A)$ and $C(p)_{MS}$ denotes the multi-set type over the base type $C(p)$.

8. $I$ is an initialization function. It is defined from $P$ into closed expressions so that:

$$\forall p \in P : [Type(I(p)) = C(p)_{MS}].$$

The distribution of tokens, called marking, in the places of a CPN determines the state of a system being modelled.

CPN models can be constructed using CPN Tools, a graphical software tool used to create, edit, simulate, and analyze models. CPN Tools has a graphical editor that allows the user to create and arrange the various Petri Net components. One of the key features of CPN Tools is that it visually divides the hierarchical components of a CPN, enhancing its readability without affecting the execution of the model. CPN Tools also provides a monitoring facility to conduct performance analysis of a system. In addition, unlike traditional discrete event systems, CPNs allow for state space based exploration and analysis, which is complementary to pure simulation based analysis. State space analysis can be used to detect system properties such as the absence of deadlocks.

The dynamic behavior of a CPN is described in terms of the firing of transitions. The firing of a transition takes the system from one state to another. A transition is enabled if the associated arc expressions of all incoming arcs can be evaluated to a multi-set, compatible with the current tokens in their respective input places, and its guard is satisfied.

CPNs are an extension of ordinary Petri Nets. Petri Nets can be used to model a wide range of various systems. Thus in a CPN model, tokens can be coded as data values of a rich set of types (called color sets) and arc inscriptions can be computed expressions and not just constants. So that's the fundamental idea of CPNs: tokens have types, and each token type has some data value associated with it. Below the fold, we'll look at how we do that and what it means.

Colored Petri Nets add a collection of extensions to the other elements of the net to take advantage of typed tokens carrying values:

Each place in the net is also assigned a data type, and can only hold tokens of its assigned type.

Incoming edges of a transition can have *conditions*: the transition is only enabled when some set of tokens from the source places satisfy the full set of conditions for its incoming edges. The conditions for the incoming edges of a transition can reference the values from other incoming edges of the same transition – so, for example, the conditions can require that the values of two tokens coming from different incoming values match.

The edges going out of a transition can have *expressions* specifying how to compute the values of tokens being produced by the transition. When a transition is successfully fired, the expressions on its outgoing edges are evaluated to produce new tokens to feed into the place at the end of the edge.

Let's look at a quick example of Figure 2 of CPN transition. Here's a very simple CPN. It's got 3 places: two of them have type (Int×String), and one has type (Int). The transition takes one token of the pair type, and one of the integer type; and it produces one token of the pair type. The edges coming into the transition declare names for the elements of the token values, and the edge leaving the transition describes how to generate the values for outgoing tokens. This little net starts with two tokens, (4,"Foo"), and (2). The transition only fires if the integer token has a value greater than or equal to one, and produces a token multiplying the two integers from the incoming token. So the transition would fire, consuming two tokens shown in the graph, and producing a token (8, "Foo") in the bottom place.

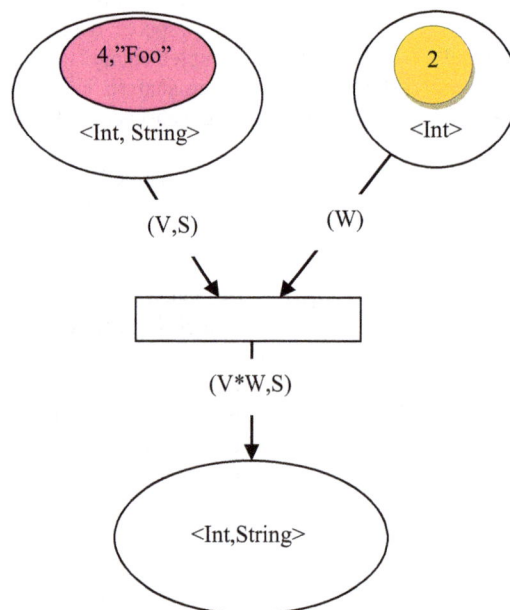

Fig.2  Example of CPN transition

For colored nets, there are a ton of variations. The basic idea is that there's some simple version of λ-calculus which is associated with elements of the CPN:

• Places are assigned lambda calculus types, which define the types of values carried by tokens that can be located in the place. At runtime, the place is a *bag* of tokens.

• Transitions are basically functions, where the incoming edges define a tuple of input parameters to the function, and the outgoing edges form a tuple of results from the function. For our example, the transition is, basically "&lamba; (v,s),w → (v*w,s)".

• The transition can have *conditions* for its firing: the condition will only be enabled for incoming token values which meet the condition. In the example, the condition is that the value coming on the edge from the integer-place must have a value greater than or equal to one. This turns the lambda function into a guarded partial function: in our example,

"&lamba; (v,s),w | w >= 1 → (v*w,s)".

Biometrics is often used by companies, governments, military, border control, hospitals, banks etc. to either verify a person's identity, for physical access control, computer log-in, welfare disbursement, international border crossing and national ID cards, e-passports, allowing access to certain building area or to identify individuals to retain information about them, i.e. criminals, forensics, etc. In automobiles, biometrics is being adopted to replace keys for keyless entry and keyless ignition [7].

The objective of a biometrical identification system is to identify individuals on the basis of physical (passive or active) features. One of the oldest and probably best known of such features is the human fingerprint. One can safely say that for a long time fingerprinting-based identification and biometrical identification have been seen as one and the same thing.

The last decade other human features have become practical, and there is now an active research community on iris-based recognition, face recognition and others [8].

Biometric identification systems were studied by O'Sullivan and Schmid [9] and Willems et al. [10]. They assumed storage of biometric enrollment sequences in the clear and determined the corresponding identification capacity. Later Turcel [11] analysed the trade-off between the capacity of a biometric identification system and the storage space (compression rate) required for the biometric templates. It should be noted that Turcel's method realizes a kind of privacy protection scheme. Recall that secrecy capacity introduced by Ahlswede and Csiszar [12] can be regarded as the amount of common secret information that can be obtained in an authentication system in which helper data are (publicly) available. Interestingly this secrecy capacity, which is equal to the mutual information between enrollment and authentication biometric sequences in the biometric setting, equals the identification capacity found by O'Sullivan and Schmid and Willems et al.

## 2. Model description

Biometrical identification in general involves two phases (Fig.3). In an enrollment phase all individuals are observed and for each individual a record is added to a database. This record contains enrollment data, i.e. a noisy version of the biometrical data corresponding to the individual. In the identification phase an unknown individual is observed again.

The resulting identification data of an unknown individual, is compared to (all) the enrollment data in the database and the system has to come up with an estimate of the individual.

Essential in this procedure is that both in the enrollment phase and in the identification phase noisy versions of the biometrical data are obtained. The actual biometrical data of each individual remains unknown.

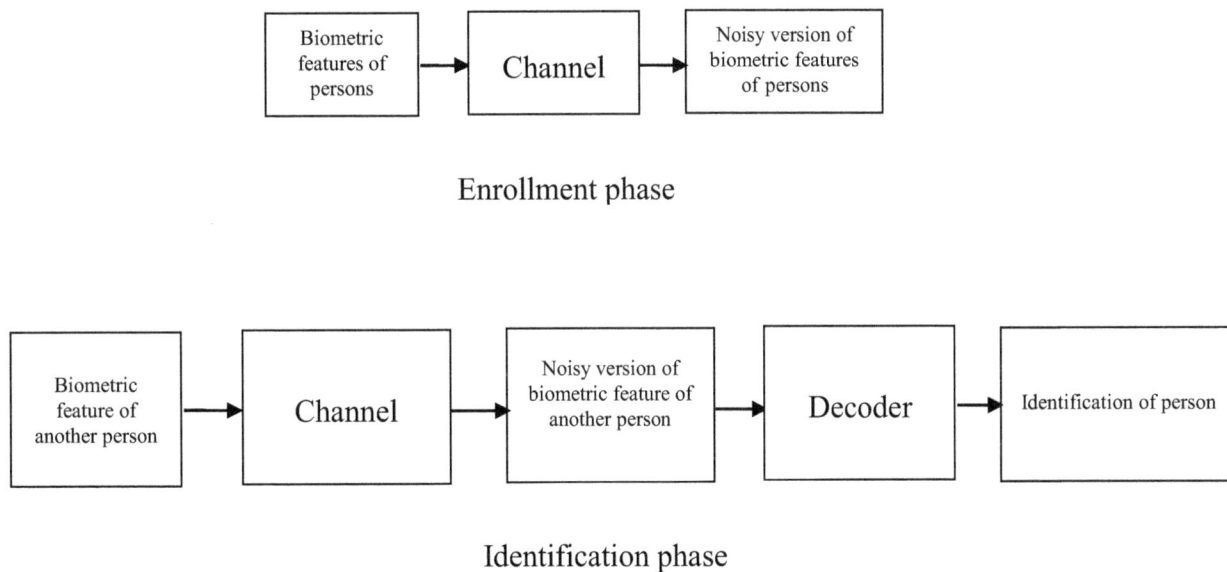

Enrollment phase

Identification phase

Fig3. The model of biometric identification system

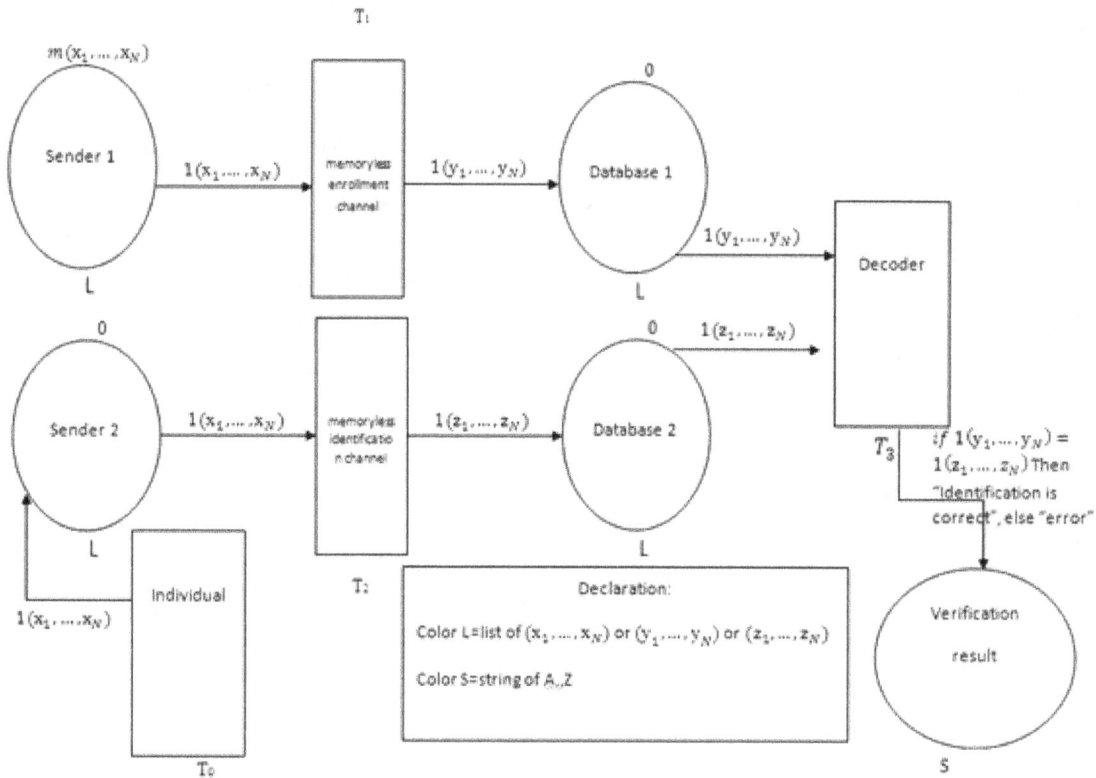

Fig.4 The model of biometric identification system by Colored Petri Net

In Fig. 4 Colored Petri Net consists of the following positions: *Sender 1, Sender2, Database1, Database2, Verification result* and the following transitions: $T_0, T_1, T_2, T_3$, which have corresponding names. *Sender1* includes N-dimensional biometric data from m-dimensional set. Element $(x_1,...x_N)$ passes through the channel to $T_1$ transition, and its corresponding encoded data $(y_1,...y_N)$ is placed in *Database1*.

After the firing of $T_0$ transition, the element of m-dimensional set is placed in *Sender2* position, after that the result of firing of $T_2$ transition, through the respective channel, it is obtained the encoded data $(z_1,...z_N)$, which is placed in the *Database2* position.

Through the firing of $T_3$ transition, it is checked the correspondence of vectors $(y_1,...y_N)$ and $(z_1,...z_N)$, gets the responce: if the identification is correct, or there is an error in the system. The arc, which is directed from $T_3$ transition to the *Verification result* position, is bound by a corresponding logical expression.

Then firing of $T_0$ transition, the next element can queue up and the cycle can be repeated again.

*Sender1, Sender2, Database1, Database2* positions that are attached to the **L** type, which is a set of N-dimensional vectors. The **S** type is attached to *Verification result* position, which is presented as a type string. The corresponding information about the types is shown in the declaration table.

### 3.  Conclusion

In the result of further studies, Colored Petri Network will be built, modelling complex Biometric Identification Systems, which will explore verification, validation, error detection and functional interactions problems in the Biometric Identification Systems.

The obtained results of the study will be used for processing of biometric data used in medical equipment, reducing time and resource consumption.

## REFERENCES

Peterson, James Lyle (1981).
Petri Net Theory and the Modelling of Systems. Prentice Hall.
ISBN 0-13-661983-5

Tadao Murata. "Petri nets: Properties, Analysis and Applications." Proc. of the IEEE, 77(4), 1989.

K. Jensen. Coloured Petri Nets: Basic Concepts, Analysis Methods and Practical Use. Springer - Verlag, Berlin, 1992.

Jensen K. Coloured Petri Nets: Basic Concepts, Analysis Methods and Practical Use. Springer, 1996. Vol. 1–3.

S. Pankanti, R. M. Bolle and A. Jain, "Biometrics-The Future of Identification", IEEE Computer, vol. 33, no.2, pp. 46-49, February, 2002.

J. D. Ullman, "Elements of ML Programming," Prentice- Hall, Upper Saddle River, 1998.

S. Pankanti, R. M. Bolle and A. Jain, "Biometrics – The Future of Identification", IEEE Computer, V33, N2, pp. 46-49, 2002.

J. A. O'Sullivan and N. A. Schmid, "Performance prediction methodology for biometric systems using a large deviations approach", IEEE Trans. On it Signal Proc., vol. 52, no. 10, pp. 3036-3045, 2004.

F. WIllems, T. Kalker, J. Goselig, and J.-P. Linnartz, "On the capacity of a biometric identification system", Intenational Symposium on Information Theory, Yokohama, Japan, p. 82, 2003.

E. Trucel, "Capacity/storage tradoff in high-dimentional identification system", IEEE International Symposium on Information Theory, Washington, USA, pp 1929-1933, 2006.

R. Ahlswede and Csiszar, "Common randomness in information theory and cryptography – Part I: Secret sharing", IEEE Trans. Information Theory, vol. IT-39, pp. 1121-1132, July 1993.

# 11

# EXAMINATION ABOUT INFLUENCE FOR PRECISION OF 3D IMAGE MEASUREMENT FROM THE GROUND CONTROL POINT MEASUREMENT AND SURFACE MATCHING

T. Anai [a], N. Kochi [a,d, *],  M. Yamada [a], T. Sasaki[a], H. Otani[b], D. Sasaki [b],
S. Nishimura [c], K. Kimoto[c], N.Yasui [c]


a General Technology Div., R&D Dept., TOPCON CORPORATION, 75-1, Hasunuma, Itabashi, Tokyo - t.anai@topcon.co.jp
b Smart Infrastructure Company, Technology Development Dept., TOPCON CORPORATION, 75-1, Hasunuma, Itabashi, Tokyo
[c] Keisoku Research Consultant Co.,Ltd.Creative design group
[d] R&D Initiative, Chuo University, 1-13-27, Kasuga, Bunkyo-ku, Tokyo, Japan

**WG IV/7 and WG V/4**

**KEY WORDS:** 3DModeling, Accuracy, Matching, Orientation, Surface measurement, UAV

**ABSTRACT:**

As the 3D image measurement software is now widely used with the recent development of computer-vision technology, the 3D measurement from the image is now has acquired the application field from desktop objects as wide as the topography survey in large geographical areas. Especially, the orientation, which used to be a complicated process in the heretofore image measurement, can be now performed automatically by simply taking many pictures around the object. And in the case of fully textured object, the 3D measurement of surface features is now done all automatically from the orientated images, and greatly facilitated the acquisition of the dense 3D point cloud from images with high precision. With all this development in the background, in the case of small and the middle size objects, we are now furnishing the all-around 3D measurement by a single digital camera sold on the market. And we have also developed the technology of the topographical measurement with the air-borne images taken by a small UAV [1~5].

In this present study, in the case of the small size objects, we examine the accuracy of surface measurement (Matching) by the data of the experiments. And as to the topographic measurement, we examine the influence of GCP distribution on the accuracy by the data of the experiments. Besides, we examined the difference of the analytical results in each of the 3D image measurement software.

This document reviews the processing flow of orientation and the 3D measurement of each software and explains the feature of the each software. And as to the verification of the precision of stereo-matching, we measured the test plane and the test sphere of the known form and assessed the result. As to the topography measurement, we used the air-borne image data photographed at the test field in Yadorigi of Matsuda City, Kanagawa Prefecture JAPAN. We have constructed Ground Control Point which measured by RTK-GPS and Total Station. And we show the results of analysis made in each of the 3D image measurement software. Further, we deepen the study on the influence of the distribution of GCP on the precision.

## 1. INTRODUCTION

As the 3D image measurement software is now widely used with the recent development of computer-vision technology, the 3D measurement from the image is now has acquired the application field from desktop objects as wide as the topography survey in large geographical areas.

However, for the measurement of higher accuracy, we must make the control point measurement on those images, as well as the geometric correction by orientation devices. When we make the topographical measurement of the air-borne images by a small UAV, often times we cannot obtain the sufficiently accurate coordinates by GPS-IMU boarded on a small UAV. Thus, the model obtained would be not of the real scale but of the relative scale. To solve this problem, we can put lots of Ground Control Points (GCP) and level up the accuracy of the orientation by measuring the GCP on the display. But at the time of natural disaster, often times it is difficult to place

enough GCP and it has not in the idealistic distribution. Besides, among the each of the software of 3D measurement, the method of orientation is different and it could create the difference in the results.

With this experiment, we made assessment separately on stereo-matching (point cloud generation) function and on the accuracy of the orientation in the control points distribution. With the experiment of surface measurement, as we wished to assess the efficiency of the stereo-matching itself, we put on the object as much feature points as possible. And for our experiment we chose a small object which is easy to assess the shape and useful for practical purposes. Actually we chose a mannequin, which has the surface and spherical shape and allowed us to obtain the point cloud with high precision by contact type 3D measuring machine. To verify the accuracy of the orientation in distribution of control points, we used UAV hovering over the testing area, where we had already measured the control points

---

\* Corresponding author.  This is useful to know for communication with the appropriate person in cases with more than one author.

by Total Station (TS) to photography and analyse. To find the effect of different ways of distributing control points, we made 3 patterns of distribution: total area, straight line and pinpoint and assessed their accuracy.

Now first we would dwell on the measuring flowchart of the each software and their features, then report on the accuracy of surface measurement (stereo-matching) as well as on the accuracy verification of different ways of the distribution of control points. And at the end we would summarize the whole system.

## 2. EXPERIMENTAL ASSESMENT

### 2.1 The flowchart of measuring process of each software

In our research, as a software which makes 3D measurement and modelling, we chose following 3 kind: Agisoft PhotoScan, Accute3D Smart3DCapture, Topcon ImageMaster UAS. PhotoScan and Smart3DCapture have sfm (structure from motion) of computer vision, as the basic principle. Whereas, ImageMaster UAS has photogrammetry as the basic principle. Each has greatly different process, method of calculation and GUI of its own. Table 1 shows the general specification of software we used.

| | Smart3DCapture | Photoscan | ImageMaster UAS |
|---|---|---|---|
| Ver. | Expert edition Ver. 3.1.0.3700 | Professional Edition Ver. 1.0.4 | Ver. 3.0 |
| Principle | Structure from motion | Structure from motion | Photogrammetry |
| GeoTag input | Exif | Exif | UAS Logger (attachment soft) |
| Camera Calibration | No (Parameters input) | Yes | Yes (necessary) |
| Orientation (Alignment) | Automatic | Automatic | Automatic/Manual |
| Control points (image) measurement | Manual | Manual/ Semi-automatic | Manual/ Semi-automatic |
| Scale Input | Ok | Yes | Yes |
| Point cloud generation (matching) | Automatic: Feature Based | Automatic: Feature  Based | Automatic: Area Based |
| Editing | No | Yes | Yes |
| Measurement | Distance (Smart 3DViewer) | Distance/Area/Volume | Distance/Area/Volume/ Contour/Line Profile |
| Ortho-photo | No | Yes | Yes |

Table 1. General specification of software

In all software manual process for the production of 3D model is required, but automatic process is possible. So, it does not demand much labour. The big difference in the software lies in the control-point-measurement function, in camera-calibration function (all software parameters input is possible), in point cloud editing function, and in 3D model measuring function. The control points setting is laborious, if the images are numerous. But this work is indispensable to obtain the result highly accurate and reliable.

### 2.1.1 Accute3D Smart3DCapture[6]: This time we used the ver.3.1 of Accute3D Smart3DCapture expert edition. Figure 1 shows its flowchart. As in the Figure, the process of Smart3DCapture is simple. But the measurement of control points is manual and much laborious, especially the images are

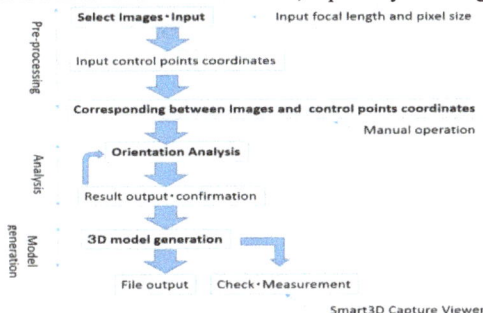

Figure 1. Measuring flowchart of Smart3DCapture

numerous as in the case of UAV. Unless we get at least 3 images the analysis is not possible. As this software perform the camera-calibration automatically, it does not have software for it, but the input of camera parameters is possible. But once the model is made, we can use other software (Smart3DCapture viewer), though limited to confirmation of the result and some simple measurement points. This software works only for minimum necessity, i.e. model production.

### 2.1.2 Agisoft PhotoScan[7]: In standard edition, there is no control points inputting function. So, we used PhotoScan professional edition ver.1.0.4. As this software is highly flexible, we can make 3D model very easily. But, in order to make the precise measurement by inputting the control points, first we have to fully understand the settings and handling order and then work on measurement. At one glance, we might feel we could make any kind of model, but if we want to measure without error, we have to very attentive. Figure 2 is the flowchart of measurement.

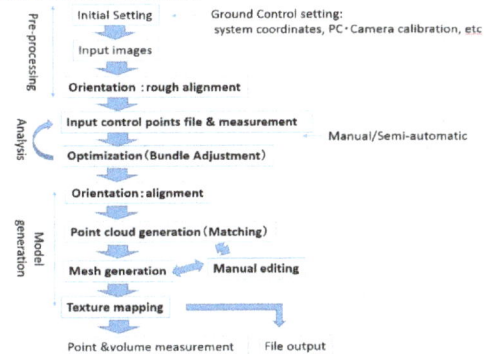

Figure 2. Measuring flowchart of PhotoScan

The measurement of control points is manual, but if the number of control points increases, the possible candidate points are automatically indicates. However, to decide the candidate points individually as one by one, the manual operation is required. The measurement result can be made and output automatically.

### 2.1.3 Topcon ImageMasterUAS[8]: As to the Topcon's ImageMasterUAS is especially defined for UAV, it has the software called Logger and Planner. And as this software is basically for the photogrammetry[1,4,5,9,10], it puts importance more on measurement, comparing with other software First, by Logger we determine the correspondence of the site with the images and information obtained by GPS mounted on UAV. By this function, we can determine, on the spot, the flight position of UAV and the distribution of the control points and we can exclude the wrong coordinate system as well as the misunderstanding of the photographing position. Figure 3 shows the flowchart of the process.

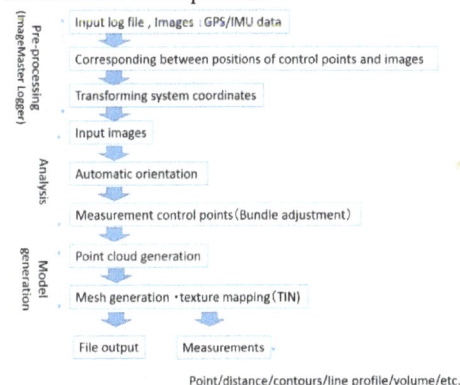

Figure 3. Measuring flowchart of the ImageMasterUAS

As to the function of control-points measurement, if we use the round target, the image measurement by image processing become feasible [9]. When the images are numerous, as in the case of UAV, the measurement of control points with no personal influence is possible. Besides, it is also labour-saving. In the control point measurement, when we measure manually several points, the automatic measurement becomes available. As the result of automatic measurement is shown in a table, we can amend it by looking at it, if necessary. And on the 3D model thus produced, we can make editing or perform the shape measurement such as the cross-section and contour lines.

## 2.2 The accuracy of surface measurement

For the assessment of accuracy on surface measurement, we used the test plane and test sphere of AIST (national institute of Advanced Industrial Science and Technology, Japan). Also, with the point cloud of the mannequin which we measured with contact type 3D measuring machine as a yardstick, we compared and assessed the point cloud of mannequin which we were measured by image measuring. The verification of these data is to make a thorough examination of the effect of the presence of pre-calibration which is the element of inner orientation in each software and also the effect of autofocus. So, we starts the test with minimum of 2 images.

Because of the limited capacity of the software: For Smart3DCapture we used at least 3 images and did not work on pre-calibration. For ImageMaster we used 2 images and worked on pre-calibration. For PhotoScan we used 2 images and more, and worked on both with and without pre-calibration. As seen in Figure 4, the measurement was done by projecting random dot pattern [10] and changing camera position we took picture with scale bar. The camera we used was Sony digital SLR ILCE(Alpha) 6000, and lens was macro-lens f 30mm.

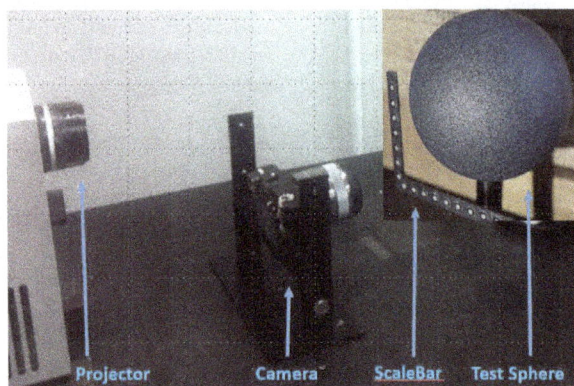

Figure 4. Scene of measurement

### 2.2.1 Test plane measurement: Figure 5 shows the measured test plane. The deviation from flatness was 2.3 micron.

Figure 5. Test plane

In this experiment the measurement resolution is 0.38mm/pixel. Assessment scale is the deviation from flatness, i.e., we have set it to be the maximum of error of flatness. The unit is mm. We checked the effect of the pre-calibration by looking at the result when it is done and when it is not done by 2 images. Also we checked whether, in the case of plural number of pictures, the

effect of autofocus exist or not by changing photo-angles in various manner. As seen from Table 2, when we made pre-calibration, the flatness of pixel resolution was obtained. So, there was no trouble. When we did not perform the pre-calibration, the result was unstable and not trustworthy.

Table 3 shows the result (deviation from flatness) of autofocus picture-taking by automatic calibration. From this result, we found out that as the number of pictures increases there is the case where the accuracy increases, but also decreases.

| | A | B | | | | B | C |
|---|---|---|---|---|---|---|---|
| | Pre-Calibration | Pre-Calibration | without Pre-Calibration | AF: 4Images | Case1 | 0.18 | 0.25 |
| | | | | | Case2 | 0.17 | 0.22 |
| Case1 | 0.38 | 0.43 (no-stability) | 0.71 | AF: 11Images | Case1 | 0.5 | 0.2 |
| Case2 | 0.31 | 0.38 | 0.66 | | Case2 | 0.47 | 0.12 |

Table 2(left). With and without pre-calibration,
Table 3(right). Autofocus
The result of test plane measurement (unit: mm)

### 2.2.2 Test sphere measurement: From the point cloud obtained by matching, we calculated the maximum diameter and minimum diameter by spherical approximation fitting. And we compared its average value with the diameter of the sphere, which was measured by the contact type 3D measuring machine. The accuracy of measuring of contact type 3D measuring machine (Zeiss Accura-J5) was 5 micron and the diameter of the sphere was 101.29mm. Table 4 shows the result when the pre-calibration was used and when it was not used by stereo photography.

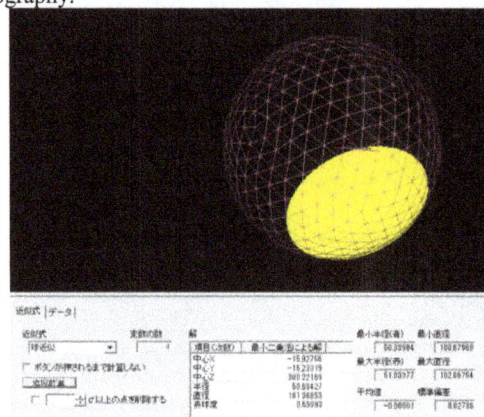

Figure 6. Spherical approximation fitting

In this experiment, the depth resolution was 0.21mm. Evidently the accuracy was better when we made pre-calibration. Table 5 shows the result of autofocus photographing. We can confirm the accuracy change by photographing. When we detect the diameter of sphere from point cloud, the size of the area where we make spherical approximation fitting (Figure 6: the area we could make matching) and even very minute error or aberration affects the result greatly. And for this reason, the assessment was difficult and had to be performed very carefully.

| | A | B | | | | B | C |
|---|---|---|---|---|---|---|---|
| | With Pre-Calibration | With Pre-Calibration | Without Pre-Calibration | AF: 5 Images | Min. diameter | 100.70 | 100.67 |
| | | | | | Max. diameter | 101.27 | 101.38 |
| Min. diameter | 100.31 | 101.18 | 107.83 | | Average. | 100.98 | 101.02 |
| Max. diameter | 101.71 | 102.44 | 109.43 | | Error | -0.31 | -0.27 |
| Average. | 101.00 | 101.81 | 108.63 | AF: 10 Images | Min. diameter | 100.68 | 101.17 |
| Error | -0.28 | 0.52 | 7.34 | | Max. diameter | 102.07 | 102.77 |
| | | | | | Average. | 101.37 | 101.97 |
| | | | | | Error | 0.08 | 0.68 |

Table 4(left). With and without pre-calibration
Table 5(right). Autofocus
The result of test spherical measurement (unit: mm)

**2.2.3    Mannequin measurement**: For the measurement of test plane and test sphere, their shape is converted into numerical values and if they are dispersed, their average is used for comparison. For this reason, in some case much noise is created and, depending on the size of the area, the numerical values fluctuate greatly. And in reality we cannot see the details of the point cloud. Thereupon, we measured at mannequin at 2014 points with contact type 3D measuring machine and same parts were correlated to each other through their point cloud. Especially we compared the difference at the area of eyes, where local variation is great. For registration we used ICP method [11] with points to areas. For photographing, we first fixed focal length and took pictures. As to A, we took 2 pictures. As to B and C, we took 5 pictures. And as shown in Figure 7, we photographed and performed the surface measurement. Figure 8 shows the distribution of errors in the result. Also Table 6 shows the quantity of errors (standard deviation: mm). In this measurement, depth resolution was 0.17mm. While from the Table 6, we cannot see the difference numerically, but from histogram we can see how much the part of the eyes, where its inclination is sharply changed, can be expressed.

Figure 7. Pictures

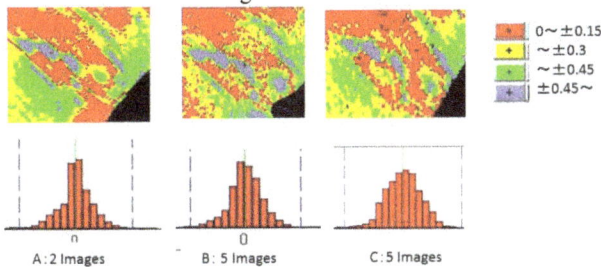

Figure 8. Distribution of errors

Furthermore, in order to investigate the influence of autofocus on the measurement, we compared the difference when we made 4 images of autofocus set up and when we added one more image to those pictures, The result is shown in the Table 7.

| | A | B | C |
|---|---|---|---|
| σ | 0.26 | 0.27 | 0.29 |
| Max | 1.62 | 2.27 | 1.34 |
| Min | -2.87 | -1.67 | -1.71 |
| Width | 4.49 | 3.94 | 3.05 |

Table 6. Distribution of errors

| | AF:4Images | | AF:5Images | |
|---|---|---|---|---|
| | B | C | B | C |
| σ | 0.27 | 0.27 | 0.24 | 0.28 |
| Max | 4.27 | 1.22 | 1.08 | 12.5 |
| Min | -1.78 | -1.81 | -2.81 | -1.59 |
| Width | 6.05 | 3.03 | 3.89 | 14.09 |

Table 7. Result of autofocus

This result shows that the standard deviation is not much, but the range of errors is widely dispersed (for example: the width of table 7 is 14.09mm). As the number of pictures and feature points became numerous and process number (feature quality) became great, average making effect became also great.

**2.2.4    Recapitulation of surface measurement:** The assessment of surface measurement required greatest prudence and care, because there are so many things are involved, such as, photographing process, feature patterns projection, orientation stabilization, wide measuring area, noise conquering etc.

However, when the number of pictures is small, definitely better result was obtained when we had performed pre-calibration. Again, when we increased the number of pictures and when we performed autofocus, the result was different in each case:

sometimes positive (became better), and at the other times negative (became worse). This indicates that as sfm works on the principle of multi-baseline stereo, the more we have the images the better becomes the resolution. But when the change of pictures is great (change of focal distance, or change of the amount of features), the system cannot absorb them all and brings out various values. Therefore, before we take up a work, we have to sturdy carefully the condition of camera and subject, because these elements affect essentially the final accuracy of the result.

**2.3    Sturdy on the distribution of control points: UAV image**

**2.3.1    Outline of the measurement**: The verification by UAV on the orientation accuracy of the distribution of control points was performed in Yadorigi test site in Kanagawa prefecture in Japan (110m x110m). On this site we had previously set up the control points and measured them by Total Station (TS) and RTK-GPS. Then we flew UAV and analysed its image data. Figure 9 shows the test site on which modelling was performed and control points for measuring. Table 8 shows the measuring conditions.

| Camera | Sony ILCE-6000 |
|---|---|
| Pixel | 6000x4000 |
| Lens | 20mm |
| Flight Height | 40m |
| Overlap | 80% |
| Baseline Length | 9m |
| Resolution | 11mm |
| Depth accuracy | 24mm |

Table 8. Measuring conditions

Figure 9. Test site

The camera used was Sony's digital SLR ILCE (Alpha)-6000 (6000x4000) with lens (f 20mm). Flight 1~4 were performed at about 40m high above the ground, overlapping about 80%, with the base length 9m. With this condition, the plane resolution is about 8mm and depth resolution is about 24mm. And the accuracy of observation of GCP was about 11mm. Figure 10 shows an example of 3D model produced.

Figure 10. Example of 3D reconstruction of topography

Analysis was performed with 6 control points chosen from those measured and the rest was used for verification point. In order to identify the error distribution we performed 4 flights and for each of the flight we made 3 different patterns: flight1234_A for total area, flight1234_B for straight line and flight1234_C for pinpoint. Figure 11 shows the distribution of control points in 3 kinds of patterns (flight1234_A, _B, _C). Big mark + is the control points, and small + is the mark for verification points.

**2.3.2 Result of measurement**: Table 9 shows the standard deviation and RMSE of the error of the coordinates x,y,z of verification points in each flight. What comes out over 40mm is written in bold. In most of cases we can measure within 40mm. However, as this value is not sufficient to grasp the result. Figure 12 shows it by contour lines after the residual of verification points were put into 3 dimension. From the pictures we can obtain the trend of the distribution of errors in the placement of control points.

**Flight pattern_ A** (control points spread in the entire area): with the minimum error at the control points, errors spread in the entire areas.

**Flight pattern_B** (control points forming a straight line): control points glow as a straight line with minimum error, and contour line is a straight line (error appear as a curved surface).

**Flight pattern_C** (control points unevenly distributed at pinpoint): here each of the control points has the least possible liability of error (center), but totally viewed, error trends to increase as it gets remote from the center, on the form of contour lines.

In the Flight pattern_A, we can see the difference in error distribution corresponding verification point among software. In the software A, error distribution is uniform for each control point, forming comparatively gentle error curved surface. Whereas, with the software B and C, while the accuracy around the control points is satisfactory, the error rapidly changes, if it gets remote from the control points. But judging from the result of the flight pattern_B and _C, while all the software A,B,C get effect of the unevenly distribution of the control points, especially in the flight pattern_B errors are numerous. The feature of the flight pattern_B is particularly manifested in the flight3_B, as control points are in the end of image, in all A,B,C the errors range from 133mm to several meters. Each of A,B,C has different value but as their errors are numerous, they are not useful for measurement. Especially the case of B in the flight3_B, other than the value Z, at a first glance looks satisfactory, but we cannot use it as measurement data in practical work. So, in this way, as we often tend to overlook, we have to be careful when we verify the result.

## 3. CONCLUSION

From our sturdy, we learned the following points.

1. In the measurement when the number of image is less than 2 images, it is indispensable to perform pre-calibration. But when it is numerous, it is not. The quality of measurement depends on the amount of the features of the object.

2. When we use autofocus, especially when the number of the images is small, we can no longer disregard the influence on the accuracy of the measurement (sudden change in focal length, photographing position, photographing angle). To make highly accurate measurement fixed focus should be used.

3. The distribution of control points has to be done evenly as a whole after having fully studied the necessary accuracy while abstaining desire for excessive quality.

4. When it is not possible to distribute evenly, it is better to distribute like encircling the area. This time, in the example of measurement, we obtained the sufficient accuracy below 2cm by encircling area.

5. When the distribution of control points becomes straight line, in the area remote from the straight line, the errors tend to become bigger by forming an inclined plane. So, we have to be careful.

6. We must avoid the distribution as much as possible, where the control points spread in pinpoint. As to the accuracy of area remote from the control points, reliability has to be carefully examined.

The measurement by images can be easily performed by anybody, because of the recent development of sfm method. And if an object has abundant features, we can easily make 3D model. As a result, it became possible to evaluate and measure the reconstructed model from many angles. But the quality of the product basically depends on the distribution of the control points and the photographing condition. To make the product useful to the purpose for which it was made (not to fail in measurement), we must examine very carefully its photographing condition and measurement condition to satisfy the necessary accuracy. After making the photographing plan, and distributing the control points, we must confirm them by simulation test. And only then, we must work on the actual photographing and measuring.

## References

[1]N.Kochi, "Photogrammetry", *Handbook of Optical Metrology, " Principles and Applications*, Yoshizawa,T.'(Ed.), Taylor and Francis, Chapter 22 (2009)

[2]M. Naumann. et. al., 2013. Accuracy Comparison of Digital Surface Models Created by Unmanned Aerial Systems Imagery and Terrestrial Laser Scanner, *ISPRS, Volume XL-1/W2, uav-g 2013* , Rostock, Germany, pp 281-286

[3]M. Bolognesi, et. al., 2014. Accuracy of Cultural Heritage 3d Models by Rpas and Terrestrial Photogrammetry, *Volume XL-5, ISPRS Technical Commission V Symposium*, 2014, Riva del Garda, Italy, pp113-119

[4]T. Anai, et. al., 2014. Aerial Photogrammetry Procedure Optimized for Micro Uav, *Volume XL-5, ISPRS Technical Commission V Symposium*, 2014, Riva del Garda, Italy, pp41-46

[5]N.Kochi, et. al., Robust surface matching by integrating edge segments, *ISPRS Annals, Volume II-5*, pp.203-210, doi:10.5194/isprsannals-II-5-203-2014, 2014.

[6]http://www.acute3d.com/ (15 Apr. 2015)

[7]http://www.agisoft.com/ (15 Apr. 2015)

[8]http://www.topcon.co.jp/positioning/products/product/3dscanner/imgmaster.html (15 Apr. 2015)

[9]N.Kochi, et. al., PC-based 3D Image Mesuring Station with Digital Camera An Example of Its Actual Application on a Historical Ruin : *ISPRS, Vol. XXXIV,Part 5/W12*, pp. 195-199. July 2003

[10]N. Kochi, et. al., "3D-Measuring-Modeling- System based on Digital Camera and PC to be applied to the wide area of Industrial Measurement", *SPIE Optical Diagnostics*, August, pp.588015-1-10 (2005)

[11]P.J.,Besl and N.D. McKay, 1992. A Method for Registration of 3-D Shapes. *IEEE Transaction on Pattern Analysis and Machine Intelligence*, vol.14, no.2, pp.239-256,

[12]M. Hess, et. al., 2014. A Contest of Sensors in Close Range 3D Imaging: Performance Evaluation with a New Metric Test Object, *Volume XL-5, ISPRS Technical Commission V Symposium*, 2014, Riva del Garda, Italy, pp277-284

[13]E. Dall'Asta, R. Roncella, 2014. A Comparison of Semiglobal and Local Dense Matching Algorithms for Surface Reconstruction, *Volume XL-5, ISPRS Technical Commission V Symposium*, 2014, Riva del Garda, Italy, pp187-194

[14]N. Haala, 2013. The Landscape of Dense Image Matching Algorithms. In: Fritsch, D. (Ed.): *Photogrammetric Week '13*, Wichmann, Berlin/Offenbach, 271-284.

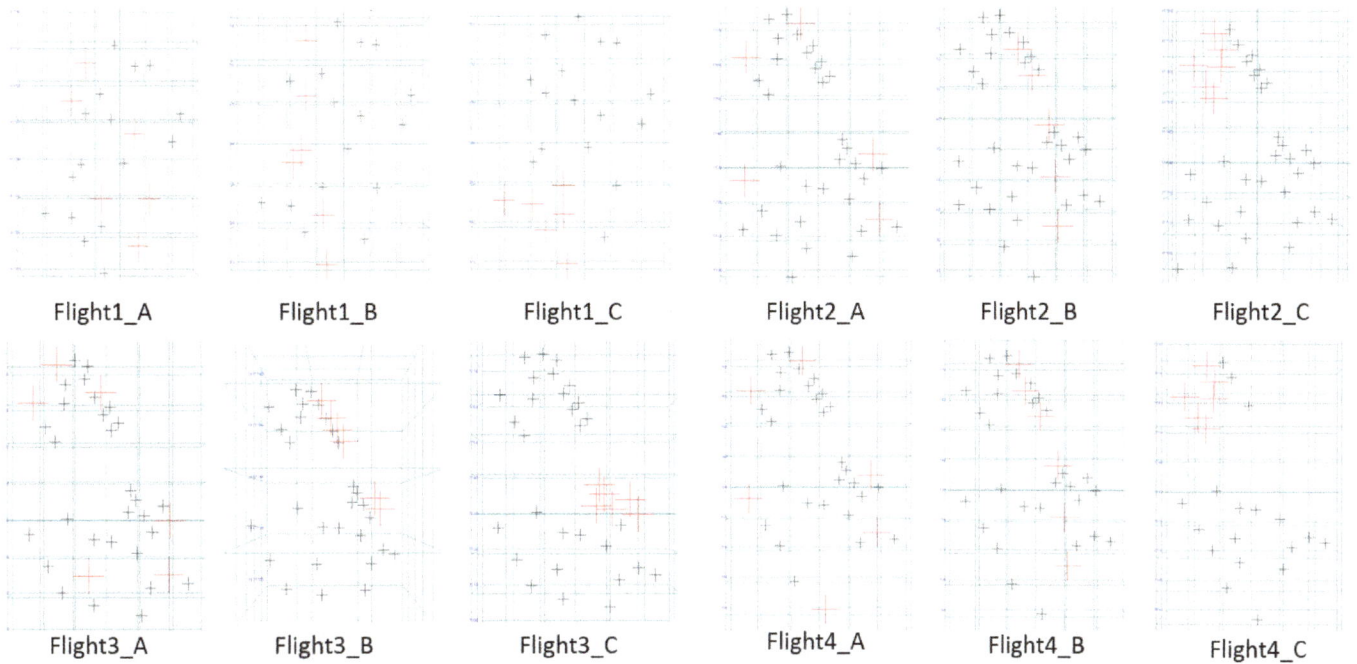

Figure 11. The distribution of control points in each flight

| Flight | Base Point Pattern | A | | | | B | | | | C | | | |
|--------|--------------------|--------|--------|--------|------|--------|--------|--------|------|--------|--------|--------|------|
| | | Sigma X | Sigma Y | Sigma Z | RMSE | Sigma X | Sigma Y | Sigma Z | RMSE | Sigma X | Sigma Y | Sigma Z | RMSE |
| Flight1 | Flight1_A | 0.008 | 0.019 | 0.014 | 0.014 | 0.013 | 0.017 | 0.006 | 0.013 | 0.026 | 0.013 | 0.022 | 0.021 |
| | Flight1_B | 0.010 | 0.018 | 0.031 | 0.021 | 0.014 | 0.016 | 0.019 | 0.016 | 0.024 | 0.014 | 0.028 | 0.023 |
| | Flight1_C | 0.031 | 0.036 | 0.020 | 0.030 | 0.011 | 0.012 | 0.015 | 0.013 | 0.021 | 0.013 | 0.023 | 0.019 |
| Flight2 | Flight2_A | 0.009 | 0.008 | 0.014 | 0.010 | 0.012 | 0.008 | 0.012 | 0.011 | 0.018 | 0.021 | 0.008 | 0.017 |
| | Flight2_B | 0.012 | 0.027 | 0.080 | 0.049 | 0.02 | 0.01 | 0.014 | 0.015 | 0.026 | 0.027 | 0.017 | 0.024 |
| | Flight2_C | 0.011 | 0.021 | 0.078 | 0.047 | 0.043 | 0.046 | 0.013 | 0.037 | 0.059 | 0.054 | 0.014 | 0.047 |
| Flight3 | Flight3_A | 0.007 | 0.007 | 0.024 | 0.015 | 0.012 | 0.009 | 0.009 | 0.010 | 0.024 | 0.024 | 0.011 | 0.021 |
| | Flight3_B | 2.993 | 4.624 | 5.042 | 4.312 | 0.04 | 0.033 | 0.133 | 0.082 | 0.052 | 0.052 | 0.273 | 0.163 |
| | Flight3_C | 0.015 | 0.012 | 0.029 | 0.020 | 0.019 | 0.006 | 0.032 | 0.022 | 0.025 | 0.028 | 0.034 | 0.029 |
| Flight4 | Flight4_A | 0.007 | 0.008 | 0.012 | 0.009 | 0.008 | 0.006 | 0.005 | 0.006 | 0.018 | 0.021 | 0.014 | 0.018 |
| | Flight4_B | 0.008 | 0.012 | 0.018 | 0.013 | 0.009 | 0.006 | 0.006 | 0.007 | 0.012 | 0.016 | 0.020 | 0.017 |
| | Flight4_C | 0.017 | 0.018 | 0.049 | 0.031 | 0.032 | 0.037 | 0.019 | 0.030 | 0.057 | 0.047 | 0.010 | 0.043 |

Table 9. Accuracy in the distribution of control points (unit m)

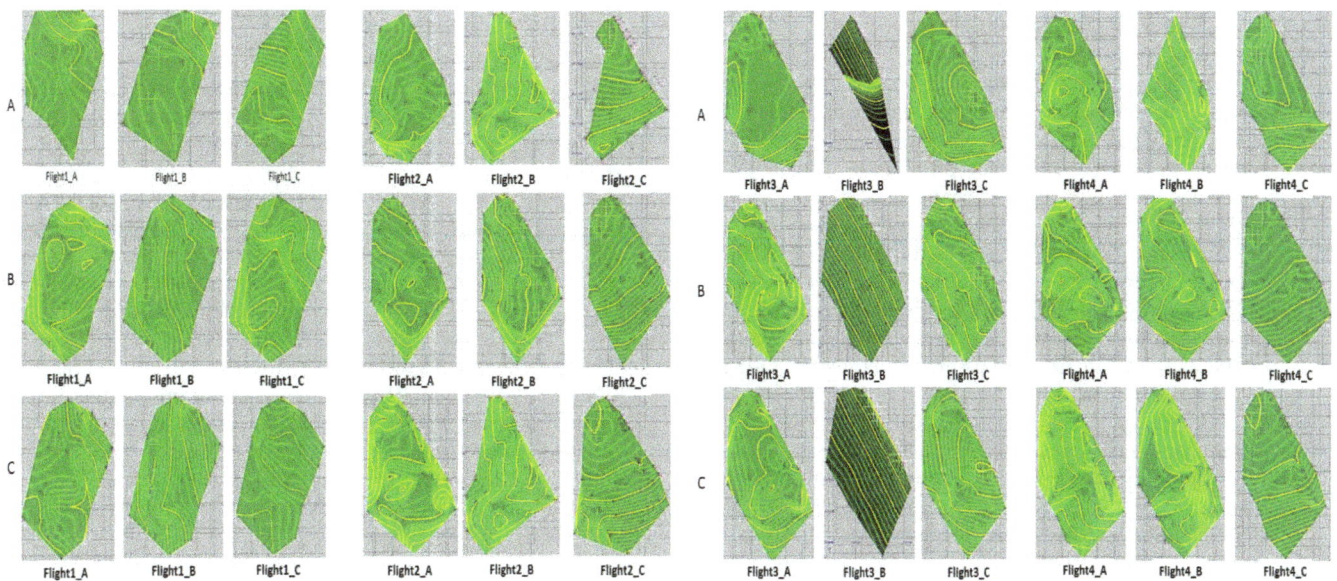

Figure 12. The error distribution in each flight

# 12

# LAND-USE AND LAND-COVER DYNAMICS MONITORED BY NDVI MULTITEMPORAL ANALYSIS IN A SELECTED SOUTHERN AMAZONIAN AREA (BRAZIL) FOR THE LAST THREE DECADES

D. Borini Alves [a*], F. Pérez-Cabello [a], M. Rodrigues Mimbrero [a],

[a] University of Zaragoza (UNIZAR),  Department of Geography and Spatial Management, and Aragon University Research Institute in Environmental Science (IUCA), Zaragoza, Spain – dborini@unizar.es, fcabello@unizar.es, rmarcos@unizar.es

**KEY WORDS:** NDVI, multitemporal dynamics, Landsat series, fire, deforestation, Brazilian Amazon.

**ABSTRACT:**

This study aims to analyse the dynamics of land-use and land-cover (LULC) in a selected southern Amazonian area (Brazil), monitoring and distinguishing trajectories in NDVI (Normalized Difference Vegetation Index) variations for the last three decades. The area, with a total of 17336 km², has been subject to significant LULC changes associated with deforestation progress and use of fire. Considering available Landsat time series, it was selected an image per year from 1984 to 2013 (path/row -231/66), at a particular period of year, atmospherically corrected using LEDAPS tools. NDVIs values were generated for each selected image. Furthermore, the images of 1984 and 2010 still underwent a classification of LULC differentiate five categories: water, forest, secondary/degraded forest, savannah/pasture and crop/bare soil. The trajectories in NDVI variation values were analysed by R software, considering intersections of classified categories. The pixels identified as forests on the images of 1984 and 2010 displayed stable trajectories of NDVI values, with average value 0.824 and coefficient of variation 3.9%. While the pixels of savannah/pasture, which was periodically affected by fire, had an average NDVI value 0.585 and coefficient of variation 15,1%. The main regressive trajectory was the transition "forest to crop/bare soil", identifying 1999 as the starting point in the drop in NDVI values, associated with an increase of the deforested areas. Therefore, the results show distinct trajectories associated with NDVIs and LULC changes that assist in better understanding the dynamics of ecological processes and the human impacts operating in the area.

## 1. INTRODUCTION

In recent decades, the southern Brazilian Amazon experienced increases in the dynamics of land use and land cover (LULC) associated with the expansion of crop and pasture areas, road and communication network construction, and population density growth (Espindola et al., 2012; Morton et al., 2006; Nepstad et al., 2001). This dynamics results in pressure on forest resources in the area due to the advance of deforestation, degradation and intensive use of fire (Aragão et al., 2008; Lima et al., 2012; Vasconcelos et al., 2013).

To monitor and better understand the dynamics of ecological processes and human impacts related to these changes in LULC, remote sensing data represents an essential source of analysis , allowing to generate systematic information in most different spatial and temporal scales (Nagendra et al., 2013). Two of the most important initiatives monitoring deforestation and fires in the Brazilian Amazon, the Assessment of Deforestation in Brazilian Amazonia (PRODES) (INPE, 2014a) and the *Queimadas* project (INPE, 2014b), have fundamental remote sensing data base to scale and generate information on the use of fire and deforestation in this area, contributing to the area management process.

Regarding remote sensing techniques, the generation of vegetation indexes calculated from the combination of spectral bands stands out (Bonham, 2013; Turner et al., 2003). Also, the NDVI (Normalized Difference Vegetation Index) is one of the most used index. NDVI relates spectral information of the red and near infrared generating a variable able to estimate quantity, quality and development of vegetation (Pettorelli et al., 2005).

To calculate this index, it is possible to count on lifting decades of satellite sensor information, which highlights the potential of the Landsat time series (Devries et al., 2015; Ding et al., 2014; Du et al., 2010; Maxwell and Sylvester, 2012; Zheng et al., 2015). This time series provides free access to an extensive gallery of relevant temporal and spatial resolution images, widely used and validated in scientific research in remote sensing.

In this context, this study aims to analyze the dynamics LULCs in a particular southern Amazonian area in Brazil, monitoring and distinguishing trajectories in NDVI variations considering the period between 1984 and 2013. We explore the continuity of Landsat series data to generate vegetation indexes that help to understand the spatial dynamics of the selected area.

### 1.1 Study area

The area covers 17336 km² and is located in the southern Brazilian Amazon (Figure 1), including cities of Amazonas state (Nova Aripuanã, Manicoré and Humaitá), Rondônia state (Machadinho d'Oeste, Cujubim and Porto Velho) and Mato Grosso state (Colniza). The main river that crosses the area is Machado River and belongs to Madeira River basin.

---

* Corresponding author.

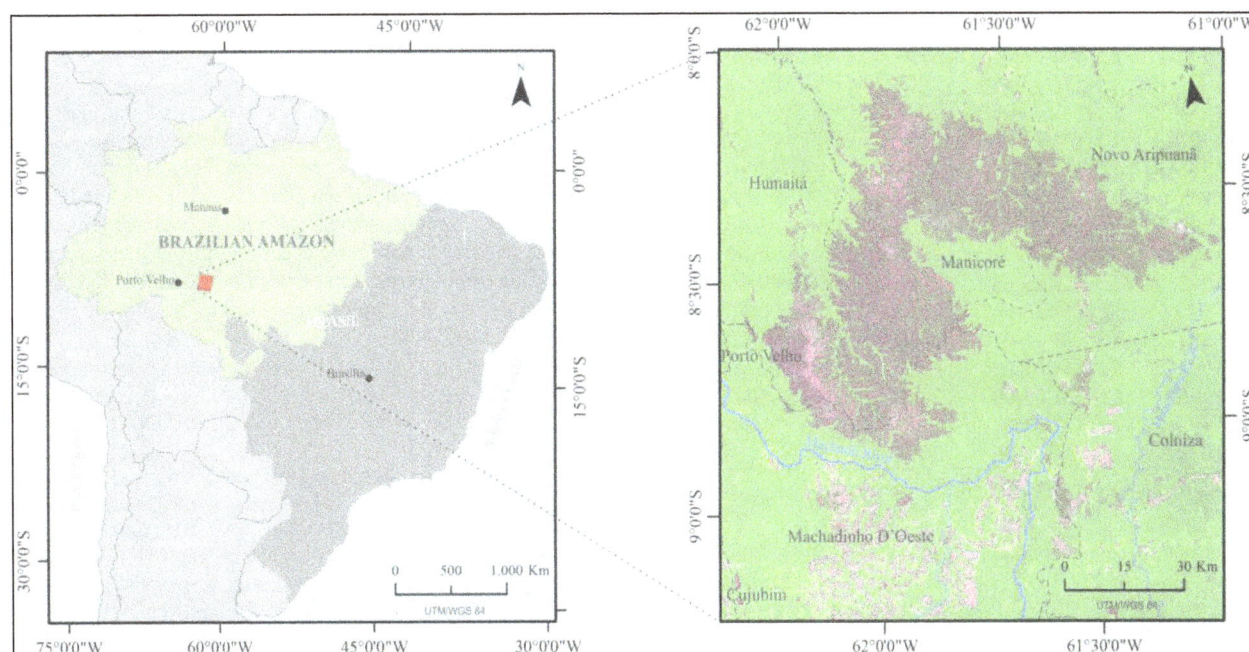

Figure 1. Study site location (red polygon) in the south of Brazilian Amazon (green polygon in the left map). In the right map, the municipal division of the study area on a clip image Landsat ETM+ (path/row-231/66, composition RGB-643) of August 28, 2013.

The study area currently consists of some regions of agro-pastoral activities, natural conservation areas (including a part of the Campos Amazônicos National Park) and indigenous lands. In zones of agro-pastoral activities, pastures with cattle creations and grain crops predominate, using fire as the main form of management.

In terms of phytogeographic composition, the study area has a savannah vegetation enclave area (Ratter et al., 2003) in a predominant Amazon biome area. The savannah area, locally identified as *cerrado* , displays grasses and shrubs (ICMBio, 2011), while the Amazon area presents tree species from rain forests, typical of the Amazon forest. Thus, this savanna-forest interface results in an area of high biological diversity.

The climatic conditions of the region have high annual average temperatures, ranging between 24°C and 28°C, and an annual rainfall of up to 2000 mm, with the presence of a dry period extending from May to October. However, variations of the dry season may happen in certain years. For example, remote sensing techniques registered, spatially and temporally, the effects of regional climate phenomena El Niño, the Pacific Decadal Oscillation and the Atlantic Multidecadal Oscillation (Marengo et al., 2011, 2008; Phillips et al., 2009).

During the dry period, hot pixel detections reveled an intensive use of fire for the management of agro-pastoral practices (Silvestrini et al., 2011). The use of fire tended to increase exponentially with the decrease of rainfall related to the dry season (Aragão et al., 2008).

This region has experienced significant LULC changes in the last decades associated with deforestation progress and use of fire as a main management instrument to the agro-pastoral practices. The main reason for these changes is the fact of being located in the midst of the agricultural frontier zone in a large area known as 'arc of deforestation' of Brazilian Amazon.

## 2. METHODOLOGY

### 2.1 Data select and preparation

Considering available Landsat time series, download at <http://glovis.usgs.gov/>, we selected an image per year (Table 1) from 1984 to 2013 (path/row - 231/66). All images belong to a particular time of year associated with the dry season in the study area, between the end of June and the end of August. The maximum daily difference of the selected images to the time series is 78 days between the images of 10/09 (2009) and 23/06 (1992).

Regarding our image selection criterion, we considered that the better pixel quality (based on the information provided by the supplier), the lower percentage cloud cover and the closer to the end of the dry season.

We gave priority to the later images on dry period seeking the maximum of the dynamic information in a given year and reducing stationary phenological effects of multitemporal analysis. This is because the closer images to the early dry season record more vigorous vegetation stages most often associated with high rainfall rates in the remaining months of the year (Kobayashi and Dye, 2005).

In order to generate a land surface reflectance for all selected images we used the software LEDAPS tools (Masek et al., 2006), which performs an atmospheric correction for the Landsat reflective bands using the MODIS/6S radiative transfer approach (Vermote et al., 1997). The algorithm runs considering an ancillary data of NCEP (National Centers for Environmental Prediction) water vapor data and TOMS (Total Ozone Mapping Spectrometer) data, included in the software distribution, with aerosols obtained from the image itself using the dark dense vegetation methodology (Kaufman et al., 1997).

| Year | Sensor/day-month | Year | Sensor/day-month |
|------|------------------|------|------------------|
| 1984 | L5-TM/04-08 | 1999 | L7-ETM+/06-08 |
| 1985 | L5-TM/07-08 | 2000 | L5-TM/15-07 |
| 1986 | L5-TM/10-08 | 2001 | L7-ETM+/11-08 |
| 1987 | L5-TM/12-07 | 2002 | L7-ETM+/27-06 |
| 1988 | L5-TM/14-07 | 2003 | L5-TM/24-07 |
| 1989 | L5-TM/01-07 | 2004 | L7-ETM+/19-08 |
| 1990 | L5-TM/04-07 | 2005 | L5-TM/30-08 |
| 1991 | L5-TM/07-07 | 2006 | L5-TM/16-07 |
| 1992 | L5-TM/23-06 | 2007 | L5-TM/04-08 |
| 1993 | L5-TM/26-06 | 2008 | L5-TM/06-08 |
| 1994 | L5-TM/15-07 | 2009 | L5-TM/10-09 |
| 1995 | L5-TM/03-08 | 2010 | L5-TM/27-07 |
| 1996 | L5-TM/05-08 | 2011 | L5-TM/30-07 |
| 1997 | L5-TM/23-07 | 2012 | L7-ETM+/09-08 |
| 1998 | L5-TM/24-06 | 2013 | L7-ETM+/28-08 |

Table 1. Selected Landsat TM/ETM+ images, path/row-231/66.

For those images that had cloud cover (1987, 1989, 1990, 1991, 1993, 1996, 1997, 2002, 2005, 2009, 2011) we generated a mask from an unsupervised classification to each image held by the ERDAS Imagine software.

We applied 3x3 pixels order filters to the ETM+ selected images dated after 2002 and which showed the failure of the SLC (Scan Line Corrector). The filters were applied six times to each image and filled the gaps using the information of the neighboring pixels.

## 2.2 Data process and analysis

Using ERDAS Imagine we generate NDVIs values for each selected image (from 1984 to 2013), relating the bands 3 and 4 of TM and ETM+ Landsat images considering the radiometric equivalence of the two instruments (Teillet et al., 2001).

We chose images of 1984 and 2010 to classify the LULC for the two most extreme dates within the time series, both from the same sensor (TM), without cloud cover and good pixel quality. We excluded the image from 2011 from the classification because it had cloud cover. Classification of images from 2012 and 2013, from the ETM + sensor, was hampered by filling the gap of SLC off.

The selected images underwent a supervised classification of LULC using the operator maximum likelihood to differentiate five categories:
* Forest (F) – areas of dense rain forest or open rain forest. The last one more associated with forest drainage channels galleries in the savannah's enclave area.
* Secondary/degraded forest (Fs) – regenerated forests or in advanced process of regeneration, as well as areas of degraded rainforests. It also includes some savanna areas with denser shrub domain.
* Savannah/pasture (SP) – areas of vegetation grasses and shrubs, largely used as areas of creation of extensive cattle on pastures.
* Crop/bare soil (CB) – includes a ranching and crop farming areas, with large harvested zones, viewed as bare soil. It also includes burned areas to crop/pastures usage.
* Water (W) –areas of rivers and small water reservoirs located in certain pastures and agricultural areas.

This classification was validated by measuring the Cohen kappa index to each classified image, comparing with a base of control points. The cloud masks were applied to the NDVI images associating a *nodata* value for these cloud pixels. The trajectories in NDVI variation values were analyzed by R

statistical software, in a sample of random points covering 3% of the area, with a total of 792411 points.

This analysis divides the main intersections of the classified categories into two groups: regressive/progressive NDVI trajectories and stable NDVI trajectories. Thanks to monitoring of the NDVI trajectories for the period between 1984 and 2013, we generated average, standard deviation and coefficient of variation for each pixel group according to the intersection of thematic classes. A break point detection was calculated for the progressive/regressive trajectories using a non-parametric approach (Pettitt, 1979).

## 3. RESULTS AND DISCUSSION

### 3.1 LULC dynamics between 1984 and 2010

The result of the classification process (Table 2) (Figure 2) allows a spatial configuration of the dynamics of the LULC in recent decades. Validation with a Cohen kappa index shows values of 0.84 and 0.86 to 1984 and 2010 classifications respectively.

Rain forest is the class that occupies the highest proportion of area in relation to others in both classified dates. In 1984 occupied 79.52% of the total area and 73.02% in 2010. The savanna/pasture areas extended 2440 $km^2$ in 1984 and gained 4.11% of the area, occupying 18.18% of the total area in 2010.

It is noteworthy that in 1984 we observed higher levels of impact of human activities within the savanna enclave area. Recently burned area for pasture management might be identified in the crop/bare soil class.

Among the LULC changes, we mainly identify the loss of forest areas associated with the advancement of agricultural areas and pastures in the southern half of the study area during the analyzed period. In quantitative terms, 563 $km^2$ of forest became crop/bare soil in 2010, and 471 $km^2$ became savannah/pasture.

The analysis records the loss of 1124 $km^2$ of forest between 1984 and 2010. This total is compatible with the data recorded by Assessment of Deforestation in Brazilian Amazonia (PRODES) (INPE, 2014a), which between 2001 and 2010 records 1069 $km^2$ of deforested areas.

Furthermore, occupation of secondary and degraded forests between 1984 and 2010 increased 1.45%. Currently, they occupy 349.34 $km^2$. In 1984, these regions were scattered in the northeast and southwest of the study area and were mainly associated with degraded zones reflecting the selective logging. In 2010, these degraded areas expanded and some areas of forest regeneration appeared thanks to the abandonment of certain areas and the creation of protected areas.

| LULC classes | Area (1984) | | Area (2010) | |
|--------------|-------------|------|-------------|------|
| | $km^2$ | % | $km^2$ | % |
| Forest | 13786.85 | 79.52 | 12660.47 | 73.02 |
| Sec./degraded forest | 97.13 | 0.56 | 349.34 | 2.01 |
| Savannah/pasture | 2440.32 | 14.07 | 3153.29 | 18.18 |
| Crop/bare soil | 938.65 | 5.41 | 1099.74 | 6.34 |
| Water | 73.75 | 0.42 | 73.86 | 0.42 |
| Total | 17336.70 $km^2$ | | | |

Table 2. LULC dynamics between 1984 and 2010.

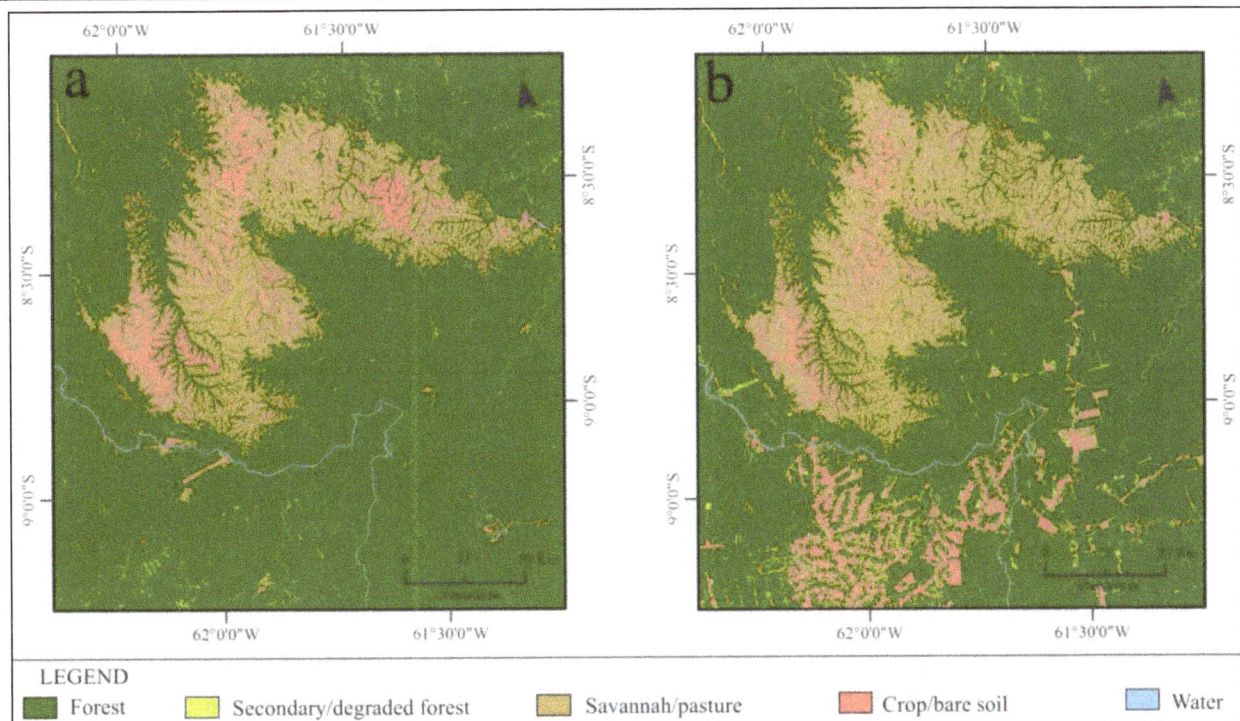

Figure 2. LULC classification of 1984 (a) and 2010 (b).

## 3.2 Monitoring NDVI trajectories (1984-2013)

In order to monitor the LULC dynamics identified in the previous step, we generate NDVI values on a yearly basis over the last three decades (1984-2013). For these trajectories, major intersections associated with the classification process were grouped as follows: NDVI stable trajectories; NDVI regressive or progressive trajectories.

We identify as stable trajectories of NDVI those pixels where no LULC was detected: 'Forest to Forest' (FF); 'Sec./degraded forest to Sec./degraded forest' (Fs-Fs); 'Savannah/pasture to Savannah/pasture' (SP-SP); and 'Crop/bare soil to Crop/bare soil' (CB-CB).

On the other hand regressive or progressive NDVI trajectories were associated to the following LULC changes: 'Forest to Savannah/pasture' (F-SP); 'Forest to Crop/bare soil' (F-CB); and 'Crop/bare soil to Forest and Sec./degraded forest' (CB-FFs).

### 3.2.1 NDVI stable trajectories: Each trajectory is located in specific interval of NDVI values (Figure 3a), according to the characteristics of its thematic category. F-F and Fs-Fs displayed their trajectory standing close to 0.80 NDVI values, linked to a good vegetation development. The average NDVI of SP-SP trajectory is slightly higher than the CB-CB, with values of 0.63 and 0.54, respectively.

We observe that the trajectories of SP-SP and CB-CB fluctuate more intensely than the FF and Fs-Fs values. In part, this variability is associated to increased sensitivity of these classes to phenological effects, as their most outstanding peaks (1987, 1994, 1998 and 2002) can also be perceived in F-F trajectory. Moreover, CB-CB and SP-SP are periodically affected by the intensive use of fire, used as a management tool for agro-pastoral activities.

These factors influence the variability in both SP-SP and CB-CB, making it even higher than F-F's. These differences are explicit when comparing F-F and SP-SP trajectories (Figure 3b), according to the reported standard deviation. F-F displayed the most stable trajectory of NDVI values, with an average value of 0.824 and a coefficient of variation of 3.9%. While the pixels of SP-SP, which was periodically affected by fire, had an average NDVI value 0.585 and coefficient of variation 15.1%.

### 3.2.2 NDVI regressive/progressive trajectories: F-CB trajectory (Figure 3c) revealed that during the first 13 years average NDVI values stand close to those from the F-F trajectory (mean values of 0.828).

After 1997 an increase in standard deviation values is observed, suggesting certain level of pressure on forested areas. In 1999, identified as break point of the trajectory (Pettitt, 1979), the average NDVI values begin to distance themselves from the F-F values. This detachment is consolidated in 2003, following gradually, provided with increased participation of CB class of pixels associated with deforestation. The average NDVI values are 0.482 in 2010, associated with the CB class.

Splitting the years into two periods 1984-2002 and 2003-2013, the average of F-CB NDVI values changed respectively from 0.821 to 0.556. This regressive trajectory is associated with 563 km$^2$ of forests that became CB in 2010, identified in the classification process.

The F-SP trajectory (Figure 3d) follows the same logic established in the F-CB transition, with a steeper decline in the second half of the review period. However, it is observed that the standard deviation of F-SP is more variable when compared to levels of F-F.

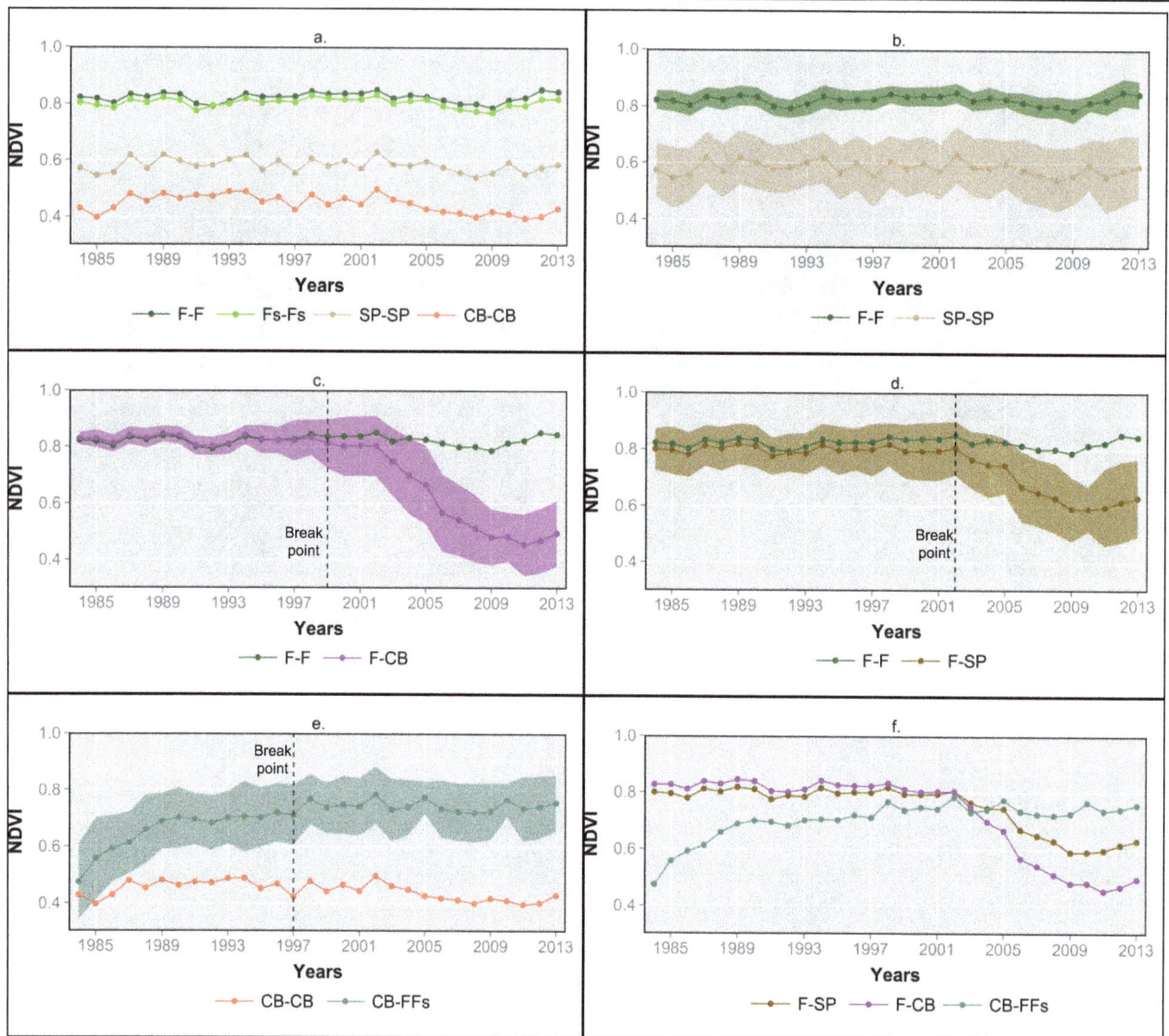

Figure 3. NDVI stable trajectories (a); Variability of F-F and SP-SP trajectories; (b); F-CB trajectory compared to F-F trajectory (c); F-SP trajectory compared to F-F trajectory (d) CB-FFs trajectory compared to CB-CB trajectory (e); NDVI regressive/progressive trajectories (f). The lines represent the average NDVI values. Maximum and minimum limits of the filled areas are the average added and subtracted to their respective standard deviations. Break points, according Pettitt (1979), are marked on the regressive/progressive trajectories.

Many pixels forest included in this theme category are associated with areas of forest-savannah interface, which contributes to the increased sample variability from the beginning of the time series. The breakpoint identified for this regressive trajectory is set in 2002, where the descents gradually expand in the following years. This regressive trajectory is associated with 471 km$^2$ of forests that became SP in 2010.

The progressive trajectory identified (Figure 3e) in the time series is associated with the combination of categories of CB and a grouping of categories of Forest and Sec./degraded forest (FFs). In this trajectory, the NDVI values start with an average value of 0.475, starting to increase gradually in the following years. The breakpoint identified is the year 1997, where the average NDVI was already performing 0.712. From 1997 until 2013, it is observed that this group of pixels keeps the average stable without reaching the levels of FF NDVI. This progressive trajectory is a minor trend within the area. Is related to the gain of 18 km$^2$ FFs in 2010, replacing areas classified as CB in 1984.

All of these regressive/progressive trajectories (Figure 3f) display information about the LULC changes operating on the study area.

## 4. CONCLUSION

The results show distinct trajectories associated with NDVIs and LULC dynamics that assist in better understanding the dynamics of ecological processes and the human impacts operating in the area.

In the stables NDVI trajectories, is possible observe different temporal behaviours depending on vegetation type. Forests have proved to be the most stable community according to its NDVI values. On the other hand, we detected high sensitivity in savannah-type vegetation (scrubs, grasslands and pastures) to disturbances such wildfire or climatic/phonological effects.

NDVI multitemporal analysis has proved to be a useful tool allowing detecting and monitoring LULC changes. In this particular case, combining NDVI monitoring and change point detection procedures, made possible to identity an increase in the deforestation process, mainly affecting rain forest communities, starting 15 years ago and becoming more intense over time.

It is worth to stand out the current potential of Landsat products since Landsat 8 satellite, launched in 2013, guarantees their continuity. This fact allows enhancing valuable multitemporal analysis and monitoring of remote areas.

## ACKNOWLEDGEMENTS

This work is supported by a grant from the CAPES Foundation (Brazil) awarded to the first author.

## REFERENCES

Aragão, L.E.O.C., Malhi, Y., Barbier, N., Lima, A., Shimabukuro, Y.E., Anderson, L., Saatchi, S., 2008. Interactions between rainfall, deforestation and fires during recent years in the Brazilian Amazonia. *Philosophical transactions of the Royal Society of London,* 363, 1779–85.

Bonham, C.D. (Colorado U., 2013. *Measurements for Terrestrial Vegetation,* 2nd ed. John Wiley & Sons, Ltd, Oxford, UK.

Devries, B., Verbesselt, J., Kooistra, L., Herold, M., 2015. Remote Sensing of Environment Robust monitoring of small-scale forest disturbances in a tropical montane forest using Landsat time series. *Remote Sensing of Environment,* 161, 107–121.

Ding, Y., Zhao, K., Zheng, X., Jiang, T., 2014. Temporal dynamics of spatial heterogeneity over cropland quantified by time-series NDVI, near infrared and red reflectance of Landsat 8 OLI imagery. *International Journal of Applied Earth Observation and Geoinformation,* 30, 139–145.

Du, H., Cui, R., Zhou, G., Shi, Y., Xu, X., Fan, W., Lü, Y., 2010. The responses of Moso bamboo (Phyllostachys heterocycla var. pubescens) forest aboveground biomass to Landsat TM spectral reflectance and NDVI. *Acta Ecologica Sinica,* 30, 257–263.

Espindola, G.M. de, Aguiar, A.P.D. de, Pebesma, E., Câmara, G., Fonseca, L., 2012. Agricultural land use dynamics in the Brazilian Amazon based on remote sensing and census data. *Applied Geography,* 32, 240–252.

ICMBio, Instituto Chico Mendes de Conservação da Bioiversidade., 2011. *Plano de Manejo* - Parque Nacional dos Campos Amazônicos, Brasília, Brazil.

INPE, Instituto Nacional de Pesquisas Espaciais, 2014a. PRODES: assessment of deforestation in Brazilian Amazonia. São José dos Campos, Brazil http://www.obt.inpe.br/prodes/index.html.

INPE, Instituto Nacional de Pesquisas Espaciais, 2014b. Queimadas project. São José dos Campos, Brazil http://www.cptec.inpe.br/queimadas/.

Kaufman, Y.J., Wald, A.E., Remer, L. a., Gao, B.-C.G.B.-C., Li, R.-R.L.R.-R., Flynn, L., 1997. The MODIS 2.1 channel-correlation with visible reflectance for use in remote sensing of aerosol. *IEEE Transactions on Geoscience and Remote Sensing,* 35, 1286–1298.

Kobayashi, H., Dye, D.G., 2005. Atmospheric conditions for monitoring the long-term vegetation dynamics in the Amazon using normalized difference vegetation index. *Remote Sensing of Environment,* 97, 519–525.

Lima, A., Silva, T.S.F., Aragão, L.E.O. e C. de, Feitas, R.M. de, Adami, M., Formaggio, A.R., Shimabukuro, Y.E., 2012. Land use and land cover changes determine the spatial relationship between fire and deforestation in the Brazilian Amazon. *Applied Geography,* 34, 239–246.

Marengo, J.A., Nobre, C.A., Tomasella, J., Oyama, M.D., Oliveira, G.S. de, Oliveira, R. de, Camargo, H., Alves, L.M., Brown, I.F., 2008. The Drought of Amazonia in 2005. *Journal of Climate,* 21, 495–516.

Marengo, J.A., Tomasella, J., Alves, L.M., Soares, W.R., Rodriguez, D.A., 2011. The drought of 2010 in the context of historical droughts in the Amazon region. *Geophysical Research Letters,* 38, 1–5.

Masek, J.G., Vermote, E.F., Saleous, N.E., Wolfe, R., Hall, F.G., Huemmrich, K.F., Gao, F., Kutler, J., Lim, T., 2006. A Landsat Surface Reflectance Dataset for North America, 1990-2000. *IEEE Geoscience and Remote Sensing Letters,* 3, 68–72.

Maxwell, S.K., Sylvester, K.M., 2012. Identification of "ever-cropped" land (1984-2010) using Landsat annual maximum NDVI image composites: Southwestern Kansas case study. *Remote Sensing of Environment,* 121, 186–195.

Morton, D.C., DeFries, R.S., Shimabukuro, Y.E., Anderson, L.O., Arai, E., Espirito-Santo, F. del B., Freitas, R., Morisette, J., 2006. Cropland expansion changes deforestation dynamics in the southern Brazilian Amazon. *Proceedings of the National Academy of Sciences of the United States of America,* 103, 14637–41.

Nagendra, H., Lucas, R., Honrado, J.P., Jongman, R.H.G., Tarantino, C., Adamo, M., Mairota, P., 2013. Remote sensing for conservation monitoring: Assessing protected areas, habitat extent, habitat condition, species diversity, and threats. *Ecological Indicators,* 33, 45–59.

Nepstad, D., Carvalho, G., Barros, A.C., Alencar, A., Capobianco, J.P., Bishop, J., Moutinho, P., Lefebvre, P., Silva Jr., U.L., Prins, E., 2001. Road paving , fire regime feedbacks , and the future of Amazon forests. *Forest Ecology and Management,* 154, 397–407.

Pettitt, A.N., 1979. A non-parametric approach to the change-point problem. *Journal of the Royal Statistical Society.* 28, 126–135.

Pettorelli, N., Vik, J.O., Mysterud, A., Gaillard, J.M., Tucker, C.J., Stenseth, N.C., 2005. Using the satellite-derived NDVI to assess ecological responses to environmental change. *Trends in Ecology and Evolution.* 20, 503–510.

Phillips, O.L., Aragão, L.E.O.C., Lewis, S.L., Fisher, J.B., Lloyd, J., López-González, G., Malhi, Y., Monteagudo, A., Peacock, J., Quesada, C.A., Heijden, G. Van Der, Almeida, S., Amaral, I., Arroyo, L., Aymard, G., Baker, T.R., Bánki, O., Blanc, L., Bonal, D., Brando, P., Chave, J., Oliveira, Á.C.A. de, Cardozo, N.D., Czimczik, C.I., Feldpausch, T.R., Freitas, M.A., Gloor, E., Higuchi, N., Jiménez, E., Lloyd, G., Meir, P., Mendoza, C., Morel, A., Neill, D.A., Nepstad, D., Patiño, S., Peñuela, M.C., Prieto, A., Ramírez, F., Schwarz, M., Silva, J., Silveira, M., Thomas, A.S., Steege, H., Stropp, J., Vásquez, R., Zelazowski, P., Dávila, E.A., Andelman, S., Andrade, A., Chao, K., Erwin, T., Fiore, A. Di, C, E.H., Keeling, H., Killeen, T.J., Laurance, W.F., Cruz, A.P., Pitman, N.C.A., Vargas, P.N., Ramírez-Angulo, H., Rudas, A., Salamão, R., Silva, N., Terborgh, J., Torres-Lezama, A., 2009. Drought Sensitivity of the Amazon Rainforest. *Science (80 )*. 323, 1344–1347.

Ratter, J.A., Bridgewater, S., Ribeiro, J.F., 2003. Analysis of the floristic composition of the Brazilian cerrado vegetation: comparison of the woody vegetation of 376 areas. *Edinburgh J. Bot Edinburgh Journal of Botany,* 57–109.

Silvestrini, R.A., Soares-Filho, B.S., Nepstad, D., Coe, M., Rodrigues, H., Assunção, R., 2011. Simulating fire regimes in the Amazon in response to climate change and deforestation. *Ecological Applications,* 21, 1573–1590.

Teillet, P., Barker, J., Markham, B., Irish, R., Fedosejevs, G., Storey, J., 2001. Radiometric cross calibration of the Landsat-7 ETM+ and Landsatt-5 TM sensors based on tandem data sets. *Remote Sensing of Environment,* 78, 39–54.

Turner, W., Spector, S., Gardiner, N., Fladeland, M., Sterling, E., Steininger, M., 2003. Remote sensing for biodiversity science and conservation. *Trends in Ecology and Evolution,* 18, 306–314.

Vasconcelos, S.S. de, Fearnside, P.M., Graça, P.M.L. de A., Dias, D.V., Correia, F.W.S., 2013. Variability of vegetation fires with rain and deforestation in Brazil's state of Amazonas. *Remote Sensing of Environment,* 136, 199–209.

Vermote, E.F., El Saleous, N., Justice, C.O., Kaufman, Y.J., Privette, J.L., Remer, L., Roger, J.C., Tanré, D., 1997. Atmospheric correction of visible to middle-infrared EOS-MODIS data over land surfaces: Background, operational algorithm and validation. *Journal of Geophysical Research,* 102, 17131.

Zheng, B., Myint, S.W., Thenkabail, P.S., Aggarwal, R.M., 2015. A support vector machine to identify irrigated crop types using time-series Landsat NDVI data. *International Journal of Applied Earth Observation and Geoinformation*, 34, 103–112.

# 13

# DOES TOPOGRAPHIC NORMALIZATION OF LANDSAT IMAGES IMPROVE FRACTIONAL TREE COVER MAPPING IN TROPICAL MOUNTAINS?

H. Adhikari [a, *], J. Heiskanen [a], E. E. Maeda [a], P. K. E. Pellikka [a]

[a] University of Helsinki, Department of Geosciences and Geography, P.O. Box 68, FI-00014, Helsinki, Finland – (hari.adhikari, janne.heiskanen, eduardo.maeda, petri.pellikka)@helsinki.fi

KEY WORDS: Landsat, Fractional tree cover, Topographic correction, Digital Elevation Model, LiDAR

**ABSTRACT:**

Fractional tree cover (Fcover) is an important biophysical variable for measuring forest degradation and characterizing land cover. Recently, atmospherically corrected Landsat data have become available, providing opportunities for high-resolution mapping of forest attributes at global-scale. However, topographic correction is a pre-processing step that remains to be addressed. While several methods have been introduced for topographic correction, it is uncertain whether Fcover models based on vegetation indices are sensitive to topographic effects. Our objective was to assess the effect of topographic correction on the accuracy of Fcover modelling. The study area was located in the Eastern Arc Mountains of Kenya. We used C-correction as a digital elevation model (DEM) based correction method. We examined if predictive models based on normalized difference vegetation index (NDVI), reduced simple ratio (RSR) and tasseled cap indices (Brightness, Greenness and Wetness) are improved if using topographically corrected data. Furthermore, we evaluated how the results depend on the DEM by correcting images using available global DEM (ASTER GDEM, SRTM) and a regional DEM. Reference Fcover was obtained from wall-to-wall airborne LiDAR data. Landsat images corresponding to minimum and maximum sun elevation were analyzed. We observed that topographic correction could only improve models based on Brightness and had very small effect on the other models. Cosine of the solar incidence angle ($cos\ i$) derived from SRTM DEM showed stronger relationship with spectral bands than other DEMs. In conclusion, our results suggest that, in tropical mountains, predictive models based on common vegetation indices are not sensitive to topographic effects.

## 1. INTRODUCTION

Landsat satellite images became available for free in 2008, which together with systematic data acquisition plan has fortified Landsat's role as a primary source of information for global land change research (e.g., Wulder et al., 2012). Landsat images also have good data consistency among Landsat missions and historical archive since 1972. The pre-processing methods of Landsat data have reached the level of maturity and pre-processed data sets have become available (e.g. Landsat Climate Data Record (CDR)) similar to moderate resolution data e.g. various MODIS data products (Masek et al., 2006). However, topographic correction remains as one of the pre-processing steps that have not been addressed globally.

Topographical effects in satellite images are caused by illumination differences between the sunlit and shaded slopes. The magnitude of the effect depends on the time of the year because of variations in the sun elevation. Hence, in the topographic normalization, the dependency of reflectance factors on topographic position is removed. Topographic correction has been shown to improve land cover mapping using object-based classification (Moreira & Valeriano, 2014) and land cover classification accuracy (Pellikka, 1996; Vanonckelen et al., 2013).

Topographic correction methods can be grouped into three categories: those based on band ratios, those based on a Hyperspherical Direction Cosine Transformation (HSDC), and those requiring digital elevation model (DEM). DEM-based methods, can be summarized as three broad types: (1) empirical methods, (2) Lambertian methods and (3) non-Lambertian methods (Gao & Zhang, 2009). The most often used methods are

Lambert cosine correction (Meyer et al., 1993), Minnaert correction (Smith et al., 1980), C-correction (Teillet et al., 1982) and Sun-canopy-sensor method (Gu & Gillespie, 1998). Several studies have demonstrated the viability of C-correction for radiometric correction of multitemporal images taken under different illumination conditions (Moreira & Valeriano, 2014; Reese & Olsson, 2011; Vanonckelen et al., 2013).

Continuous fields of vegetation attributes can be estimated using a multitude of predictors derived from Landsat images. However, the topographic effect on different types of predictors can be different. For example, canopy cover, fractional tree cover (Fcover) and leaf area index (LAI) are typically estimated based on vegetation indices such as Normalized Difference Vegetation Index (NDVI) and Reduced Simple Ratio (RSR) (Majd et al., 2013; Korhonen et al., 2013; Wu, 2011).

NDVI is a simple index used to accentuate vegetation from imagery containing reflectance in the red and NIR portions of the spectrum. RSR is an empirical SWIR modification to the simple ratio (SR) vegetation index (Brown et al., 2000). Both indices attempt to depress background reflectance and improve the accuracy in extracting vegetation information from remotely sensed data.

Tasseled Cap transformation (TC) compress spectral data into bands associated with the physical characteristics of scene (Crist, 1985). TC indices (Brightness, Greenness and Wetness) have been used, for example, for forest disturbance detection (Healey et al., 2005; Jin & Sader 2005; Skakun et al. 2003) and forest classification (Dymond et al., 2002).

The correction of satellite data for illumination differences due to

---

topography requires a DEM. Currently, there are several sources of global DEMs, such as SRTM and ASTER DEM. SRTM DEM at 30 m resolution was made recently globally available (https://lta.cr.usgs.gov/SRTM1Arc), improving its applicability for topographic correction of Landsat data. In addition to the global DEMs, many areas are supplemented by regional DEMs based on topographic map data. However, viability of different DEMs for topographic correction have been rarely assessed.

Several authors have compared topographic correction methods for Landsat data (e.g., Hantson & Chuvieco, 2011; Moreira & Valeriano, 2014; Vanonckelen et al., 2013). However, these assessments are typically focusing on the removal of the topographic effect, instead of assessing how the accuracy of output products is affected. Although some studies have used topographic correction as a step to improve the land cover classification accuracy, it has been only rarely tested if topographic correction improves the accuracy of continuous variables (Törmä & Härmä, 2003), such as Fcover. Furthermore, classification is typically based on individual bands, but not vegetation indices, which are commonly used in Fcover, LAI and biomass mapping.

In this study, our objective was to assess the effect of topographic correction on the accuracy of Fcover predictions based on Landsat images with and without topographic correction. In particular, we aimed to examine if predictive models based on NDVI, RSR and TC indices are affected differently, and if results depend on the source of DEM.

## 2. MATERIAL AND METHODS

### 2.1 Study area

The Taita Hills (3°25′S, 38°20′E) are located in the northernmost part of the Eastern Arc Mountains in southeastern Kenya (Figure 1). The area is characterized by distinct topographical variation and the mountainous hills raise up to 2200 m a.s.l. from the Tsavo Plains at 600–900 m a.s.l. Taita Hills are considered as one of the world's most important biodiversity hotspots. However, due to the favorable climate and edaphic conditions, the indigenous cloud forests of the Taita Hills have suffered substantial deforestation and degradation due to agriculture, grazing and logging of forest for firewood and charcoal manufacturing since the early 1960′s (Pellikka et al., 2009).

### 2.2 Landsat images

Landsat surface reflectance (CDR) is a product of the Landsat Ecosystem Disturbance Adaptive Processing System (LEDAPS) at the National Aeronautics and Space Administration (NASA) Goddard Space Flight Center. The images have been calibrated to radiance, converted to top-of-atmosphere reflectance and then atmospherically corrected using the MODIS/6S methodology (Masek et al., 2006).

| Sensor | Date | Sun Elevation | Sun Azimuth |
|---|---|---|---|
| Landsat 7 ETM+ | 29.9.2013 | 64.08° | 89.79° |
| Landsat 7 ETM+ | 25.6.2013 | 51.35° | 45.62° |

Table 1. Details of Landsat images used in this study.

The Taita Hills lies on the south east quarter of the Landsat image (WRS path 167 and row 62). Landsat scenes with cloud cover

less than 30% in the lower right quarter in 2013 were downloaded and two images less affected by cloud, shadow and missing data due to Scan line corrector (SLC) off and corresponding to the minimum and maximum sun elevation were used in this study. Both images were acquired at 7:32 UTC time in the morning and image processing level was L1T.

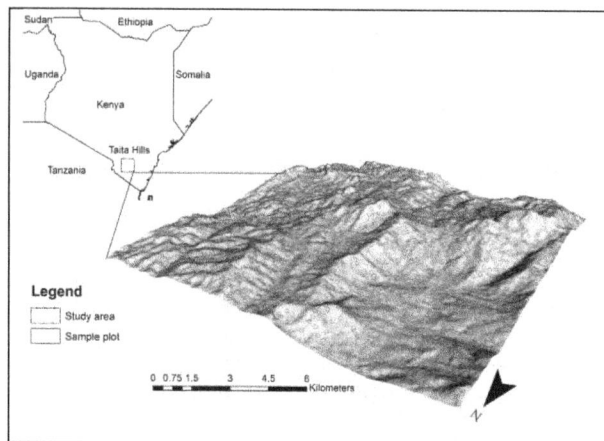

Figure 1. Location of the study area and hillshade view of the topography.

### 2.3 Digital elevation models

DEM used for topographic correction were obtained from various sources. Advanced Spaceborne Thermal Emission and Reflection Radiometer (ASTER) Global Digital Elevation Model (GDEM) has been generated from the stereoscopic ASTER satellite images. ASTER GDEM is available at the resolution of 30 m. The ASTER GDEM covers land surfaces between 83°N and 83°S. It is a product of Ministry of Economy, Trade, and Industry (METI) and the National Aeronautics and Space Administration (NASA) (https://www.jspacesystems.or.jp/ersdac/GDEM/E/4.html).

Shuttle Radar Topography Mission (SRTM) 1 arc-second global elevation data has been available since 23 September 2014 (https://lta.cr.usgs.gov/SRTM1Arc). SRTM provides DEM for 80% of the earth's surface (all land areas between 60° N and 56° S). SRTM is a joint project between the National Geospatial-Intelligence Agency (NGA) and NASA. The DEM has been filled to remove small artifacts.

All Landsat images, ASTER and SRTM DEM were downloaded from the United States Geological Survey (USGS) Earth Explorer platform (http://earthexplorer.usgs.gov/).

A regional DEM (TOPO DEM) for the Taita Hills was created using the contour lines of the topographic maps at 1:50 000 scale produced by the Survey of Kenya (Pellikka et al., 2004). DEM has a pixel size of 20 m × 20 m, and it was resampled to 30 m × 30 m pixel size similar to the Landsat images.

### 2.4 Airborne LiDAR data and fractional tree cover

We used Fcover (1–canopy gap fraction) derived from airborne LiDAR as a reference data. Previous studies have shown that LiDAR provides canopy cover and gap fraction estimates with an accuracy comparable to the field measurements (Korhonen et al., 2011; Heiskanen et al., in press). Furthermore, wall-to-wall LiDAR data provided a large sample size and the use of random sampling scheme in the logistically difficult terrain.

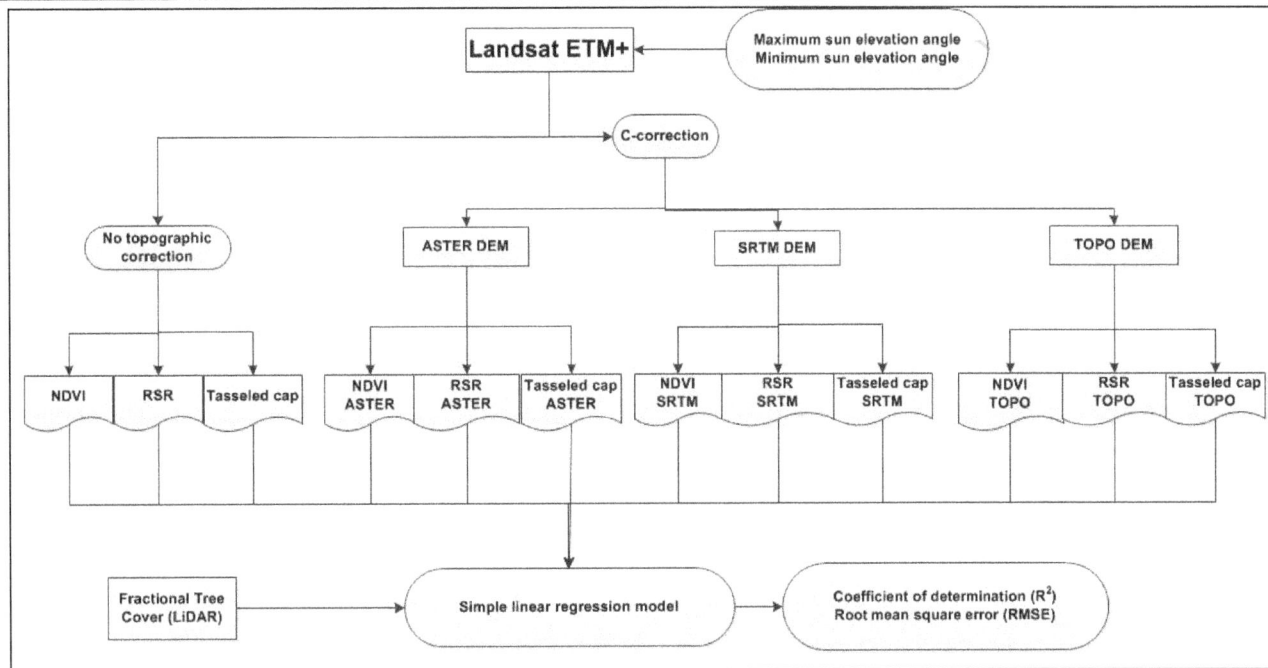

Figure 2. Overview of the methodology

The discrete return LiDAR data was acquired 4–5 February 2013 for 10 km × 10 km area (Table 2). Data were pre-processed by the data vendor (Topscan Gmbh) and delivered as georeferenced point cloud in UTM/WGS84 coordinate system with ellipsoidal heights. Vendor also filtered ground returns by using Terrascan software (Terrasolid Oy). Furthermore, we filtered buildings and powerlines. Then, the ground returns were used to generate DEM at 1 m cell size. We also removed the overlap between the adjacent flight lines based on minimum scan angle using lasoverage tool in LAStools software (rapidlasso GmbH).

| Parameter | Value |
|---|---|
| Date of acquisition | 4–5 February, 2013 |
| Sensor | Optech ALTM |
| Mean flying height (m AGL) | 760 |
| Flying speed (knots) | 116–126 |
| Pulse rate (kHz) | 100 |
| Scan rate (Hz) | 36 |
| Maximum scan angle (degrees) | 16 |
| Pulse density (pulses m$^{-2}$) | 9.6 |
| Return density (returns m$^{-2}$) | 11.4 |
| Maximum number of returns per pulse | 4 |
| Beam divergence at $1/e^2$ (mrad) | 0.3 |
| Footprint diameter (cm) | 23 |

Table 2. Characteristics of the LiDAR data.

FUSION software (McGaughey, 2014) was used for computing a proxy of Fcover. We extracted LiDAR returns for 90 m × 90 m sample plots corresponding to 3 × 3 pixel windows of Landsat ETM+ images. Heiskanen et al. (in press) found that all echo cover index (ACI) (e.g., Morsdorf et al., 2006) gave an unbiased estimate of canopy gap fraction using the same LiDAR data. Hence, we used ACI as a proxy of Fcover. ACI was computed as:

$$ACI\ (\%) = 100\ \frac{\sum All_{canopy}}{\sum All} \quad (1)$$

where, *All* refers to the returns of all return types (i.e. single, first, intermediate and last returns) and *canopy* refers to the returns

from the forest canopy. Laser returns with height less than 1.5 m from the ground were considered as ground in order to exclude understory vegetation from Fcover.

### 2.5 Topographic correction of Landsat-data

C-correction is a semi-empirical topographic correction method, which consists of a modified cosine correction with the empirical parameter $c_\lambda$ (Reese & Olsson, 2011). $c_\lambda$ (Eq. 4) is calculated for every band ($\lambda$) separately based on the linear relationship (Eq. 3) between the spectral data and the cosine of the solar incidence angle (*cos i*) (Eq. 2). Each band were topographically corrected using Eq. 5.

$$\cos i = \cos sz \times \cos sl + \sin sz \times \sin sl \times \cos\ (az - as) \quad (2)$$

$$\rho_{\lambda,t} = b_\lambda + m_\lambda \times \cos i \quad (3)$$

$$c_\lambda = \frac{b_\lambda}{m_\lambda} \quad (4)$$

$$\rho_{\lambda,n} = \rho_{\lambda,t}\ \frac{\cos\ sz + c_\lambda}{\cos\ i + c_\lambda} \quad (5)$$

where,

$i$ = solar incidence angle with respect to surface normal
$sz$ = solar zenith angle
$sl$ = slope
$az$ = solar azimuth angle
$as$ = aspect
$\rho_{\lambda,t}$ = topographically influenced (t) reflectance of band $\lambda$
$b_\lambda$ = intercept of linear regression
$m_\lambda$ = slope of linear regression
$c_\lambda$ = c-factor calculated for every band separately
$\rho_{\lambda,n}$ = topographically normalized (n) reflectance of band $\lambda$

*cos i*, slope and aspect were calculated from each DEM. Resolution of each DEM was made similar to Landsat image. Areas with a slope ≤ 2% were not included in the parameter estimation. Land cover stratification based on NDVI has been successfully used to consider the land cover dependency of $c_\lambda$

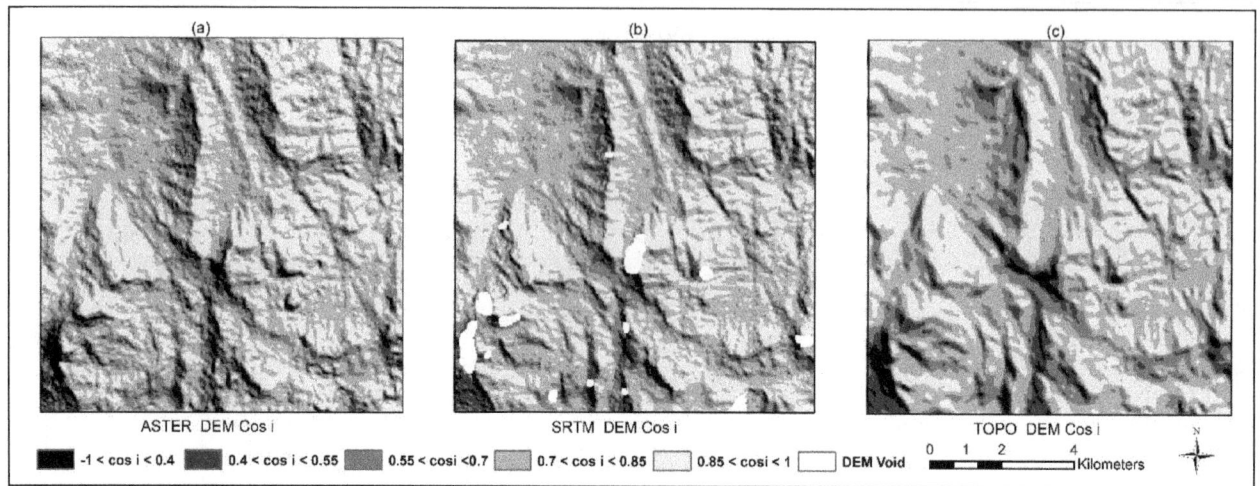

Figure 3. Five *cos i* classes based on (a) ASTER DEM, (b) SRTM DEM and (c) TOPO DEM.

(Hantson & Chuvieco, 2011; McDonald et al., 2000). The study area was divided into three NDVI classes. Class 1 (NDVI < 0.4) covered urban areas, bare areas, grasslands and agriculture lands, Class 2 (NDVI 0.4–0.6) covered secondary forest, plantation and regeneration and Class 3 (NDVI 0.6–1) covered indigenous forest. Each class was further subdivided into five sub classes (-1–0.4, 0.4–0.55, 0.55–0.70, 0.70–0.85, 0.85–1.0) based on *cos i* (Figure 3). 500 samples from each sub class, i.e. 2500 samples for each class and 7500 samples for each image were selected for linear regression between *cos i* and spectral bands. $c_\lambda$ was calculated for each NDVI class and each band.

## 2.6 Predictor variables for fractional tree cover

NDVI, RSR and TC indices (Brightness, Greenness and Wetness) were used as predictor variables. NDVI combines information only from NIR and Red bands, but RSR includes additional SWIR bands to reduce the effects of background reflectance (Brown et al., 2000). Since Landsat ETM+ images were in surface reflectance, tasseled cap coefficients for Landsat TM reflectance factor data were used.

$$NDVI = \frac{(NIR-R)}{(NIR+R)} \tag{6}$$

$$RSR = \frac{NIR}{R} \times \frac{(SWIR1max - SWIR1)}{(SWIR1max - SWIR1min)} \tag{7}$$

$$Brightness = 0.2043 \times B + 0.4158 \times G + 0.5524 \times R + 0.5741 \times NIR + 0.3124 \times SWIR1 + 0.2303 \times SWIR2 \tag{8}$$

$$Greenness = -0.1603 \times B - 0.2819 \times G - 0.4934 \times R + 0.7940 \times NIR - 0.0002 \times SWIR1 - 0.1446 \times SWIR2 \tag{9}$$

$$Wetness = 0.0315 \times B + 0.2021 \times G + 0.3102 \times R + 0.1594 \times NIR - 0.6806 \times SWIR1 - 0.6109 \times SWIR2 \tag{10}$$

where, B =blue band (450–520 nm), G = green band (520–600 nm), R= red band (630–690 nm), NIR= NIR band (770–900 nm), SWIR1 =SWIR band (1550–1750 nm) and SWIR2 = SWIR band (2090–2350 nm). $SWIR1_{max}$ and $SWIR1_{min}$ were defined as 99% and 1% points in the cumulative histogram of the SWIR1 respectively. NDVI, RSR and TC indices were calculated from topographically corrected and non-topographically corrected surface reflectance values of Landsat ETM + spectral bands.

Finally, 2000 random sample plots (size of 90 m × 90 m corresponding to 3 pixels × 3 pixels windows of Landsat image) were chosen for extracting average NDVI, RSR, Brightness, Greenness and Wetness from Landsat image and for computing Fcover from LiDAR data. Only sample plots without any missing data were retained for the further analysis. Hence, the number of sample plots was reduced to around 1500.

## 2.7 Regression analysis

We used simple linear regression to model relationships between Fcover and NDVI, RSR, and TC indices. The results were evaluated based on coefficient of determination ($R^2$) and root mean square error (RMSE).

## 3. RESULTS

The various predictor variables explained variation in Fcover differently (Figure 4). Out of all predictors, the highest $R^2$ and the lowest RMSE were obtained using RSR. NDVI and Greenness showed the weakest relationships with Fcover. NDVI, RSR, Greenness and Wetness showed positive correlation with Fcover as they are affected by the amount of vegetation, while Brightness had negative correlation as it increases with the higher amount of open soil and lower vegetation cover.

The strength of the relationship also varied between the images acquired under different illumination conditions (Table 3). NDVI, RSR, Brightness, Greenness and Wetness explained more variation in Fcover when extracted from the image with higher sun elevation angle (29.9.2013). For example, topographically uncorrected Brightness indices explained only 37.5% of variation in Fcover when extracted from the lower sun elevation angle image in comparison to 45.2% when extracted from the higher sun elevation angle image. NDVI was also affected by the time of image acquisition. This was consistent for all the predictor variables.

The effect of topographic correction on Fcover regression models was not consistent for all predictors (Figure 4 and Table 3). The models based on Brightness and Wetness showed some improvement in terms of $R^2$ and RMSE but other models were not significantly affected, particularly when based on the larger solar elevation angle image. In the case of NDVI, there was no improvement at all due to topographic correction in either image.

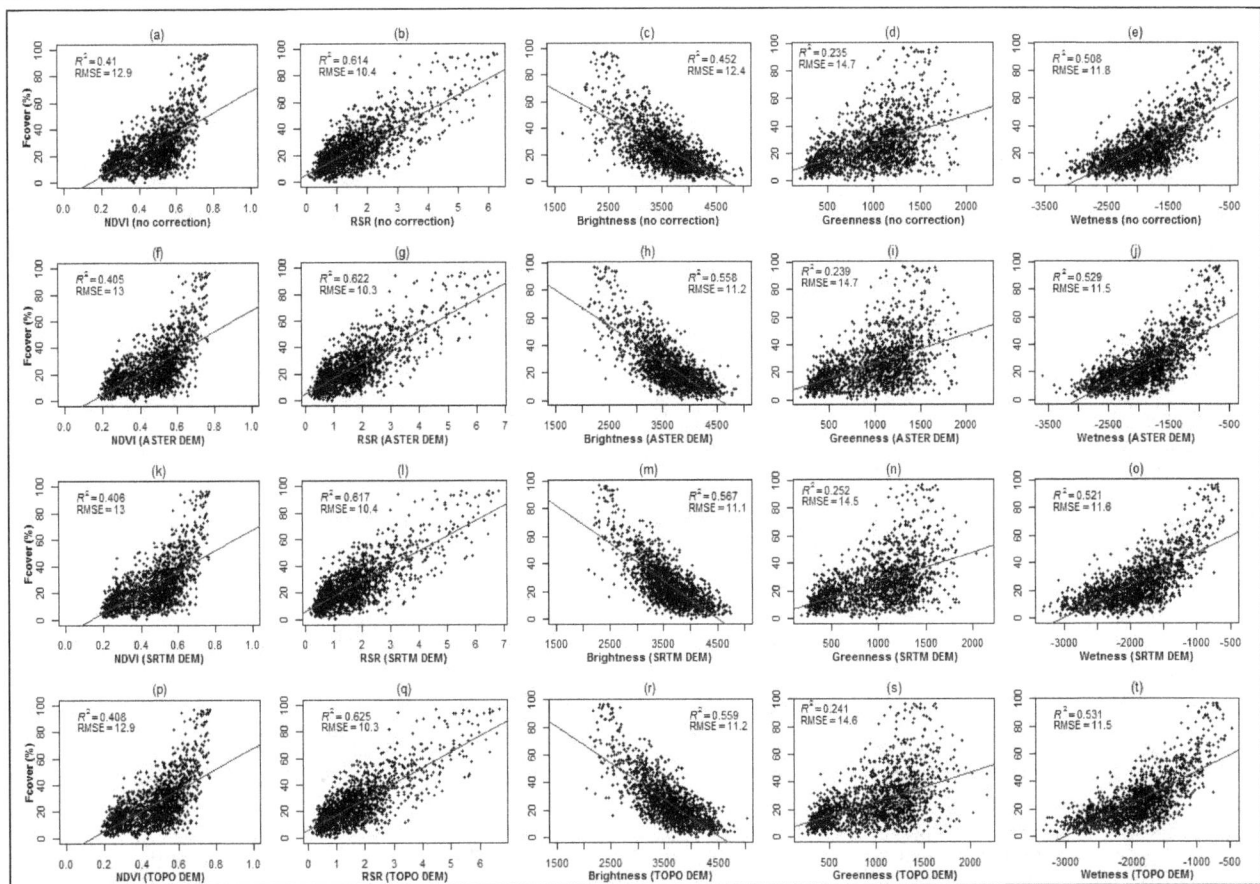

Figure 4. Simple linear regression between Fcover and NDVI, RSR and TC indices before topographic correction (first row) and after topographic correction using ASTER DEM (second row), SRTM DEM (third row) and TOPO DEM (fourth row). The image was taken at 29.9.2013 with a solar elevation angle of 64.08°.

The source of DEM also affected the results in different ways (Figure 4, Table 3), although differences were small like improvements in the model fit due to topographic correction. However, SRTM DEM performed better for Brightness and Greenness indices which were most affected by the topographic correction. Furthermore, when estimating c-factors by regression, the relationships between *cos i* and spectral bands were the strongest in terms of $R^2$ (Table 4) when derived from SRTM DEM.

## 4. DISCUSSION

According to our result RSR had the strongest linear relationship with Fcover obtained from LiDAR data. NDVI showed weak relationship with Fcover. These results are similar to those obtained by Korhonen et al. (2013) from a boreal forest site. However, our results differ from Majd et al. (2013) and Wu (2011) who found strong relationship between Fcover and NDVI in pecan orchards and savannas ecosystem. Weak relationship between NDVI and Fcover can be explained by the heterogeneous landscape in the study area and saturation of NDVI as the study area contains also patches of mountain rain forests.

However, it is difficult to say if this is due to weaker topographic effects or a result of variations in vegetation phenology. Image with lower sun elevation match better in terms of phenology to LiDAR acquisition.

Furthermore, our results demonstrate that topographic correction has very small or no effect on Fcover models. In the case of NDVI, no improvement was observed. Therefore, topographic correction might not be necessary for achieving the best Fcover prediction accuracy in tropical mountains that have relatively high solar elevation angles. In particular, models based on ratio based vegetation indices seem to be not affected by topographic correction. The fact that NDVI is less sensitive to topographic conditions has also been demonstrated by Matsushita et al. (2007). However, with multiplicative indices (e.g. Brightness), topographic correction is likely to have considerable effect.

Even though artifacts could be clearly observed in the SRTM DEM, it performed well in comparison to other sources of DEM, in agreement with similar previous studies (Vanonckelen et al., 2013; Balthazar et al., 2012). Most noticeably, SRTM DEM provided better $R^2$ in the C-correction than regional DEM based on the best topographic maps of the study area.

More comprehensive assessments on the effect of topographic correction in modelling vegetation attributes is necessary. Uncertainties related to the use of LiDAR data as a source of reference data needs to be further analyzed. Optimally, field data would be used but LiDAR data allowed us to sample Fcover from all topographical conditions with large sample size. Furthermore, we only considered one topographic correction method (C-correction) and results could be different when other topographic correction methods are used. However, C-correction has usually performed well in comparison to other DEM-based topographic correction methods. Furthermore, similar comparisons would be needed across larger latitudinal range because topographic effects increase with increasing latitude and decreasing solar elevations.

| Predictor variables | 25.6.2013 | | 29.9.2013 | |
|---|---|---|---|---|
| | RMSE | $R^2$ | RMSE | $R^2$ |
| NDVI | **13.5** | **0.356** | 12.9 | 0.410 |
| NDVI$_{ASTER}$ | 13.5 | 0.355 | 13.0 | 0.405 |
| NDVI$_{SRTM}$ | 13.5 | 0.355 | 13.0 | 0.406 |
| NDVI$_{TOPO}$ | 13.5 | 0.355 | 12.9 | 0.408 |
| RSR | 10.8 | 0.588 | 10.4 | 0.614 |
| RSR$_{ASTER}$ | 10.6 | 0.605 | 10.3 | 0.622 |
| RSR$_{SRTM}$ | 10.6 | 0.600 | 10.4 | 0.617 |
| RSR$_{TOPO}$ | **10.4** | **0.616** | **10.3** | **0.625** |
| Brightness | 14.3 | 0.375 | 12.4 | 0.452 |
| Brightness$_{ASTER}$ | 12.5 | 0.450 | 11.2 | 0.558 |
| Brightness$_{SRTM}$ | **11.8** | **0.508** | **11.1** | **0.567** |
| Brightness$_{TOPO}$ | 12.5 | 0.450 | 11.2 | 0.559 |
| Greenness | 15.3 | 0.171 | 14.7 | 0.235 |
| Greenness$_{ASTER}$ | 15.3 | 0.180 | 14.7 | 0.239 |
| Greenness$_{SRTM}$ | **15.1** | **0.192** | **14.5** | **0.252** |
| Greenness$_{TOPO}$ | 15.3 | 0.170 | 14.6 | 0.241 |
| Wetness | 12.6 | 0.438 | 11.8 | 0.508 |
| Wetness$_{ASTER}$ | **11.9** | **0.500** | 11.5 | 0.529 |
| Wetness$_{SRTM}$ | 11.9 | 0.497 | 11.6 | 0.521 |
| Wetness$_{TOPO}$ | 12.0 | 0.496 | **11.5** | **0.531** |

Table 3. Summary of the linear regression results between different predictor variables and Fcover, and two images. The predictors with subscripts were topographically corrected using either ASTER, SRTM and TOPO DEM. Values in bold are the highest $R^2$ and lowest RMSE. All the models were statistically highly significant (p < 0.0001).

| Image and DEM | | $R^2$ | | | | | |
|---|---|---|---|---|---|---|---|
| | | B | G | R | NIR | SWIR1 | SWIR2 |
| 25.6.2013 | ASTER | 0.15 | 0.26 | 0.29 | 0.33 | 0.37 | 0.33 |
| | SRTM | **0.24** | **0.42** | **0.43** | **0.57** | **0.58** | **0.50** |
| | TOPO | 0.16 | 0.28 | 0.31 | 0.37 | 0.42 | 0.36 |
| 29.9.2013 | ASTER | 0.23 | 0.29 | 0.32 | 0.32 | 0.36 | 0.32 |
| | SRTM | **0.28** | **0.42** | **0.43** | **0.55** | **0.54** | **0.45** |
| | TOPO | 0.26 | 0.37 | 0.38 | 0.48 | 0.44 | 0.38 |

Table 4. The mean $R^2$ of three NDVI classes obtained from the linear regression between *cos i* from different DEMs and each spectral bands from the two images. Values in bold are the highest $R^2$.

## 5. CONCLUSIONS

We assessed the effect of topographic correction on the accuracy of Fcover predictions based on Landsat images. Several sources of DEM were evaluated when performing the topographic correction. Among the tested indices, RSR provided the best regression model for Fcover. There were no major difference between the linear models based on topographically corrected or non-corrected NDVI and RSR, which indicates that NDVI and RSR are relatively robust against topographical effects. Therefore, in tropical mountain environments, studies based on NDVI or RSR do not necessarily need to consider the effect of topography. However, topographic correction should be considered for lower sun elevation angle images if models are developed using TC Brightness or Wetness. Finally, SRTM was

shown to be the best DEM source for topographic corrections in the tested conditions.

## ACKNOWLEDGEMENTS

The authors acknowledge Building Biocarbon and Rural Development in West Africa (BIODEV) project funded by the Ministry for Foreign Affairs of Finland. Mr. Adhikari holds a grant from CIMO of Finland (Center for International Mobility). Dr. Eduardo Maeda is currently funded by a research grant from the Academy of Finland.

## REFERENCES

Balthazar, V., Vanacker, V., & Lambin, E. F., 2012. Evaluation and parameterization of ATCOR3 topographic correction method for forest cover mapping in mountain areas. *Int. J. Appl. Earth Obs. Geoinf.* (18), pp. 436–450.

Brown, L., Chen, J. M., Leblanc, S. G., & Cihlar, J., 2000. A shortwave infrared modification to the simple ratio for LAI retrieval in boreal forests: An image and model analysis. *Remote Sens. Environ.*, 41 (1), pp.16-25.

Crist, E. P., 1985. A TM tasseled cap equivalent transformation for reflectance factor data. *Remote Sens. Environ.*, 17, pp. 301-306.

Dymond, C.C., Mladenoff, D.J., & Radeloff, V.C., 2002. Phenological differences in Tasseled Cap indices improve deciduous forest classification. *Remote Sens. Environ., 80*, pp. 460-472.

Gao, Y.N., & Zhang, W.C., 2009. A simple empirical topographic correction method for ETM+ imagery. *Int. J. Remote Sens.*, 30 (9), pp. 2259–2275.

Gu, D., & Gillespie, A., 1998. Topographic normalization of Landsat TM images of forest based on subpixel sun-canopy-sensor geometry. *Remote Sens. Environ.*, 64, pp. 166-175.

Hantson, S., & Chuvieco, E., 2011. Evaluation of different topographic correction methods for Landsat imagery. *Int. J. Appl. Earth Obs. Geoinf., 13*, pp. 691-700.

Healey, S.P., Cohen, W.B., Zhiqiang, Y., & Krankina, O.N., 2005. Comparison of Tasseled Cap-based Landsat data structures for use in forest disturbance detection. *Remote Sens. Environ., 97*, pp. 301-310.

Heiskanen, J., Korhonen, L., Hietanen, J., Pellikka, P. K. E., in press. Airborne LiDAR for estimating canopy gap fraction and leaf area index of tropical montane forests. *Int. J. Remote Sens.*

Jin, S., & Sader, S.A., 2005. Comparison of time series tasseled cap wetness and the normalized difference moisture index in detecting forest disturbances. *Remote Sens. Environ., 94*, pp. 364-372.

Korhonen, L., Korpela, I., Heiskanen, J., Maltamo, M., 2011. Airborne discrete-return LiDAR data in the estimation of vertical canopy cover, angular canopy closure and leaf area index. *Remote Sens. Environ.* 115(4), pp. 1065-1080.

Korhonen, L., Heiskanen, J., & Korpela, I., 2013. Modelling lidar-derived boreal forest canopy cover with SPOT 4 HRVIR data. *Int. J. Remote Sens.*, *34*, pp. 8172-8181.

Majd A. M. S., Bleiweiss, M. P., DuBois, D., & Shukla, M. K., 2013. Estimation of the fractional canopy cover of pecan orchards using Landsat 5 satellite data, aerial imagery, and orchard floor photographs. *Int. J. Remote Sens.*, 34(16), pp. 5937-5952.

Masek, J. G., Vermote, E. F., Saleous, N. E., Wolfe, R., Hall, F. G., Huemmrich, K. F., Gao, F., Kutler, J., & Lim, T. K., 2006. A Landsat Surface Reflectance Dataset for North America, 1990–2000. *IEEE Geosci. Remote. Sens. Let.*, 3(1). pp. 68-72.

Matsushita, B., Yang, W., Chen, J., Onda, Y., & Qiu, G., 2007. Sensitivity of the Enhanced Vegetation Index (EVI) and Normalized Difference Vegetation Index (NDVI) to Topographic Effects: A Case Study in High-Density Cypress Forest. *Sensors* 7, pp. 2636-2651.

McDonald, E.R., Wu, X., Caccetta, P.A., & Campbell, N.A., 2000. Illumination correction of Landsat TM data in south east NSW. Paper Presented at: Proceedings of the Tenth Australasian Remote Sensing Conference.

McGaughey, R. J., 2014. FUSION/LDV: Software for LIDAR Data Analysis and Visualization. United States Department of Agriculture, Forest Service, Pacific Northwest Research Station.

Meyer, P., Itten, K. I., Kellenberger, T., Sandmeier, S., & Sandmeier, R., 1993. Radiometric corrections of topographically induced effects on Landsat TM data in an alpine environment. *ISPRS J. Photogramm. Remote, Sens.*, 484, pp. 17–28.

Moreira, E.P., & Valeriano, M.M., 2014. Application and evaluation of topographic correction methods to improve land cover mapping using object-based classification. *Int. J. Appl. Earth Obs. Geoinf.*, *32*, pp. 208-217.

Morsdorf, F., Kötz, B., Meier, E., Itten, K. I., Allgöwer, B., 2006. Estimation of LAI and fractional cover from small footprint airborne laser scanning data based on gap fraction. *Remote Sens. Environ.* 104(1), pp. 50-61.

Pellikka, P., 1996. Illumination compensation for aerial video image to improve land cover classification accuracy in the mountains. *Can. J. Remote Sens.* 22:4, pp. 368-381.

Pellikka, P., Clark, B., Hurskainen, P., Keskinen, A., Lanne, M., Masalin K., Nyman-Ghezelbash, P., & Sirviö, T., 2004. Land use change monitoring applying geographic information systems in the Taita Iiills, SE-Kcnya. In the Proceedings of the 5th African Association of Remote Sensing of Environment Conference, 17-22 Oct. 2004, Nairobi, Kenya.

Pellikka, P., Lötjönen, M., Siljander, M., & Lens, L., 2009. Airborne remote sensing of spatiotemporal change 1955-2004 in indigenous and exotic forest cover in the Taita Hills, Kenya. *Int. J. Appl. Earth Obs. Geoinf.*, 11, pp. 221 -232.

Reese, H., & Olsson, H., 2011. C-correction of optical satellite data over alpine vegetation areas: A comparison of sampling strategies for determining the empirical c-parameter. *Remote Sens. Environ.*, 115, pp. 1387–1400.

Skakun, R.S., Wulder, M.A., & Franklin, S.E., 2003. Sensitivity of the thematic mapper enhanced wetness difference index to detect mountain pine beetle red-attack damage. *Remote Sens. Environ.*, *86*, pp. 433-443.

Smith, J. A., Lin, T. L., & Ranson, K. J., 1980, The Lambertian assumption and Landsat data. *Photogramm. Eng. Remote Sens.*, 46, pp.1183–1189.

Teillet, P. M., Guindon, B., & Goodenough, D. G., 1982. On the slope-aspect correction of multispectral scanner data. *Can. J. Remote Sens.* 8(2), pp. 84–106.

Törmä, M., & Härmä, P., 2003. Topographic correction of landsat ETM-images in Finnish Lapland. Geoscience and Remote Sensing Symposium, IGARSS '03. Proceedings. IEEE International, 6, pp. 3629-3631.

Vanonckelen, S., Lhermitte, S., & Van Rompaey, A., 2013. The effect of atmospheric and topographic correction methods on land cover classification accuracy. *Int. J. Appl. Earth Obs. Geoinf.*, *24*, pp. 9-21.

Wu, W., 2011. Derivation of tree canopy cover by multiscale remote sensing approach. *ISPRS - Int. Arch. Photogramm. Remote Sens. Spa. Inf. Sci.*, XXXVIII-4 (W25), pp. 142-149.

Wulder, M. A., Masek, J. G., Cohen, W. B., Loveland, T. R., & Woodcock, C. E., 2012. Opening the archive: How free data has enabled the science and monitoring promise of Landsat. *Remote Sens. Environ.*, 122, pp. 2-10.

# 14

# CLOUD PHOTOGRAMMETRY FROM SPACE

K. Zakšek [a, *,] A. Gerst [b], J. von der Lieth [a], G. Ganci [c], M. Hort [a]

[a] University of Hamburg, CEN, Institute of Geophysics, Bundesstr. 55, 20146 Hamburg, Germany – klemen.zaksek@uni-hamburg.de, jost.lieth@uni-hamburg.de, matthias.hort@uni-hamburg.de
[b] ESA, European Astronaut Centre, Linder Höhe, 51147 Köln, Germany
[c] INGV, Sezione di Catania, Piazza Roma, 2, 95125 Catania, Italy – gaetana.ganci@ct.ingv.it

**KEY WORDS:** photogrammetry, cloud top height, volcanic ash, SEVIRI, MODIS, ISS

**ABSTRACT:**

The most commonly used method for satellite cloud top height (CTH) compares brightness temperature of the cloud with the atmospheric temperature profile. Because of the uncertainties of this method, we propose a photogrammetric approach. As clouds can move with high velocities, even instruments with multiple cameras are not appropriate for accurate CTH estimation. Here we present two solutions. The first is based on the parallax between data retrieved from geostationary (SEVIRI, HRV band; 1000 m spatial resolution) and polar orbiting satellites (MODIS, band 1; 250 m spatial resolution). The procedure works well if the data from both satellites are retrieved nearly simultaneously. However, MODIS does not retrieve the data at exactly the same time as SEVIRI. To compensate for advection in the atmosphere we use two sequential SEVIRI images (one before and one after the MODIS retrieval) and interpolate the cloud position from SEVIRI data to the time of MODIS retrieval. CTH is then estimated by intersection of corresponding lines-of-view from MODIS and interpolated SEVIRI data. The second method is based on NASA program Crew Earth observations from the International Space Station (ISS). The ISS has a lower orbit than most operational satellites, resulting in a shorter minimal time between two images, which is needed to produce a suitable parallax. In addition, images made by the ISS crew are taken by a full frame sensor and not a push broom scanner that most operational satellites use. Such data make it possible to observe also short time evolution of clouds.

## 1. INTRODUCTION

Information on height of clouds independent of their origin (natural or anthropogenic aerosol clouds, meteorological clouds) is important in different research fields. Cloud top height (CTH) is especially interesting for meteorologists. In some countries, like Midwest USA, a lot of effort is dedicated towards observation of convective clouds (Cumulonimbus) that might develop in so called super cells. Tops of such clouds can reach heights of over 20 km making them a source of dangerous severe weather, such as hail, heavy precipitation, or tornadoes (Heymsfield et al., 1983). Furthermore, the top cloud height is relevant for climatology, because the height is related to the amount of long-wave radiation that is emitted to space. Computing the heights for a part of the available data archives may extract some interesting climatological trends. Another important application is monitoring of aerosols produced in forest fires or industrial accidents (for instance explosions in chemical factories, oil refineries or power plants).

In this contribution, we focus on the volcanic ash heights because of the huge economic loss of the 2010 Eyjafjallajökull volcanic eruption. We knew already before that volcanic eruptions can have a significant impact on air traffic; between 1953 and 2009 Guffanti et al. (2010) report 128 encounters between aircraft and volcanic ash worldwide. The International Air Transport Association (IATA) stated that the total loss for the airline industry as a result of the airspace closure during the eruption of Eyjafjallajökull was around €1.3 billion (BBC News, 21/04/10): over 95,000 flights had been cancelled all across Europe during the six-day travel ban (BBC News, 21/04/10), with later figures suggesting 107,000 flights

cancelled during an 8 day period, accounting for 48% of total air traffic affecting roughly 10 million passengers (Bye, 2010). The 2010 Eyjafjallajökull volcanic eruption on Iceland was not particularly large compared to the 1991 Pinatubo eruption (~20 times more erupted material) or the 1815 Tambora eruption (~500 times more). However, it totally paralysed the air traffic in Europe because of our inability to make an exact prediction of the volcanic ash dispersion. State of the art ash dispersion models are very sophisticated but the accuracy of their predictions is limited by the unacceptably low quality information on the eruption (Zehner, 2010). The crucial parameter is the ash cloud height. This is detailed in the study of Heinold et al. (2012) showing that emplacement heights of the ash into the atmosphere result in significantly different ash transport patterns. The reason for that is the wind field that can strongly vary with height. Significant differences in the wind velocity or significant differences of the wind direction are possible across height intervals of less than 500 m. Therefore, the ash dispersion models, in order to provide a reliable forecast, require ash cloud top height at a very high accuracy.

In this contribution we first give an overview of CTH estimation from space (section 2). Then we show two methods for CTH estimation. In section 3 we describe a method based on simultaneous observations from polar orbiting satellite and geostationary satellite. In section 4 we show an example of CTH estimation from photos made by ISS astronauts. Finally in section 5, we discuss the methodology and its further possible development.

---

\* Corresponding author

## 2. STATE OF THE ART IN CTH ESTIMATION

Clouds can be observed from the ground by common weather radar or by atmospheric lidar that received increasing attention. Although these are excellent tools, all ground based measurements are restricted to their low spatial and temporal availability. Compared to ground or airborne based observations satellite remote sensing provides global observations. In the table 1, we briefly review different satellite remote sensing techniques for aerosol / meteorological clouds height retrieval.

| Methodology | Pros / Cons |
| --- | --- |
| Lidar and radar | + very high vertical resolution and accuracy |
| (Carn et al., 2009; Karagulian et al., 2010; Prata et al., 2015; Stohl et al., 2011) | − too long revisit time (16 days) and only nadir observations from currently operational instruments (lidar CALIOP, radar CPR) |
| Radio occultation | + high resolution in lower troposphere |
| (Kursinski et al., 1997; Solheim et al., 1999) | − globally available only about 2000 times per day |
| Backward trajectories modelling | + possible estimate even for clouds drifted away from the source |
| (Eckhardt et al., 2008; Oppenheimer, 1998; Tupper et al., 2004) | − requires wind field data for a large area; homogenous wind field results in high uncertainty of the source height |
| Brightness temperature | + easy to apply, possible with instruments having a short revisit time |
| (Genkova et al., 2007; Oppenheimer, 1998; Prata and Grant, 2001; Tupper et al., 2004) | − requires atmospheric profile and emissivity of the cloud; assumption of thermal equilibrium; problems around tropopause |
| $O_2$ A-band absorption | + high accuracy |
| (Dubuisson et al., 2009) | − requires high spectral resolution data (not available on many satellites→long revisit time); good performance only over dark surface; requires radiative transfer modelling (slow); daytime only |
| $CO_2$ absorption | + good performance also by semi-transparent clouds |
| (Chang et al., 2010; Richards et al., 2006) | − accurate only in the high levels of troposphere; problems around tropopause |
| Shadow length | + easy to apply; requires no additional data |
| (Glaze et al., 1989; Prata and Grant, 2001) | − possible only during daytime; retrieves the height of the cloud horizontal edge and not its top |
| Stereoscopy | + high accuracy; requires no additional data; based on geometry→no problems in the case of ash reaching the stratosphere |
| (Genkova et al., 2007; Hasler, 1981; Prata and Turner, 1997; Scollo et al., 2012; Virtanen et al., 2014; Zakšek et al., 2013) | − requires simultaneous retrieval of data from two different viewpoints |

Table 1. Comparisons of satellite methods for aerosol / meteorological cloud top height retrieval

The accuracy of the listed methods depends on the sensor's spatial resolution and the cloud's height (Genkova et al., 2007). The best estimates using lidar have an accuracy better than 200 m but they have a revisit time of 16 days. The operational height estimates based on $CO_2$ absorption are available several times a day but with an accuracy worse than 1000 m (Holz et al., 2008). Therefore, the state of the art satellite measurements of the ash cloud top height do not provide adequate accuracy and temporal availability at the same time.

Stereoscopy (last line of table 1) can be considered as the optimal technique for cloud height observations if two images are made simultaneously from two different viewpoints. The stereoscopy is a classic photogrammetric technique that is optimal for retrieval of 3D form if the observed object does not change between retrievals.

Several attempts with multi-angle instruments or instruments in different orbits have already been applied for CTH estimation (Genkova et al., 2007; Hasler, 1981; Nelson et al., 2013; Prata and Turner, 1997; Scollo et al., 2012; Virtanen et al., 2014; Zakšek et al., 2013). The wind velocities on high altitudes, however, can reach even 100 m/s. This means that unless the observations are made exactly at the same time, the results of stereoscopic analysis will contain systematic errors. It is possible to apply the appropriate correction if accurate wind field is known, but this is often not the case.

For instance, it takes almost 80 s for the SLSTR instrument on-board Sentinel 3 (to be launched in fall 2015) between the nadir and backward view (inclined for 55° back from nadir). If a cloud moves at 180 km/h it moves in this time 4000 m. This can lead into a parallax error of maximal the same value of 4000 m (depends on the direction of the wind in relationship to the satellite track) causing a height error of maximal 2.8 km.

Therefore, the clue to accurate use of stereoscopy is in observations from two different viewpoints simultaneously. This is technically still not possible – for that we would need a constellation of at least two satellites flying in formation and observing the same area on the ground at the same time.

A good alternative is using geostationary satellites (section 3). They retrieve data with high frequency. This makes possible to acquire an image from a geostationary satellite nearly simultaneously with an image from another orbit.

Another alternative for reducing the time gap between two images is using lower orbit (section 4). Instruments in lower orbit (300–500 km) are closer to Earth, thus they move faster. In addition because the orbit height is lower, the baseline between two satellites can be shorter as well.

## 3. SIMULTANEOUS STEREOSCOPIC OBSERVATIONS FROM SEVIRI AND MODIS

Here we describe a photogrammetric method based on the parallax observations between data retrieved from satellites in geostationary orbit and polar orbiting satellites. We use a combination of Moderate-resolution Imaging Spectroradiometer (MODIS) aboard Terra and Aqua satellites (polar orbit) and Spinning Enhanced Visible and InfraRed Imager (SEVIRI) aboard Meteosat Second Generation (MSG) satellites in a geostationary orbit. The described method has already been tested for the ash cloud Eyjafjallajökull eruption in April 2010 (Zakšek et al., 2013).

The proposed method of CTH estimation consists of three main steps. In the first step we aggregate MODIS data to SEVIRI spatial grid. The second step is automatic image matching. In the third step, lines of sight connecting observed points of both satellites are generated; the intersection points of SEVIRI and MODIS lines of sight are then used to estimate CTH.

### 3.1 Data pre-processing

To be able to perform automatic image matching it is necessary to pre-process data so that MODIS and SEVIRI datasets are comparable. In the previous retrievals of meteorological cloud top height (Hasler, 1981) both images from GOES were projected to a standard map projection. We decided to leave SEVIRI data in its own grid system. MODIS data have much better spatial resolution, thus they can be projected to the SEVIRI grid system without significantly influencing the resulting accuracy.

In addition, the geolocation has to be adjusted. SEVIRI's geolocation is according to our experience, often false by a pixel or two. This was confirmed also by an independent study (Aksakal, 2013). Thus we used coastlines to automatically align MODIS and SEVIRI data.

### 3.2 Image matching

The goal of the image matching is to accurately identify point pairs between two satellite images. This might be difficult if the images are not retrieved by the same instrument. The problem involves different resolutions, different viewing geometries, and different instruments response functions. In addition, the appearance of the same object in two different images might contain a large illumination variation, and thus the local descriptors of the same feature point are different. A number of automatic image matching approaches have been proposed to solve these issues. Here we used the same procedure as already described by e.g. Scambos et al. (1992) or Prata and Turner (1997). We computed cross correlation (eq. 1) between a reference subset around each pixel within a moving window analysis:

$$CI = \frac{\sum\limits_{i=1-n_{C1}/2, j=1-n_{L1}/2}^{i=n_{C1}/2-1, j=n_{L1}/2-1}\left(DNr_{i,j} - \mu r\right)\cdot\left(DNm_{i,j} - \mu s\right)}{\sqrt{\sum\limits_{i=1-n_{C1}/2, j=1-n_{L1}/2}^{i=n_{C1}/2-1, j=n_{L1}/2-1}\left(DNr_{i,j} - \mu r\right)^2}\cdot\sqrt{\sum\limits_{i=1-n_{C1}/2, j=1-n_{L1}/2}^{i=n_{C1}/2-1, j=n_{L1}/2-1}\left(DNm_{i,j} - \mu s\right)^2}} \quad (1)$$

where     $CI$ – the correlation index between subsets
$DNm_{i,j}$ and $DNr_{i,j}$ – digital numbers of the moving window and reference subset
$\mu m$ and $\mu r$ – the mean values of reference subsets and the moving window set
$i$ and $j$ – shifts between the central pixels of the reference subset and the moving window

Results of image matching depend on the size of the search area and moving window. A large moving window can detect large features but it usually fails to detect small features. In contrast, a small moving window detects small features but generates a lot of noise in the results. The appropriate optimization is image matching over image pyramids. We consider image pyramids as a multi-resolution representation of the original image (Anderson et al., 1984). Each higher pyramid is merely a regridded lower pyramid. Image matching is first done on coarse pyramids and the measured shifts are then used to initialize image matching on the original data.

### 3.3 CTH estimation from a pair of SEVIRI images and one MODIS image

Because MODIS and SEVIRI times of retrieval are usually not simultaneous, there is always a time gap between them. As the plume can move during this time, we use a pair of sequential SEVIRI images – one before and one after MODIS retrieval. Therefore, image matching has to run twice to find matching points in all three images. The effect of possible advection of the eruption cloud between the MODIS and the SEVIRI images is considered for each pixel triple: the coordinates of a virtual SEVIRI pixel are interpolated from position of both SEVIRI pixels to the time of MODIS retrieval (fig. 1).

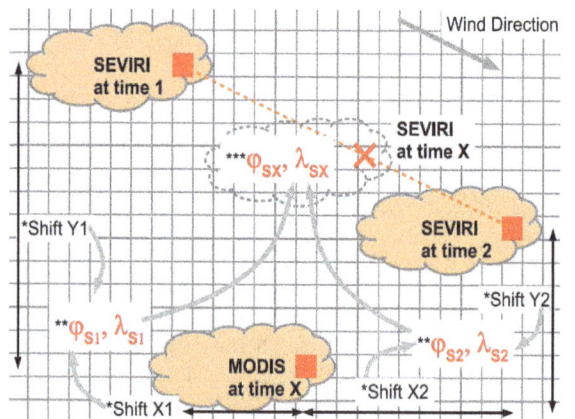

Figure 1. The procedure of determining the position of a cloud in SEVIRI image at the time of MODIS retrieval. * Shifts in column and line direction are estimated twice by automatic image matching between MODIS (retrieved at time X) and SEVIRI (retrieved at times 1 and 2). ** Estimated geographic cloud's positions are observed by SEVIRI at times 1 and 2. *** Interpolated geographic position of the plume as SEVIRI would observe it at times X corresponding to MODIS retrieval.

In a Cartesian coordinate system we can define lines connecting coordinates of the virtual SEVIRI pixels with the position of the MSG satellite ("SEVIRI lines") and corresponding lines connecting coordinates of the MODIS pixels with the position of the Terra/Aqua satellite ("MODIS lines"). The solution of the following linear system gives the intersection of the line pair:

$$\begin{bmatrix} x \\ y \\ z \end{bmatrix}_M + t_M \cdot \begin{bmatrix} v_x \\ v_y \\ v_z \end{bmatrix}_M = \begin{bmatrix} x \\ y \\ z \end{bmatrix}_S + t_S \cdot \begin{bmatrix} v_x \\ v_y \\ v_z \end{bmatrix}_S \qquad (2)$$

where    $[x,y,z]_M$ and $[x,y,z]_S$ – the positions of the MODIS aboard Terra/Aqua and SEVIRI aboard MSG

$[v_x,v_y,v_z]_M$ and $[v_x,v_y,v_z]_S$ – direction vectors of MODIS and SEVIRI lines

$t_M$ and $t_S$ are – unknowns defining the point of intersection.

The system in eq. 2 is over determined, thus it can be solved by a least-square technique. The geocentric Cartesian coordinates of the intersection are then converted back from geocentric Cartesian to the geographic coordinate system: longitude, latitude, height above ellipsoid – i.e. CTH.

MODIS and SEVIRI lines never intersect because the data are not continuous but discrete pixels. The lines rather pass each other. Thus the eq. 2 really provides just the pair of closest points on the corresponding lines. CTH can then be estimated from one of these two points or as their average. Such a procedure makes also possible to estimate the intersection quality. It can be described by the distance between MODIS and SEVIRI lines; if it is small, the accuracy of CTH is high.

## 3.4  Etna Ash Cloud 8 September 2011

Following the sunrise on 8 September, Etna produced a series of ash emissions (fig. 2) followed by increased intensity and frequency of the explosions. At about 06:30 GMT, the activity passed from Strombolian into a pulsating lava fountain, accompanied by increasing amounts of volcanic ash. The paroxysm totally ceased around 08:45 GMT.

Figure 2. Eruption of Mount Etna as seen from the airport of Catania; note its height (Etna's peak is 3350 m high). Photo is the courtesy of S. Scollo, INGV Catania).

Fig. 3 shows SEVIRI and MODIS data combined into RGB image. The first example is based on 15 min data from MSG2 (fig. 3 above) and the second one on 5 min data from MSG1 (fig. 3 below). The volcanic ash cloud (actually all elevated objects) is coloured because of the wind and the parallax. Sea is white because the colours are inverted (in visible images is sea

normally dark). The yellow part of the cloud corresponds to MODIS data only, cyan to the first SEVIRI image and purple to the second SEVIRI image. The other colours, like dark blue corresponds to the mixture of images – in this case first and second SEVIRI images.

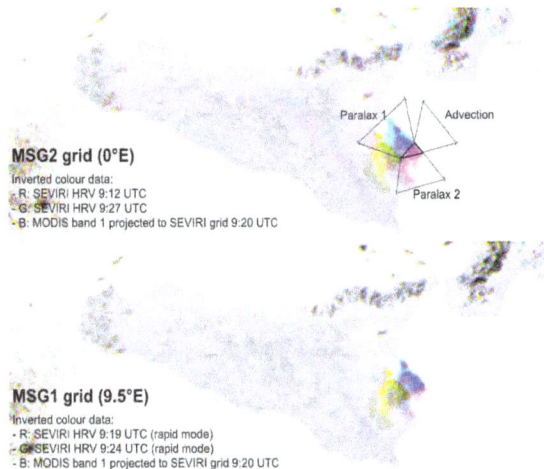

Figure 3. SEVIRI and MODIS data combined into RGB image; above data from 15 min SEVIRI retrieval on MSG2, below data from 5 min SEVIRI retrieval on MSG1.

The colours make it possible to observe the parallax between MODIS and both SEVIRI images: for the south-eastern "corner" of the cloud we show corresponding points on all three images (fig. 3 above). They are connected with lines corresponding to the parallax between MODIS and first SEVIRI image (parallax 1), MODIS and second SEVIRI Image (parallax 2) and effect of advection between both SEVIRI images (advection).

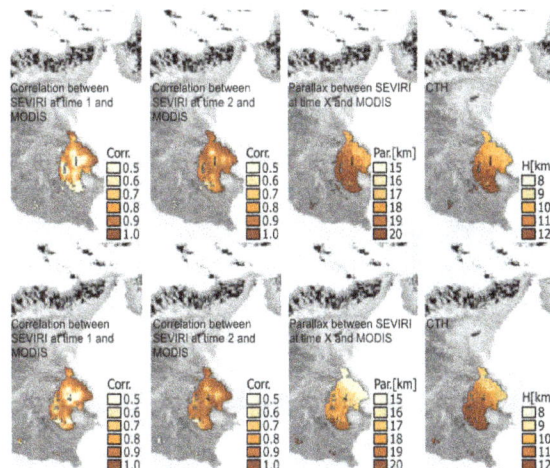

Figure 4. Cross correlation between MODIS and both SEVIRI images, estimated parallax and corresponding CTH of the ash cloud from Etna; above data from 15 min retrieval from MODIS and SEVIRI on MSG2, below data from 5 min retrieval from MODIS and SEVIRI on MSG1.

The 15 min standard SEVIRI retrieval resulted in larger parallax and also larger CTH (fig. 4). The correlation index was as expected higher for the 5 min SEVIRI retrieval, which means that these results are of lower uncertainty than results based on 15 min data.

## 4. STEREOSCOPIC OBSERVATIONS FROM A LOW ORBIT

A low orbit (height of 300–500 km) has many advantages over higher orbits when it comes to cloud photogrammetry. Because of its lower height and higher speed, the instruments can reach a suitable baseline between two positions within some seconds. This significantly reduces the influence of wind. In addition, a lower orbit usually means also a higher spatial resolution of data, resulting into higher level of details. This can provide more texture that is necessary for a reliable image matching.

Low orbits were not used very often in the past. These orbits have some limitations for other fields of remotes sensing. Its main disadvantage is its narrower swath, which increases also revisit time. Low orbits have gained on importance in the recent years. The reason for that is increased number of launches of small satellites. The philosophy of such satellites is their cost-effectiveness. The most expensive post by a small satellite mission is its launch. As the price of launch depends also on the orbit height, most of small satellites are launched into low orbits.

### 4.1 Crew Earth Observations

Besides small satellites, into a low orbit was positioned also the International Space Station (ISS). ISS does not carry many sophisticated earth observation instruments. But NASA started some years ago a programme named Crew Earth Observations. Within this programme, the astronauts take photos of the Earth. NASA made available these images for scientific purpose. It is also possible to make an acquisition request of some specific area. The co-author of this paper, Alexander Gerst, was a crew member of ISS missions 40/41. As an acknowledged volcanologist he was asked to provide photos of volcanic clouds. But there was "unfortunately" no major eruption during his mission. He still managed to provide photos of Zhupanovsky volcano (Kamchatka, Russia) during its activity in September 2014 (fig. 5).

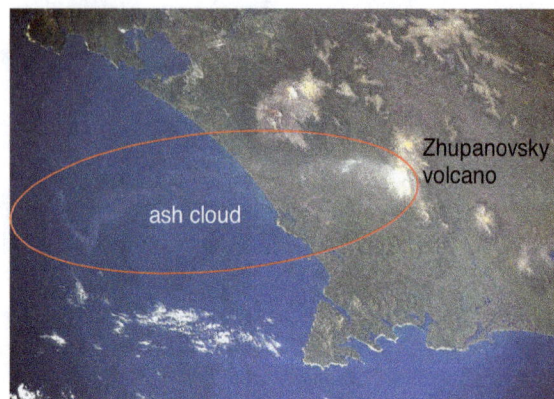

Figure 5. A photo (number ISS041e000162) of Zhupanovsky volcano and it ash cloud on Septemebr 10 2014 at 23:11 UTC; it is a courtesy of the Earth Science and Remote Sensing Unit, NASA Johnson Space Center.

### 4.2 ISS images pre-processing

Here we have to point out that the used camera contains a typical full-frame sensor. Usual satellite instruments contain a push broom scanner (along track scanner). In a push broom sensor, a line of sensors arranged perpendicular to the flight direction of the spacecraft is used. Different areas of the surface are imaged as the spacecraft flies forward. This means, that each point on earth is scanned only once by such an instrument. But a full frame sensor can take even a video of an area, which is a significant advantage of full-frame sensors over push broom technology.

The main difference between the ISS crew images and data retrieved by classical satellite instruments is that the ISS images are usual photos and not calibrated data. Because the JPG files available in Gateway to Astronaut Photography of Earth (http://eol.jsc.nasa.gov/) contain not enough of radiometric details, especially if a cloud is transparent, we requested RAW files. They contain for each band 14 bit data, which is a huge improvement over 8 bit data given in each JPG channel. We wrote our own code that converts Nikon NEF file to a TIFF file with 16 bit per channel.

Furthermore, the images are not geolocated. In the metadata of each image is given the location of ISS as the image was taken. But to estimate CTH a described by eq. 2 we require also the coordinates of each pixel. An average pixel size in fig. 5 is ~50 m. We have manually georeferenced the image using coastlines and then projected data to UTM projection (zone 57N) in 200 m spatial resolution (fig. 6). It was necessary to choose only points with zero elevation for geolocation, so the higher objects preserved their parallax.

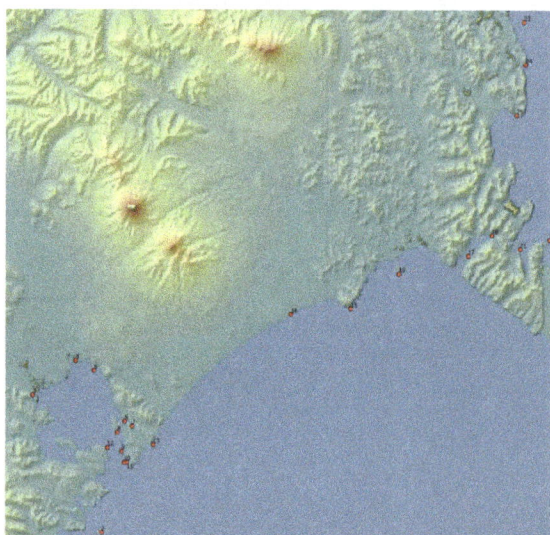

Figure 6. Points (red dots) selected for geolocation of ISS images. Zhupanovsky is in the figure above, in the middle left you can see also Koryaksky and Avachnisky volcanoes.

For the case study we selected a pair of images ISS041e000162 and ISS041e000164. The time gap between both images was 11.65 s. The ISS moved within this time almost for 130 km. Considering its height was at that time 419 km, the baseline was appropriately long for a reliable image matching at 200 m resolution data. The georeferenced data (only green channel that showed in the case study the largest contrast within the clouds) are shown in fig. 7.

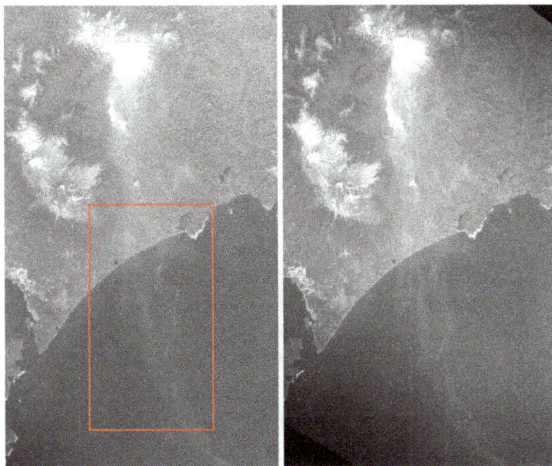

Figure 7. Geolocated green channel of the images numbered ISS041e000162 (left) and ISS041e000164 (right). The red rectangle shows the extent of results in fig. 8.

We ran basically the same procedure as described in sections 3.2 and 3.3. The position of ISS was retrieved from the web archive http://www.isstracker.com/historical and interpolated to the exact time the image was taken. The relative time between images is given by a hundred of a second. The absolute time accuracy can be as bad as several seconds, because the camera is not synchronized with the GPS. The time on the camera is always set manually. To account for this we have an iterative solution: we change the time of the first image was taken so long that the estimated height of some recognizable mountain peaks agrees with their true elevation.

### 4.3 Zhupanovsky Ash Cloud 10 September 2014

After 54 years of inactivity, the volcano began erupting on October 23, 2013 and again in 2014, continuing into 2015. Here we present only results for 10 September 2014. For the further processing we decided to focus only on data over sea providing a homogenous background. We have ignored the data in the coastal region because of the turbid water that has approximately the same reflectance as the ash over the sea.

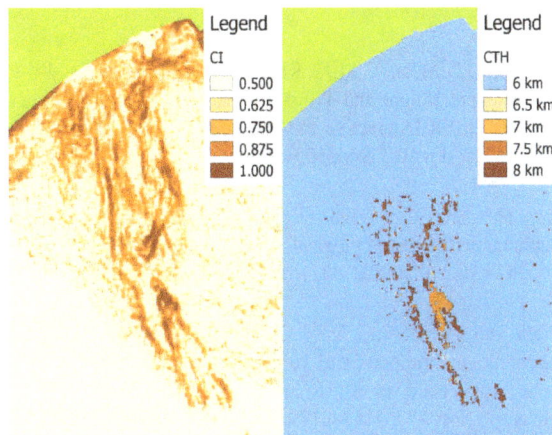

Figure 8. Correlation index over the sea on the left and CTH for pixels that passed the internal accuracy control on the right. The extent of this image is shown with the red rectangle on fig. 7.

The correlation index between both images was very high (fig. 8 left), in most parts of the cloud around 90%, which shows that this procedure is capable of producing good results even over transparent plume. In the end we have filtered out the data having lower correlation, or larger distance between lines of sight and the pixels that remained have height between 7 and 8 km (fig. 8 right). Global Volcanism Program (2014) reported satellite estimation of ash height to be up to 4 km during 9–11 September but it does not provide any source. In this time ash clouds drifted about 1000 km due South.

## 5. DISCUSSION AND CONCLUSIONS

We presented two innovative ways of CTH estimation. Both use essentially the same methodology but different instruments. The first method has been already validated (Zakšek et al., 2013) – the results for 2010 Eyjafjallajökull eruption had an accuracy of ~600 m. But here we showed that the same methodology can be used also on small clouds that might be still developing. In the case that clouds are still developing in vertical direction, it is obviously better to use SEVIRI data retrieved each 5 min instead of 15 min.

The results based on ISS images have not been validated because we did not have any independent data of this area for the same time. The approximate validation could be possible by comparing the standard MODIS cloud product made on 11 September at 00:45 UTC, but for an accurate comparison is the time gap between ISS and MODI retrieval too large (Terra flew ~90 min later over Zhupanovsky than ISS; see also http://earthobservatory.nasa.gov/NaturalHazards/view.php?id=84386).

For conclusion, we can take a look at advantages and disadvantages of stereoscopic observations more closely. The most important advantage of the proposed method is its independence of physical assumptions. For its use we do not need emissivity of the cloud, it does not matter whether a cloud is close or even above the tropopause, etc. Secondly, the method is perhaps not as fast as CTH estimation from brightness temperature but it is still much faster than some methods depending on the atmospheric transport modelling or radiative transfer modelling. And finally, the method is very accurate, especially if subpixel image matching is applied. An alternative to the image matching is selection of typical points in the ash clouds, like centre of ash mass (personal communication with Stefano Corradini). This procedure has not been practised so far but it is most likely the best option for accurate selection of corresponding points over transparent clouds.

The accuracy of image matching depends mostly on appropriate texture in data. This can be problematic especially with transparent clouds, where the texture of the background becomes dominant. An additional parameter that influences accuracy is the geolocation accuracy of the input images. This should be always (automatically) checked and corrected using independent GIS layers. The greatest issue of the cloud photogrammetry is simultaneous data retrieval from different points. We have here presented only a nearly simultaneous retrieval, but it performed well in both case studies.

However, to overcome this "disadvantage" we would need at least two satellites separated by an appropriate baseline. They should be following each other in the same orbit. Such a pair would provide truly simultaneous observations that could lead into CTH heights with accuracy of ~200 m.

## ACKNOWLEDGEMENTS

The research has been supported by grants from the German Science Foundation (DFG) number ZA659/1-1.

The SEVIRI L1.5 data were obtained through the online archive of EUMETSAT images (Earth Observation Portal). The MODIS L1B data were obtained through the online Data Pool at the NASA Land Processes Distributed Active Archive Center (LP DAAC), USGS/Earth Resources Observation and Science (EROS) Center, Sioux Falls, South Dakota (https://lpdaac.usgs.gov/get_data). ISS images were obtained through the online Gateway to Astronaut Photography of Earth (http://eol.jsc.nasa.gov/); they are courtesy of the Earth Science and Remote Sensing Unit, NASA Johnson Space Center. The RAW images were provided by ESA and NASA.

## REFERENCES

Aksakal, S.K., 2013. Geometric Accuracy Investigations of SEVIRI High Resolution Visible (HRV) Level 1.5 Imagery. Remote Sens. 5, 2475–2491. doi:10.3390/rs5052475

Anderson, C.H., Bergen, J.R., Burt, P.J., Ogden, J.M., 1984. Pyramid Methods in Image Processing. RCA Eng. 29, 33–41.

Bye, B.L., 2010. Volcanic Eruptions: Science And Risk Management [WWW Document]. URL http://www.science20.com/planetbye/volcanic_eruptions_scienc e_and_risk_management-79456 (accessed 2.26.13).

Carn, S.A., Pallister, J.S., Lara, L., Ewert, J.W., Watt, S., Prata, A.J., Thomas, R.J., Villarosa, G., 2009. The Unexpected Awakening of Chaitén Volcano, Chile. Eos 90, 205–206. doi:200910.1029/2009EO240001

Chang, F.-L., Minnis, P., Lin, B., Khaiyer, M.M., Palikonda, R., Spangenberg, D.A., 2010. A modified method for inferring upper troposphere cloud top height using the GOES 12 imager 10.7 and 13.3 μm data. J. Geophys. Res. 115, D06208. doi:10.1029/2009JD012304

Dubuisson, P., Frouin, R., Dessailly, D., Duforêt, L., Léon, J.-F., Voss, K., Antoine, D., 2009. Estimating the altitude of aerosol plumes over the ocean from reflectance ratio measurements in the O2 A-band. Remote Sens. Environ. 113, 1899–1911. doi:10.1016/j.rse.2009.04.018

Eckhardt, S., Prata, A.J., Seibert, P., Stebel, K., Stohl, A., 2008. Estimation of the vertical profile of sulfur dioxide injection into the atmosphere by a volcanic eruption using satellite column measurements and inverse transport modeling. Atmospheric Chem. Phys. 8, 3881–3897. doi:10.5194/acp-8-3881-2008

Genkova, I., Seiz, G., Zuidema, P., Zhao, G., Di Girolamo, L., 2007. Cloud top height comparisons from ASTER, MISR, and MODIS for trade wind cumuli. Remote Sens. Environ. 107, 211–222. doi:10.1016/j.rse.2006.07.021

Glaze, L.S., Francis, P.W., Self, S., Rothery, D.A., 1989. The 16 September 1986 eruption of Lascar volcano, north Chile: Satellite investigations. Bull. Volcanol. 51, 149–160. doi:10.1007/BF01067952

Global Volcanism Program, 2014. Zhupanovsky: Weekly Re ort [WWW Document]. URL

http://www.volcano.si.edu/volcano.cfm?vn=300120#September 2014

Guffanti, M., Casadevall, T.J., Budding, K., 2010. USGS Data Series 545: Encounters of Aircraft with Volcanic Ash Clouds: A Compilation of Known Incidents, 1953–2009 [WWW Document]. Data Ser. 545. URL http://pubs.usgs.gov/ds/545/ (accessed 2.26.13).

Hasler, A.F., 1981. Stereographic Observations from Geosynchronous Satellites: An Important New Tool for the Atmospheric Sciences. Bull. Am. Meteorol. Soc. 62, 194–212. doi:10.1175/1520-0477(1981)062<0194:SOFGSA>2.0.CO;2

Heinold, B., Tegen, I., Wolke, R., Ansmann, A., Mattis, I., Minikin, A., Schumann, U., Weinzierl, B., 2012. Simulations of the 2010 Eyjafjallajökull volcanic ash dispersal over Europe using COSMO–MUSCAT. Atmos. Environ. 48, 195–204. doi:10.1016/j.atmosenv.2011.05.021

Heymsfield, G.M., Blackmer, R.H., Schotz, S., 1983. Upper-Level Structure of Oklahoma Tornadic Storms on 2 May 1979. I: Radar and Satellite Observations. J. Atmospheric Sci. 40, 1740–1755. doi:10.1175/1520-0469(1983)040<1740:ULSOOT>2.0.CO;2

Holz, R.E., Ackerman, S.A., Nagle, F.W., Frey, R., Dutcher, S., Kuehn, R.E., Vaughan, M.A., Baum, B., 2008. Global Moderate Resolution Imaging Spectroradiometer (MODIS) cloud detection and height evaluation using CALIOP. J. Geophys. Res. Atmospheres 113, D00A19. doi:10.1029/2008JD009837

Karagulian, F., Clarisse, L., Clerbaux, C., Prata, A.J., Hurtmans, D., Coheur, P.F., 2010. Detection of volcanic SO2, ash, and H2SO4 using the Infrared Atmospheric Sounding Interferometer (IASI). J. Geophys. Res. Atmospheres 115, D00L02. doi:201010.1029/2009JD012786

Kursinski, E.R., Hajj, G.A., Schofield, J.T., Linfield, R.P., Hardy, K.R., 1997. Observing Earth's atmosphere with radio occultation measurements using the Global Positioning System. J. Geophys. Res. Atmospheres 102, 23429–23465. doi:10.1029/97JD01569

Nelson, D.L., Garay, M.J., Kahn, R.A., Dunst, B.A., 2013. Stereoscopic Height and Wind Retrievals for Aerosol Plumes with the MISR INteractive eXplorer (MINX). Remote Sens. 5, 4593–4628. doi:10.3390/rs5094593

Oppenheimer, C., 1998. Review article: Volcanological applications of meteorological satellites. Int. J. Remote Sens. 19, 2829. doi:10.1080/014311698214307

Prata, A.J., Grant, I.F., 2001. Retrieval of microphysical and morphological properties of volcanic ash plumes from satellite data: Application to Mt Ruapehu, New Zealand. Q. J. R. Meteorol. Soc. 127, 2153–2179. doi:10.1002/qj.49712757615

Prata, A.J., Turner, P.J., 1997. Cloud-top height determination using ATSR data. Remote Sens. Environ. 59, 1–13. doi:10.1016/S0034-4257(96)00071-5

Prata, A.T., Siems, S.T., Manton, M.J., 2015. Quantification of volcanic cloud-top heights and thicknesses using A-train observations for the 2008 Chaitén eruption. J. Geophys. Res. Atmospheres 2014JD022399. doi:10.1002/2014JD022399

Richards, M., Ackerman, S.A., Pavolonis, M.J., Feltz, W.F., 2006. Volcanic ash cloud heights using the MODIS CO2-slicing algorithm. Presented at the 12th Conference on Aviation Range and Aerospace Meteorology, Atlanata, USA.

Scambos, T.A., Dutkiewicz, M.J., Wilson, J.C., Bindschadler, R.A., 1992. Application of image cross-correlation to the measurement of glacier velocity using satellite image data. Remote Sens. Environ. 42, 177–186. doi:10.1016/0034-4257(92)90101-O

Scollo, S., Kahn, R.A., Nelson, D.L., Coltelli, M., Diner, D.J., Garay, M.J., Realmuto, V.J., 2012. MISR observations of Etna volcanic plumes. J. Geophys. Res. 117, D06210. doi:10.1029/2011JD016625

Solheim, F.S., Vivekanandan, J., Ware, R.H., Rocken, C., 1999. Propagation delays induced in GPS signals by dry air, water vapor, hydrometeors, and other particulates. J. Geophys. Res. Atmospheres 104, 9663–9670. doi:10.1029/1999JD900095

Stohl, A., Prata, A.J., Eckhardt, S., Clarisse, L., Durant, A., Henne, S., Kristiansen, N.I., Minikin, A., Schumann, U., Seibert, P., Stebel, K., Thomas, H.E., Thorsteinsson, T., Tørseth, K., Weinzierl, B., 2011. Determination of time- and height-resolved volcanic ash emissions for quantitative ash dispersion modeling: the 2010 Eyjafjallajökull eruption. Atmospheric Chem. Phys. 11, 5541–5588. doi:10.5194/acpd-11-5541-2011

Tupper, A., Carn, S., Davey, J., Kamada, Y., Potts, R., Prata, F., Tokuno, M., 2004. An evaluation of volcanic cloud detection techniques during recent significant eruptions in the western "Ring of Fire." Remote Sens. Environ. 91, 27–46. doi:10.1016/j.rse.2004.02.004

Virtanen, T.H., Kolmonen, P., Rodríguez, E., Sogacheva, L., Sundström, A.-M., de Leeuw, G., 2014. Ash plume top height estimation using AATSR. Atmos Meas Tech 7, 2437–2456. doi:10.5194/amt-7-2437-2014

Zakšek, K., Hort, M., Zaletelj, J., Langmann, B., 2013. Monitoring volcanic ash cloud top height through simultaneous retrieval of optical data from polar orbiting and geostationary satellites. Atmospheric Chem. Phys. 13, 2589–2606.

Zehner, C., 2010. Monitoring volcanic ash from space, in: Monitoring Volcanic Ash from Space. Presented at the ESA-EUMETSAT workshop on the 14 April to 23 May 2010 eruption at the Eyjafjöjull volcano, South Iceland, Frascati, Italy. doi:10.5270/atmch-10-01

# 15

# National Scale Monitoring, Reporting and Verification of Deforestation and Forest Degradation in Guyana

P. Bholanath[a], K. Cort[b]

[a] Guyana Forestry Commission, Planning and Developmnet Division, 1 Water Street, Kingston, Georgetown, Guyana – project.coordinator@forestry.gov.gy
[b] Guyana Forestry Commission, Manager, Forest Area Assessment Unit, 1 Water Street, Kingston, Georgetown, Guyana – kerryanne.cort@gmail.com

**KEY WORDS:** Deforestation, Forest Degradation, Landsat, RapidEye, MRV System, and persistent cloud.

**ABSTRACT:**

Monitoring deforestation and forest degradation at national scale has been identified as a national priority under Guyana's REDD[+] Programme. Based on Guyana's MRV (Monitoring Reporting and Verification) System Roadmap developed in 2009, Guyana sought to establish a comprehensive, national system to monitor, report and verify forest carbon emissions resulting from deforestation and forest degradation in Guyana. To date, four national annual assessments have been conducted: 2010, 2011, 2012 and 2013.

Monitoring of forest change in 2010 was completed with medium resolution imagery, mainly Landsat 5. In 2011, assessment was conducted using a combination of Landsat (5 and 7) and for the first time, 5m high resolution imagery, with RapidEye coverage for approximately half of Guyana where majority of land use changes were taking place. Forest change in 2013 was determined using high resolution imagery for the whole of Guyana. The current method is an automated-assisted process of careful systematic manual interpretation of satellite imagery to identify deforestation based on different drivers of change. The minimum mapping unit (MMU) for deforestation is 1 ha (Guyana's forest definition) and a country-specific definition of 0.25 ha for degradation.

The total forested area of Guyana is estimated as 18.39 million hectares (ha). In 2012 as planned, Guyana's forest area was re-evaluated using RapidEye 5 m imagery. Deforestation in 2013 is estimated at 12 733 ha which equates to a total deforestation rate of 0.068%. Significant progress was made in 2012 and 2013, in mapping forest degradation. The area of forest degradation as measured by interpretation of 5 m RapidEye satellite imagery in 2013 was 4 352 ha. All results are subject to accuracy assessment and independent third party verification.

## 1. Introduction

Historical deforestation in Guyana has been very low (0.02% to 0.079% $yr^{-1}$ over the past 22 years), but this trend may change in the future as deforestation increases to meet growing demands for agriculture, timber, minerals, and human settlements. Guyana is therefore considered to be a high forest cover low emission/deforestation rate (HFLE/D) country, with forests covering approximately 85% of the country (forest area of 18.5 million hectares) and containing an estimated 19.5 billion tons (or Gt) of $CO_2$ in live and dead biomass pools.

In addition to being one of Guyana's most valuable natural assets, these forests are suitable for logging and agriculture, and are underlain with significant mineral deposits. Mining has been the primary driver of deforestation in Guyana, accounting for approximately 60% of all deforestation between 1990 and 2009 and more than 90% of deforestation between 2009 and 2012. Other drivers include forestry infrastructure, agriculture, and other infrastructure.

The Joint Concept Note (JCN) between the Government of Guyana and the Government of the Kingdom of Norway identifies the stepwise and progressive development of the Guyana Monitoring Reporting and Verification System (MRVS) based on REDD+ Interim Indicators and reporting requirements. The intention is that these interim measures will be phased out as the full-fledged MRVS is established. Guyana's MRVS, which is composed of the Forest Area

Assessment System and the Forest Carbon Monitoring System (FCMS) form the link between historical assessments and current/future assessments, enabling consistency in the data and information to support the implementation of REDD+ activities.

The initial steps of the MRVS allowed for a historical assessment of forest cover to be completed, key database integration to be fulfilled and for interim/intermediate indicators of emissions from deforestation and forest degradation to be reported for subsequent periods. The basis for comparison of the area-based interim measures is the 2009 September Benchmark Map. Four annual Forest Area Assessments have been completed so far. The first reporting period (termed Year 1) is set from 01 October 2009 to 30 September 2010 with the second reporting period (Year 2) covering 01 October 2010 to 31 December 2011, a fifteen (15) month period. The Year 3 and Year 4 reports both cover the 2013 and 2014 calendar years, respectively.

The transition from medium resolution (30 m) Landsat to high resolution RapidEye images (5 m pixel resolution) has increased the opportunity to better delineate and detect land use change. The analysis is subject to independent audit, firstly by the accuracy assessors University of Durham (UoD) and secondly by the project verifiers Det Norske Veritas (DNV).

An accompanying and closely connected programme of work is being implemented by Guyana Forestry Commission (GFC), with the assistance of a specialist firm (Winrock International) is the development of a national forest carbon measurement system and related emission factors. This programme will establish national carbon conversion values, expansion factors, wood density and root/shoot ratios as necessary. The MRVS details the methods required to quantify the changes in forest cover and changes in forest carbon stocks in Guyana, develop driver-specific emission factors by forest strata, and monitor emissions from land cover/land use change over time based on a variety of management activities.

This paper provides a summarised description of the work undertaken to complete the annual forest area assessments.

## 1.1 Country Description

The total land area for Guyana is 21.1 million hectares (ha) and spans from 2 to 8° N and 57 to 61° W. Guyana shares common borders with three countries: to the north-west - Venezuela, the south-west - Brazil, and on the east - Suriname. Guyana's 460 km coastline faces the Atlantic on the northern part of the South American continent. The coastal plain is only about 16 km wide but is 459 km long.

It is dissected by 16 major rivers and numerous creeks and canals for irrigation and drainage. The main rivers that drain into the Atlantic Ocean include the Essequibo, Demerara, Berbice, and Corentyne. These rivers have wide mouths, mangroves, and longitudinal sand banks so much associated with Amazonia, and mud flows are visible in the ocean from the air. The geology in the centre of the country is a white sand (*zanderij*) plateau lying over a crystalline plateau penetrated by intrusions of igneous rocks which cause the river rapids and falls.

## 1.2 Land Eligible under Guyana's LCDS

The Low Carbon Development Strategy (LCDS) outlines a national programme that aims to protect and maintain its forests in an effort to reduce global carbon emissions and at the same time attract resources to foster growth and development along a low carbon emissions path. Under the Memorandum of Understanding (MOU) between Guyana and Norway, not all land is included in Guyana's LCDS, only lands under the ownership of the State. Tenure classifications in Guyana were changed in 2013 to include protected areas along with State Land, State Forest and Amerindian Land. This change meant that Iwokrama Forest Reserve and Kaieteur National Park are now amalgamated into the new single class termed 'Protected Areas' for technical classification although still separate for administrative purposes.

## 1.3 Establishing Forested Area

Land classified as forest follows the definition as outlined in the Marrakech Accords (UNFCCC, 2001). Guyana has elected to classify land as forest if it meets the following criteria:

☐ Tree cover of minimum 30%
☐ At a minimum height of 5 m
☐ Over a minimum area of 1 ha.

In summary, the process used to define the national forest cover involved:

☐ Determination of the 1990 forest area using medium resolution satellite images (Landsat) by excluding non-forest areas (including existing infrastructure) as at 1990.

☐ From this point forward accounting for forest to non-forest land use changes that have occurred between 1990 and 2010 using a temporal series of satellite data.

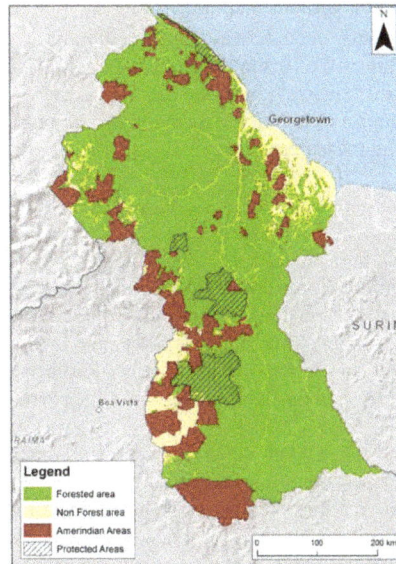

Figure 1: Land classes of Guyana

The 2010 Interim Measures report estimated that as at the benchmark period (30 September 2009) the total forest area that met the above definition was 18.39 million ha (± 0.41 million ha). This figure was further verified by the University of Durham (UoD) with an indicative accuracy of (97.1%).

The 2012 (Year 3) assessment used a forest area (including State Land, State Forest and Amerindian Villages) of 18.50 million ha as the starting point. The increase in forest area resulted from the re-analysis of the 1990 forest / non-forest classification. These boundaries were updated using 5 m satellite imagery. This was a necessary change in order to ensure the delineation of mapped change events are at a consistent resolution with the updated forest / non-forest boundary. This means that historical change was included in the reported forest area figures until year two. From year three forward, the analysis does not take into account historical change mapped from Landsat. This entails comparing different analyses based on imagery of significantly different resolution. To generate a truly comparative figure, a full 'back cast' analysis of historical change events at the updated RapidEye resolution would be necessary. This is a comprehensive exercise and would essentially entail an extensive long term analysis of all historical mapping periods, with reference to all historical imagery.

## 2. Monitoring & Spatial Datasets

The datasets used for the change analysis have evolved over time. This progression is outlined as follows:
☐ 1990 to 2000 – Landsat 30 m
☐ 2001 to 2005 – Landsat 30 m
☐ 2006 to 2009 September - Landsat 30 m
☐ 2009 – 2010 October (Year 1) - Landsat 30 m and DMC 22 & 32 m
☐ 2010- 2011 December (Year 2) Landsat 30 m and RapidEye 5 m
☐ 2012 December (Year 3) RapidEye 5 m supplemented as necessary by Landsat 5 & 7

☐ 2013 December (Year 4) RapidEye 5 m supplemented as necessary by Landsat 8

It is worth noting that currently there are very few operational medium resolution satellite systems that are freely available, or that obtain images frequently enough to allow national reporting of change. To reduce the risk of inadequate coverage GFC has invested in the tasking of an individual satellite data provider. The overall aim is to improve operational methods and to phase out or replace the interim measures.

## 2.1 RapidEye

The RapidEye constellation consists of five satellites which have been providing high resolution multi-spectral images since the start of RapidEye's commercial operations in February 2009. RapidEye holds imagery in an online image archive, and is also available to be tasked to cover specific areas. RapidEye provides both '1B' and '3A' 5 m resolution products.

The decision to commission this coverage was to ensure national coverage at a resolution high enough to capture forest change and degradation activities. The coverage also allows for robust estimates of change – as required for the national MRVS. GFC has tasked the RapidEye constellation to provide a countrywide coverage of Guyana.

Since 2012 GFC has progressively improved the positional accuracy of the RapidEye image base. This process initially involved co-registering the RapidEye 'image swaths' to match the existing Geo-Cover base map. The updated tie points were then returned to RapidEye and used to correct 2013 (Year 4) image coverage.

In 2014 RapidEye updated the positional accuracy over Guyana using control points derived from VHR (Very High Resolution) Digital Globe imagery. In the West of Guyana an offset of up to 30 m is observed. This is due to the steep topographic relief and change in the UTM zone to 20 N.

It is proposed for Year 5 that the GFC team update and improve the existing base maps using RapidEye's improved 3A ortho-corrected product. The revised basemap will be used as a reference from the next reporting period onwards.

For the analysis a higher priority is placed on images acquired at the end Year 4 reporting period, with the majority of images acquired in November 2013. Due to the typically cloudy nature of satellite imagery over Guyana multiple scenes over the same location are required. Nearly all areas have three separate images covering each footprint. Supplementary to the RapidEye acquisition, 30 m Landsat 8 data is also analysed. Wall to wall coverage of Landsat imagery for Guyana has been downloaded from the United States Geological Survey (USGS) online catalogue.

## 2.2 Landsat

Landsat 8 imagery launched on 11 February 2013 also provides temporal coverage over Guyana. This imagery is archived and is freely available and can be sourced from either the United States Geological Survey (USGS) or National Institute for Space Research (INPE) Brazil. Imagery sourced through USGS comes processed as "L1T" or terrain corrected (using SRTM 90 m DTM), whereas INPE imagery typically does not.

Landsat acquires images over the same area every 16 days. The Landsat Data Continuity Mission Landsat 8 provides a source of freely available imagery at 30 m resolution. The sensor collects 11 spectral bands from visible (~0.5μm) to thermal (~12μm) wavelengths.

## 2.3 Accuracy Assessment Datasets

The purpose of the Accuracy Assessment (AA) is to provide an assessment of the quality of the GFC's mapping of land cover land use change across Guyana. It is established practice that data used for accuracy assessment be either an independent interpretation of the same datasets used for the change mapping or, if available, higher resolution data.

Currently, there are no commercially available satellites capable of supplying imagery of sufficiently high spatial resolution with appropriate revisit frequency on a national scale. The accuracy assessment conducted for Year 2 (2011-12) noted that a pixel size of at least 1-2 m is needed to identify forest degradation resulting from human infrastructure.

As part of a continuous improvement process GFC and Indufor Asia Pacific have developed an operational method that captures high-resolution aerial imagery using a highly portable aerial multispectral imaging system. The camera system (provided by GeoVantage) is a flexible unit that can be installed quickly and easily on to various models of light aircraft. The resolution of the images captured across Guyana ranged from about 25 to 60 cm, a resolution capable of identifying forest degradation with some certainty.

The strategy employed uses the imaging system to capture high-quality image data at sites pre-determined by a stratified random sample that covers the majority of Guyana. The full sample coverage is achieved by including the RapidEye images over areas where it is not possible to safely operate a small aircraft.

The locations of these transects were provided to Indufor by the independent accuracy assessment team from Durham University, UK. Individual image frames acquired over the sample site locations were stitched together to form a mosaic. The mosaics obtained from the system were then delivered to the accuracy assessment team for analysis. The system is versatile enough to operate at low altitude (2000 ft) which increases flexibility in cloudy conditions.

In Year 3 the Accuracy Assessment involved the collection of 143 sample units randomly selected from primary sampling units. The accuracy assessment in Year 3 was carried out primarily using GeoVantage aerial imagery. Therefore in order to generate the best possible change reference dataset a repeat coverage of aerial imagery was acquired for Year 4.

It is recognised that there are practical and operational difficulties in generating an identical dataset with perfect overlap between Years 3 and 4. For example, there will be areas where GeoVantage data are missing or cannot be collected in areas where long-range flights with a light aircraft are not feasible or safe. In such cases the best available RapidEye data were selected and reinterpreted. Where possible the RapidEye data were used in parts of the *low risk stratum* where human access is particularly limited and there is no history of logging or mining.

Figure 2: Comparative resolution of RapidEye and aerial imagery

## 3. Image Processing

The image processing steps have been automated using an ENVI 4.7 custom batch processing tool created on the IAP toolbar. The user can select to perform the following processing:

- Create tiles from swaths
- Convert Domain Number to Reflectance
- Perform Dark Object Subtraction
- Produce an Enhanced Vegetation Index

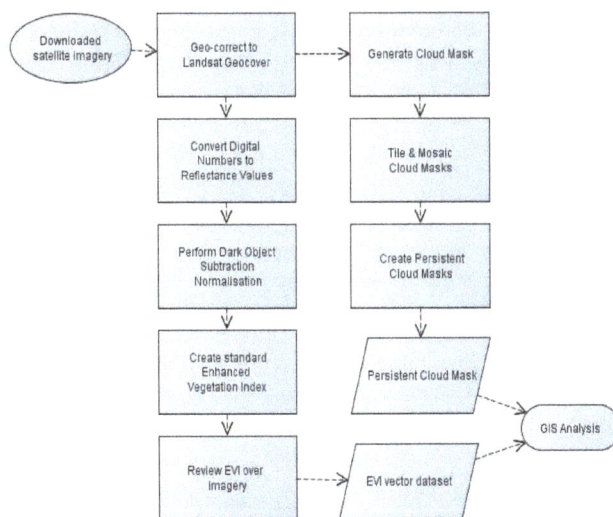

Figure 3: Image processing flow diagram © The Guyana Forestry Commission and Indufor 2013

### 3.1 Image Geo-correction

To ensure consistency across datasets all imagery is geo-referenced to a base mosaic image which was generated from data provided in MrSid format by the Global Land Cover Facility (GLCF). The GLCF holds a global set of regional images which are divided into tiles and overlap each other seamlessly at their edges. This ensures consistency between images of a similar type, and also between different image types and resolutions. All satellite images are co-registered to the 2005 Landsat Geocover base map. Accurate co-registration is important to ensure that changes detected in future time periods are valid and not simply artifacts caused by inaccurate co-registration. Mismatches should be less than one Geocover pixel (<14.25 m).

### 3.2 Image Normalisation

Radiometric normalisation is a recommended image processing practise to ensure the radiometric values within images obtained over different time periods and by different sensors are calibrated to common reference values. There are many methods applied for the normalisation of images that perform either a relative correction to a single scene or an absolute correction to standard reflectance units.

For practical purposes based on the project timeline, the number of RapidEye images to process, the generally high level of clouds per image and the availability of atmospheric correction data, the dark subtraction radiometric normalisation method implemented in ENVI was chosen. Each scene is evaluated and the band minimum Digital Number (DN) values were automatically selected from each scene and subtracted from all pixels within the scene with the assumption the band minimum values are dark targets that are only influenced by atmospheric scattering.

The method of change detection adopted uses a combination of automated (calculation of vegetation indices) and manual interpretation and editing. The objective of the approach was to use a vegetation index to delineate areas of forest and non-forest. Identified areas of non-forest within the forest mask represent potential areas of forest change (i.e. deforestation or degradation). The delineated non-forest areas were input into a GIS and used as an ancillary layer in the Year 4 change analysis mapping.

The key to differentiating forest from non-forest is to link the reflectance properties of the vegetation to its structure. Several vegetation indices exist that enhance non-forest detection as described by *Asner (1998)*.

For this work the Enhanced Vegetation Index (EVI) as described in *Huete et al.* (1997) was favoured over other vegetation indices as it includes the blue reflectance. The strength of the EVI is in its ratio concept which provides a correction for soil background signal and reduces atmospheric influences, including aerosol scattering. This is particularly relevant given the lack of any aerosols, water vapour, and ozone concentrations to correct atmospheric conditions.

The EVI is calculated with the following equation as presented and described in *Huete et al 2002*,

$$\text{EVI} = G \frac{\rho_{\text{NIR}} - \rho_{\text{red}}}{\rho_{\text{NIR}} + C_1 \times \rho_{\text{red}} - C_2 \times \rho_{\text{blue}} + L} \qquad (1)$$

where $G$ is the gain factor, $\rho$ are atmospherically corrected or partially atmosphere corrected (Rayleigh and ozone absorption) surface reflectances, $L$ is the canopy background adjustment that addresses nonlinear, differential NIR and red radiant transfer through a canopy, and $C1$, $C2$ are the coefficients of the aerosol resistance term, which uses the blue band to correct for aerosol influences in the red band. The coefficients adopted in the EVI algorithm are, $L=1$, $C1=6$, $C2 = 7.5$ and $G = 2.5$.

The EVI values range from 0 to 1 with low values indicating non-vegetative surfaces and those closer to 1 representing closed canopy forest. The same approach was successfully applied to separate forest and non-forest components for the 1990-2010 period[1].

The method has also been widely discussed in the scientific literature. *Deng et.al. (2007)* found that EVI was effective in vegetation monitoring, change detection, and in assessing seasonal variations of evergreen forests. The automated change detection process produces a vector layer delineating the potential areas of non-forest. The vector layer is subsequently input into the GIS for review, editing and attribution.

### 3.3 Persistent Cloud Mask

One potential issue is detection of change in areas of sporadic and persistent cloud. In areas of sporadic cloud (i.e. where at least one period is clear) the change was attributed to the relevant change period. If areas are under persistent cloud cover then it is not possible to evaluate the area for change.

The impact of cloud was assessed by generating cloud masks for each RapidEye and Landsat image to identify those areas of persistent cloud. Coincident pixels that are cloudy in all time periods are defined as persistent cloud coincident pixels that are nodata in all time periods are defined as persistent no data.
The masks were generated by a simple band threshold approach and edited to remove areas of non-forest. The cloud mask does not identify cloud shadow so it provides only a broad estimate of cloud coverage.

The analysis showed that for Year 4 less than 0.2% of the land area was persistently covered in cloud. The distribution of the cloud is quite scattered and located over the country most notably in the SE and NW of the country as shown on Figure 2.

Figure 4: Persistent Cloud Cover © The Guyana Forestry Commission and Indufor 2013

---

[1]The independent accuracy assessment conducted in 2011 reported the accuracy of the forest and non-forest mapping to be 99%.

### 4. Spatial Mapping of Land Cover Change

The GIS-based monitoring system is designed to map change events in the year of their occurrence and then monitor any changes that occur over that area each year. The process developed aims to enable areas of change (>1 ha) to be tracked spatially through time, by driver (i.e. mining, infrastructure and forestry). The approach adopted seeks to provide a spatial record of temporal land use change across forested land (commensurate to an Approach 3). The mapping process involves a systematic review of each 24 x 24 km tile, divided into 1 km x 1 km tiles at a resolution of 1:8000.

The EVI vector outputs from the change detection process are edited as required to delineate new change events. Change is attributed with the acquisition date of the pre and post change image, driver of change event, and resultant land use class. A set of mapping rules has been established that dictate how each event is classified and recorded in the GIS.

The input process is standardised through the use of a customised GIS tool which provides a series of pre-set selections that are saved as feature classes. The mapping process is divided into mapping and quality control (QC). The QC team operates independently to the mapping team and is responsible for reviewing each tile as it is completed.

Potentially there is some overlap between drivers as the exact cause of the forest change can be difficult to determine. This is particularly relevant when deciding on the driver of road construction when mining and forestry areas use the same access routes. Supplementary GIS layers are also included in the decision-making process to reduce this uncertainty. The decision based rules are outlined in the mapping guidance documentation. This documentation held at GFC provides a comprehensive overview of the mapping process and rules. The two types of change events mapped are deforestation and forest degradation.

### 4.1 Deforestation

Formally, the definition of deforestation is summarised as the long-term or permanent conversion of land from forest use to other non-forest uses (GOFC-GOLD, 2010). An important consideration is that a forested area is only deemed deforested once the cover falls and remains below the elected crown cover threshold (30% for Guyana). In Guyana's context forest areas under sustainable forest management (SFM) that adhere to the forest code of practice would not be considered deforested as they have the ability to regain the elected crown cover threshold.

The five historic anthropogenic change drivers that lead to deforestation include:

1. Forestry (clearance activities such as roads and log landings)
2. Mining (ground excavation associated with small, medium and large scale mining)
3. Infrastructure such as roads (included are forestry and mining roads)
4. Agricultural conversion
5. Fire (all considered anthropogenic and depending on intensity and frequency can lead to deforestation).

In Year 4, a new driver 'settlements' has been added to the driver matrix. It allows the team to describe human settlement driven change such as new housing developments.

## 4.2 Degradation

There is still some debate internationally over the definition of forest degradation. A commonly adopted definition outlined in IPCC (2003) report is:

"*A direct human-induced long-term loss (persisting for X years or more) of at least Y% of forest carbon stocks [and forest values] since time T and not qualifying as deforestation or an elected activity under Article 3.4 of the Kyoto Protocol*".

The main sources of degradation are identified as:

☐ Harvesting of timber (reported since 2011 using the Gain Loss Method)
☐ Shifting cultivation (prototype method developed in 2012)
☐ Fire
☐ Associated with mining sites and road infrastructure

In Guyana forest degradation around deforestation sites is unique, with the main contributors being the opening of roads linked to new infrastructure, and degradation mainly associated with mining activity - which is dynamic.

The method development was supported by field inspections that measured the stock changes caused by degradation. The field assessment involved the establishment of field transects 20 m in width from the edge of deforestation events. The field measurements suggest that infrastructure-related degradation is restricted to the immediate area around the deforestation site.

Interpretation of the images showed that the forest cover returns to an intact state inside 40 m from the deforested event. Beyond this point it is possible to identify forest disturbances provided the disturbances are large enough (>100 m$^2$) and the vegetation is disturbed to the point where the soil is exposed.

Further image coverages obtained in Years 3 and 4 indicate that degraded forest areas are either in transition to a state of deforestation or are only temporary in nature. It is also important to consider the possibility that historical mining sites may be re-entered or areas of small-scale prospecting extended. To ensure these activities are captured in the MRVS, the mapping team revisits all areas identified in preceding assessments (post 2011) using high-resolution imagery and update areas if changes have occurred.

### 4.2.1 Monitoring and Updating Degradation Event:

When updating an historical forest degradation event there are 3 possible updates that can be applied to any one polygon, these are:–
The historical polygon still shows signs of degradation over the area in the recent imagery. In this case it appears there is still activity, and the area shows minimal canopy evidence of reforestation or abandonment of the area.

    -Driver - Existing historical driver
    -EndLUC - Degraded Forest by type

The polygon no longer shows any visible sign of degradation. The forest canopy has closed and it appears as though the area is now free of anthropogenic activity.
    - Driver - Reforestation
    -EndLUC - Degraded Forest by type

The polygon has been deforested.
    -Driver - As appropriate according to the mapping guide
    -EndLUC - As appropriate according to the mapping guide

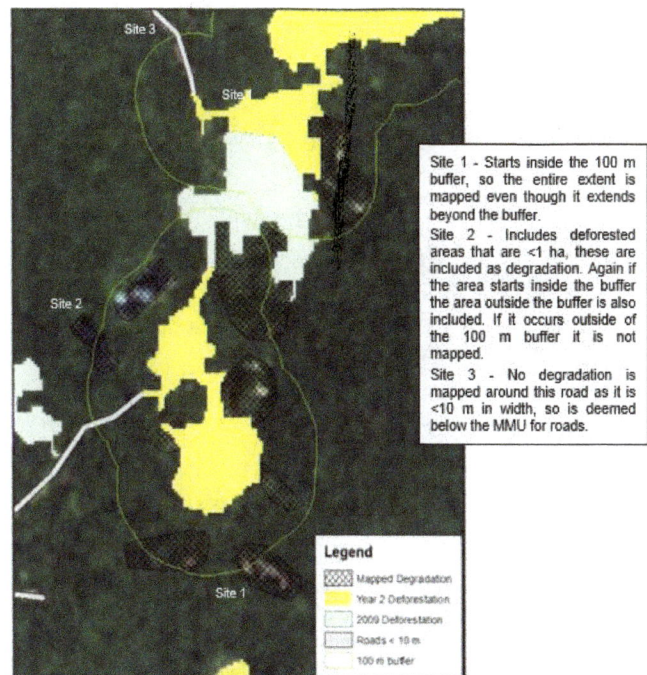

Site 1 - Starts inside the 100 m buffer, so the entire extent is mapped even though it extends beyond the buffer.
Site 2 - Includes deforested areas that are <1 ha, these are included as degradation. Again if the area starts inside the buffer the area outside the buffer is also included. If it occurs outside of the 100 m buffer it is not mapped.
Site 3 - No degradation is mapped around this road as it is <10 m in width, so is deemed below the MMU for roads.

Legend
▨ Mapped Degradation
☐ Year 2 Deforestation
☐ 2009 Deforestation
☐ Roads < 10 m
☐ 100 m buffer

Figure 5: Change mapping © The Guyana Forestry Commission and Indufor 2013

### 4.2.2. Monitoring Shifting Cultivation:

In Year 4, for the first time new shifting agriculture areas are reported under forest degradation. An evaluation of methods for detecting and mapping of areas under shifting cultivation has been undertaken. There are currently no best practice methodologies for doing this, especially on a national scale.

An appropriate detection and mapping methodology has been developed and operationalised in the year 4 analysis. The method adopted allows the calculation of the area which tracks newly cleared areas >0.25 ha. This is much smaller than the 1 ha minimum mapping unit (MMU) applied to deforestation. Shifting agriculture has been sub-categorised into:

- Pioneer; which consists of newly cut consists of newly cut areas which were seen as high forest in the previous year. All available evidence suggests these areas have not historically been degraded or anthropogenically affected. They tend to occur around the fringes of historical rotational shifting cultivation areas. A 100% carbon loss is assumed here as the pre change landcover was high forest.

- Rotational shifting cultivation consists of historically degraded and impacted areas. All available evidence suggests these areas are in various states of succession from newly burnt areas to late successional secondary forest areas. They tend to occur around the areas of long term human habitation. Field work is required to determine a carbon value/emission factor for these systems, as they are technically 'forest remaining forest'.

Further work is required to confirm the emission and removal factors for areas under shifting cultivation. Once calculated these can be linked to the spatial representation. This will enable a calculation of the carbon stock change to be included in the MRV.

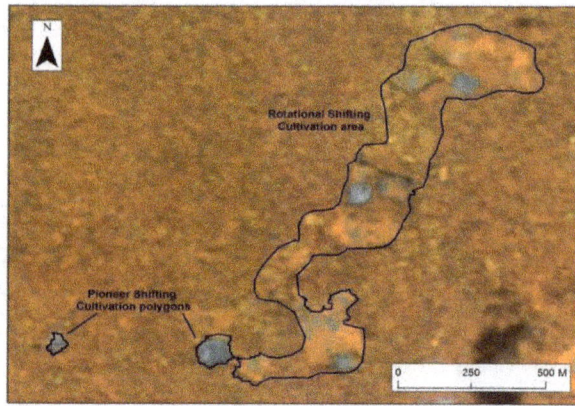

Figure 6: Shifting cultivation example © The Guyana Forestry Commission and Indufor 2013

## 5. Forest Area Analysis

Based on the initial 1990 forest area, the forest cover change for the 1990-2009 period is estimated at 0.41% (i.e.<1%). As with Year 1, the FAO (1995) equation as cited in Puyravaud (2003) has been used to calculate the annual rate of change. Puyravaud (2003) suggests an alternative to this equation, but at low rates of deforestation the two are essentially the same.

$$q = \left(\frac{A_2}{A_1}\right)^{1/(t_2-t_1)} - 1 \qquad (2)$$

Whereby the annual rate of change (%/yr or ha/yr) is calculated by determining the forest cover $A_1$ and $A_2$ at time periods $t_1$ and $t_2$.

If the 1990-2009 period is annualised this represents an average rate of change of about 3 800 ha/yr-1 which is equivalent to a deforestation rate of - 0.02%/ yr.

From this point the deforestation increased for the Year 1 period to 0.06% and has remained at a similar level for Year 2 (0.054%). The rate is in fact lower (0.043%) if the change is expressed as an annual rate rather than presented for the entire Year 2 period.

In Year 3 the deforestation rate increased relative to previous years to 0.079%, but in Year 4 a decrease has occurred to 0.068%

Overall, Guyana's Year 4 deforestation rate is low when compared to the rest of South America, which according to the FAO 2010 forest resource assessment (FRA) is tracking at an annual deforestation rate of -0.41%/yr[12].

The trend shows that deforestation rates have increased since 1990 and peaked in 2012. From 2009-10 onwards the deforestation rate has fluctuated between 0.054% and 0.079%. A decline in deforestation compared to 2012 is observed in 2013.

Significant progress was made in Years 3 and 4, in mapping forest degradation. The main cause of degradation in Year 4 continues to be mining which accounts for 68% of all degradation mapped. This is expected as mining also accounts for the largest area of deforestation. The established trend is that forest degradation impacts are largely detected around mining

areas. The remaining contributors to degradation are from newly established shifting agriculture areas (18%), fire (9%), roading construction and settlements (3%), and forestry related activities such as degradation during road formation (~1%).

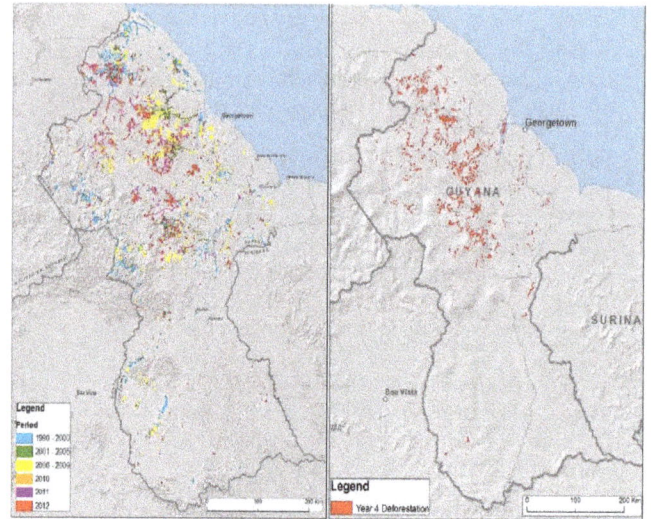

Figure 7: Historical and year 4 forest change © The Guyana Forestry Commission and Indufor 2013

| Driver | Historical Period | | | Year 1 | Year 2 2010-11 (15 months) | | Year 3 2012 | | Year 4 2013 | |
|---|---|---|---|---|---|---|---|---|---|---|
| | 1990 to 2000 | 2001 to 2005 | 2006 to 2009 | 2009-10 | Deforestation | Degradation | Deforestation | Degradation | Deforestation | Degradation |
| | Area (ha) | | | | | | | | | |
| Forestry (includes forestry infrastructure) | 6 094 | 8 420 | 4 784 | 294 | 233 | 147 | 240 | 113 | 330 | 85 |
| Agriculture (permanent) | 2 030 | 2 852 | 1 797 | 513 | 52 | N/A | 440 | 0 | 424 | N/A |
| Mining (includes mining infrastructure) | 10 843 | 21 438 | 12 624 | 9 384 | 9 175 | 5 287 | 13 516 | 1 629 | **11 251 | 2 955 |
| Infrastructure | 590 | 1 304 | 195 | 64 | 148 | 5 | 127 | 13 | 278 | 112 |
| Fire (deforestation) | 1 708 | 235 | | 32 | 58 | 28 | 184 | 208 | 96 | 395 |
| Settlements | | | | | | | | | 23 | 20 |
| Year 4 Shifting Agriculture | | | | | | | | | | 765 |
| Year 2 forest degradation converted to deforestation | | | | | | | 148 | | 67 | N/A |
| Year 3 forest degradation converted to deforestation | | | | | | | | | 200 | N/A |
| Amalia Falls development (Infrastructure Roads) | | | | | 225 | | | | 64 | 20 |
| Area Change | 21 267 | 34 249 | 19 400 | 10 287 | 9 891 | 5 467 | 14 655 | 1 963 | 12 733 | 4 352 |
| Area Change for Year 4 without Shifting Agriculture | | | | | | | | | | 3 587 |
| Total Forest Area of Guyana | 18 473 394 | 18 452 127 | 18 417 878 | 18 398 478 | 18 388 190 | | 18 502 531 | | 18 487 876 | |
| Total Forest Area of Guyana Remaining | 18 452 127 | 18 417 878 | 18 398 478 | 18 388 190 | 18 378 299 | | 18 487876 | | 18 475 143 | |
| Period Deforestation (%) | 0.01% | 0.04% | 0.02% | 0.056% | 0.054% | | 0.079% | | 0.068% | |

**Forestry infrastructure accounts for the full total of deforestation from forestry activities.

**Mining Infrastructure accounts for 918 ha in 2013 out of the total deforestation driven by mining of 11 518 ha, when Year 2 & 3 transitional areas are taken into account.

***Amalia Falls Development has been split from other infrastructure driven change for reporting purposes.

Table 1: Forest change area by period and driver from 1990 to 2013

## 6. Present and Future Development Areas

Guyana has established a robust MRVS that is able to spatially account for the area of deforestation and degradation with confidence. It is envisaged that the reference measure as well as the interim performance indicators will only apply while aspects of the MRVS are being developed and will be phased out and replaced by a full forest carbon accounting system as methodologies are proven. The future focus is to enhance the MRVS to ensure it keeps abreast of international best practice guidance, new datasets, processes and routines. Specifically these developments include:

Development of a second reporting framework aligned to the IPCC Land Use, Land-Use Change and Forestry (LULUCF) template for annual assessments. This is based on the IPCC 2003 GPG tabular format. The LULUCF area change has been reported formally for the first time in November 2014.

-Further sub-division of the non-forest area into the relevant IPCC classes. In preceeding reports this area has been presented as 'non-forest'.

-Development of methodology and guidelines for mapping and monitoring shifting cultivation.

-Development of relevant emission factors for degradation due to forest harvesting activities. Further work is on-going to determine appropriate emission factors for other forms of forest degradation.

-Integration of carbon measurements with spatial datasets to create activity-specific emission factors for degradation and shifting cultivation. This work is in on-going collaboration with Winrock International.

-Alignment of the Community MRV (CMRV) to facilitate integration with the national MRVS.

## 7. Acknowledgements

In addition to GFC, a number of agencies and individuals have assisted in providing inputs into the MRVS programme. GFC would like to recognise Indufor Asia Pacific for the technical contribution and guidance. We would like to acknowledge the support of the Ministry of Natural Resources and the Environment and the Office of Climate Change for their strategic guidance. The continued support and oversight of the members of the Multi-Stakeholder Steering Committee of the LCDS are also especially acknowledged. The GFC team would also like to acknowledge the following colleagues for their support; Winrock International for work on the forest carbon monitoring system, Conservation Iinternational for their role in supporting the implementation of this, as well as other aspects of the Guyana MRVS, the project team of Global Canopy Programme, North Rupununi District Development Board and the North Rupununi communities, Kanashen Village, World Wildlife Fund, Iwokrama and other partners working on the CMRV Project, Guiana Shield Facility and UNDP for supporting work under the MRVS and other partners.

## 8. References

Asner G.P., 1998, Biophysical and Biochemical Sources of Variability in Canopy Reflectance, *Remote Sensing of Environment*, 64:234-253.

Brown, S. and Braatz, B. 2008. *Methods for estimating CO2 emissions from deforestation and forest degradation. Chapter 5 in GOFC-GOLD. Reducing greenhouse gas emissions from deforestation and degradation in developing countries: a sourcebook of methods and procedures for monitoring, measuring and reporting.GOFC-GOLD* Report version COP 13-2. GOFC-GOLD Project Office, Natural Resources Canada, Alberta, Canada.

COP 7 29/10 - 9/11 2001 MARRAKESH, MOROCCO MARRAKESH ACCORDS REPORT (www.unfccc.int/cop7) *FAO Forest Resource Assessment, 2010* http://foris.fao.org/static/data/fra2010/FRA2010_Report_1oct2010.pdf

Deng, F., Su, G., & Liu, C. (2007).Seasonal variation of MODIS vegetation indices and their statistical relationship with climate over the subtropic evergreen forest in Zhejiang, China.IEEE *Geoscience and Remote Sensing Letters*, 4(2), 236–240.

GOFC-GOLD Sourcebook 2010. *A sourcebook of methods and procedures for monitoring and reporting anthropogenic greenhouse gas emissions and removals caused by deforestation, gains and losses of carbon stocks in forests remaining forests, and forestation GOFC-GOLD*. Report version COP16-1, (GOFC-GOLD Project Office, Natural Resource Canada, Alberta, Canada).

Hansen, M.C., Stehman, S.V., Potapov, P.V., Loveland, T.R., Townshed, J.R.G., DeFries, R.S., Pittman, K.W., Arunarwati, B., Stolle, F., Steininger, M.K., Carroll, M. and DiMiceli, C. Copyright © The Guyana Forestry Commission and Indufor 64 2008. Humid tropical forest clearing from 2000 to 2005 quantified by using multitemporal and multi-resolution remotely sensed data. *PNAS* 105(27):9439-9444.

Huete, A.R., H. Liu, K. Batchily, and W. van Leeuwen, 1997. A Comparison of Vegetation Indices Over a Global Set of TM Images for EOS-MODIS. *Remote Sensing of Environment* 59(3):440-451.

IPCC *Report on Definitions and Methodological Options to Inventory Emissions from 15 Direct Human-induced Degradation of Forests and Devegetation of Other Vegetation Types, 2003* (http://www.ipcc.ch/publications_and_data/publications_and_data_reports.htm#2)

Watt, P.J., Haywood, A.H., 2007. *Mapping Forest Clearfelling using MODIS Satellite Data. Contract Report 38A08772.* New Zealand Ministry for the Environment.

# 16

# TOPOLOGICAL 3D MODELING USING INDOOR MOBILE LIDAR DATA

M. Nakagawa [a,*], T. Yamamoto [a], S. Tanaka [a], M. Shiozaki [b], T. Ohhashi [b]

[a] Dept. of Civil Engineering, Shibaura Institute of Technology, Tokyo, Japan - mnaka@shibaura-it.ac.jp
[b] Nikon Trimble Co., Ltd., Tokyo, Japan - (shiozaki.makoto, ohhashi.tetsuya)@nikon-trimble.net

**Commission IV / WG 7**

**KEY WORDS:** Indoor mobile mapping, Point cloud, Point-based rendering

**ABSTRACT:**

We focus on a region-based point clustering to extract a polygon from a massive point cloud. In the region-based clustering, RANSAC is a suitable approach for estimating surfaces. However, local workspace selection is required to improve a performance in a surface estimation from a massive point cloud. Moreover, the conventional RANSAC is hard to determine whether a point lies inside or outside a surface. In this paper, we propose a method for panoramic rendering-based polygon extraction from indoor mobile LiDAR data. Our aim was to improve region-based point cloud clustering in modeling after point cloud registration. First, we propose a point cloud clustering methodology for polygon extraction on a panoramic range image generated with point-based rendering from a massive point cloud. Next, we describe an experiment that was conducted to verify our methodology with an indoor mobile mapping system in an indoor environment. This experiment was wall-surface extraction using a rendered point cloud from some viewpoints over a wide indoor area. Finally, we confirmed that our proposed methodology could achieve polygon extraction through point cloud clustering from a complex indoor environment.

## 1. INTRODUCTION

Point-cloud clustering is an essential technique for modeling massive point clouds acquired with a terrestrial laser scanner or mobile laser scanner in an indoor environment, as shown in Figure 1.

Figure 1. Point cloud clustering: colored point cloud (left image) and clustered point cloud (right image)

There are three clustering approaches in point-cloud clustering: model-based clustering (Boyko et al. 2011), edge-based clustering (Jiang et al. 1999), and region-based clustering (Vosselman et al. 2004). Model-based clustering requires CAD models to estimate simple objects or point clusters from the point cloud. In 3D industrial modeling, standardized objects, such as pipes and parts, are prepared as CAD models in advance. However, the CAD model preparation approach is unsuitable for modeling unknown objects. On the other hand, edge-based and region-based clustering are often used to model unknown objects (Tsai et al. 2010). These approaches also focus on geometrical knowledge (Pu, et al. 2009) and 2D geometrical restrictions, such as the depth from a platform (Zhou, et al. 2008) and discontinuous point extraction on each scanning plane from the mobile mapping system (Denis et al. 2010) to extract simple boundaries and features in urban areas.

Point-cloud data acquired in urban areas and indoor environments often include many complex features with unclear boundaries. Moreover, general 3D modeling using terrestrial and mobile data uses point-cloud data acquired from many viewpoints and view angles. Thus, for integrated point-cloud data taken by terrestrial or mobile LiDAR, the application of the abovementioned restrictions for extracting features is difficult. Additionally, viewpoints for range image rendering are limited to data-acquisition points.

We focus on the region-based point clustering to extract a polygon from a massive point cloud. In region-based clustering, Random Sample Consensus (RANSAC) (Schnabel et al. 2007) is a suitable approach for estimating surfaces. However, local workspace selection is required to improve a performance in a surface estimation from a massive point cloud. Moreover, with conventional RANSAC, it is hard to determine whether a point lies inside or outside a surface.

In this paper, we propose a method for panoramic rendering-based polygon extraction from indoor mobile LiDAR data. Our aim is to improve region-based point-cloud clustering in indoor modeling. First, we propose a point-cloud clustering methodology for polygon extraction on a panoramic range image generated with point-based rendering from a massive point cloud. Next, we describe an experiment that was conducted to verify our methodology with an indoor mobile mapping system. Finally, we confirm that our proposed methodology can achieve polygon extraction through point-cloud clustering from a complex indoor environment.

## 2. METHODOLOGY

Figure 2 shows our proposed methodology. It consists of: (1) viewpoint decision for point-based rendering; (2) point-based rendering; (3) normal vector clustering for surface estimation; (4) point-cloud interpolation using a rectangular template; and (5) point tracing.

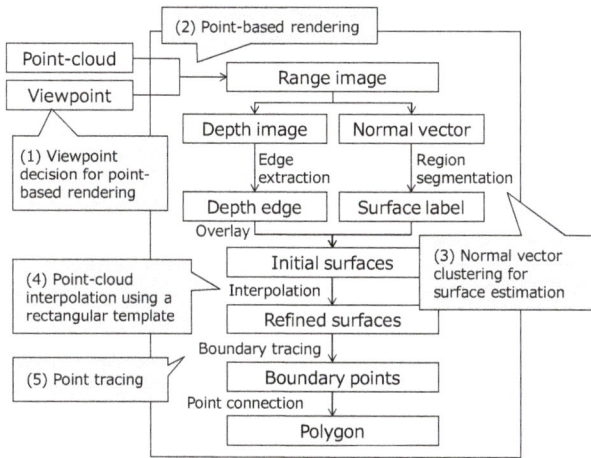

Figure 2. The five components of processing flow

## 2.1 Viewpoint decision for point-based rendering

Viewpoints for point-based rendering are selected in point-cloud data through two steps. In the first step, an orthobinary image is generated from the point cloud to represent a rough floor surface as a viewpoint candidate. In the next step, the orthoimage is eroded with morphology processing to generate a viewpoint candidate network. Intersections on the network are selected as the viewpoints for point-based rendering.

## 2.2 Point-based rendering

Point-cloud visualization has two issues. The first is the near-far problem caused by distance differences between the viewpoint and the scanned points. The second is the transparency effect caused by rendering hidden points among near-side points. These effects degrade the quality of a point-cloud visualization. Splat-based ray tracing (Linsen et al. 2007) is a methodology that generates a photorealistic curved surface on a panoramic view using normal vectors from point-cloud data. The long time period required for surface generation in the 3D work space is a problem. Furthermore, the curved-surface description is inefficient in representing urban and natural objects as Geographical Information System data. Thus, we have applied a point-based rendering application with a simpler filtering algorithm (Nakagawa 2013) to generate panoramic range images from a random-point cloud. The processing flow of point-based rendering is described in Figure 3.

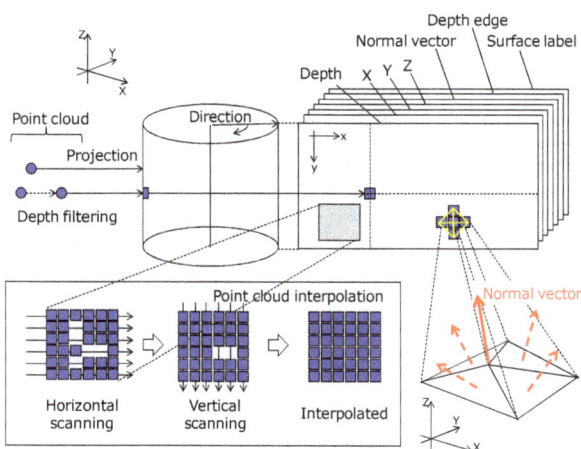

Figure 3. Point-based rendering

First, the point cloud is projected from 3D space to panorama space. This transformation simplifies viewpoint translation, filtering, and point-cloud browsing. The panorama space can be represented by a spherical, hemispherical, cylindrical, or cubic model. Here, the cylindrical model is described for wall modeling. The measured point data are projected onto a cylindrical surface, and can be represented as range data. The range data can preserve measured point data such as a depth, X, Y, Z, and some processed data in the panorama space in a multilayer style. Azimuth angles and relative heights from the viewpoint to the measured points can be calculated using 3D vectors generated from the view position and the measured points. When azimuth angles and relative heights are converted to column counts and row counts in the range data with adequate spatial angle resolution, a cylindrical panorama image can be generated from the point cloud.

Second, the generated range image is filtered to generate missing points in the rendered result using distance values between the viewpoint and objects. Two types of filtering are performed in the point-based rendering. The first is a depth filtering with the overwriting of occluded points. The second is the generation of new points in the no-data spaces in the range image. New points are generated with the point tracking filter developed in this study.

Moreover, a normal vector from each point is estimated in the range image. Normal vector estimation is often applied to extract features in point-cloud processing. Generally, three points are selected in the point cloud to generate a triangle patch for normal vector estimation. Mesh generation is the basic preprocessing step in this procedure. In 2D image processing, the Delaunay division is a popular algorithm. It can also be applied to 3D point-cloud processing with millions of points (Chevallier et al. 2011). However, using the Delaunay division, it is hard to generate triangle patches for more than hundreds of millions of points without a high-speed computing environment (Fabio 2003) (Böhm et al. 2006). Thus, we focused on our point-cloud rendering, which restricts visible point cloud data as a 2D image. A closed point detection and topology assignment can be processed as 2D image processing, as shown in the lower right image in Figure 2.

The processing flow of normal vector estimation is described below. First, a point and its neighbors in the range image are selected. Second, triangulation is applied to these points as vertexes to generate faces. Then, the normal vector on each triangle is estimated using 3D coordinate values of each point. In this research, an average value of each normal vector is used as the normal vector of a point, because we used the point cloud taken from a laser scanner that presents difficulties for measuring edges and corners clearly. These procedures are iterated to estimate the normal vectors of all points.

## 2.3 Normal vector clustering for surface estimation

Normal vectors of all points are grouped to detect regions in a range image as a point-cloud classification. The accuracy of point-cloud classification can be improved with several approaches such as the Mincut (Golovinskiy et al. 2009), Markov network-based (Shapovalov et al. 2011), and fuzzy-based (Biosca et al. 2008) algorithms. However, in this study, we improved the accuracy with point-cloud interpolation and point tracking. Thus, we applied multilevel slicing as a simple algorithm to classify normal vectors.

Moreover, building knowledge is used as a restriction in the normal vector and point-cloud classification. In general, walls in a room and building consist of parallel and orthogonal planes. Thus, four clusters in a horizontal direction are enough to detect walls in a general indoor environment. Although cylindrical surfaces are divided into some clusters, these surfaces can be reconstructed using surface merging. The processing flow of normal vector clustering with restrictions is described below. First, stronger peaks are extracted from a histogram of normal vectors. More than one strong peak is required to detect seed points in each approximate 90° change in horizontal direction. Next, boundaries of clusters are generated from the peaks of the histograms. Then, the normal vectors and point clouds are grouped into four clusters. Finally, initial 3D surfaces are estimated from the grouped normal vectors and point cloud.

This classification detected boundaries of point clusters with the same normal vectors. The point-cloud clustering methodology for extracting the intersection of planes as ridge lines requires appropriate initial values such as curvature, fitting accuracy and distances to closed points (Kitamura et al. 2010). However, our approach can extract boundaries from a point cloud without these parameters.

### 2.4 Point-cloud interpolation with a rectangular template

Estimated 3D initial surfaces are refined in a point-cloud interpolation procedure. In general, it is difficult to trace the boundaries of the initial surfaces because of holes and jaggy boundaries. Therefore, point-cloud interpolation is applied as a refinement of the initial surfaces in this procedure.

When flat and cylindrical surfaces are projected into a range image based on a cylindrical model, these surfaces are represented as rectangles with the following two restrictions. The first restriction is that points have the same $X$- and $Y$- coordinate values along the $y$-direction in the range image. The second restriction is that the points have the same $Z$-coordinate values along the $x$-direction in the range image. Based on these restrictions, point interpolation is applied along the $x$- and $y$- directions in the range image, as shown in Figure 4.

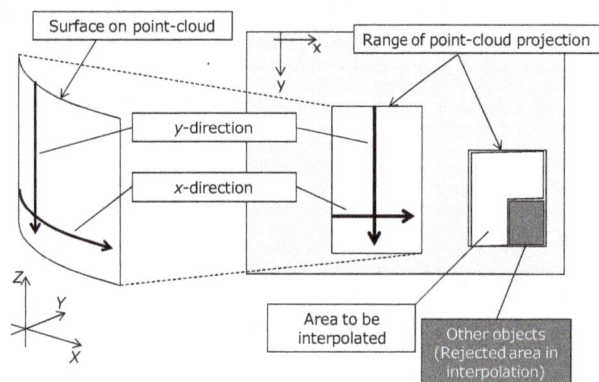

Figure 4. Point-cloud interpolation with a rectangular template in a range image

The point interpolation is as follows. First, a rectangular template is fitted to projected points in a range image. Next, missing points are detected in the rectangular template. Finally, the missing points are interpolated using neighboring points. When other objects exist in a rectangular template, the overlapped area is excluded from point interpolation.

### 2.5 Point tracing

Boundaries of features can be estimated from the refined surfaces in a range. Moreover, 3D polygons can be extracted with topology estimation using these boundaries in the range image. In this procedure, a point tracing is required to connect points in 3D space along the boundary, as shown in Figure 5.

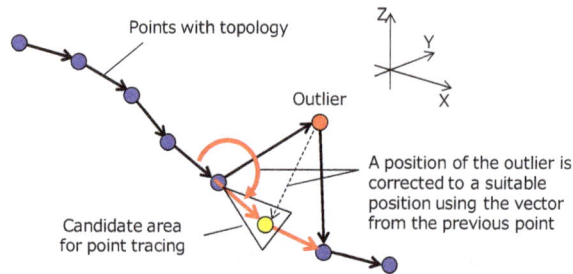

Figure 5. Point tracing

In general, least squares fitting and polynomial fitting are applied to extract straight and curved lines from points. When the point cloud includes noise, RANSAC is a suitable approach to estimate a feature. However, these approaches require a decision whether straight lines or curved lines are to be extracted before the fitting procedure. In this paper, we wish to extract polygons with a combination of straight and curved lines. Thus, we propose point tracing based on the region-growing approach to extract complex geometry as follows. First, a topology of points is estimated in a range image. When a polyline or polygon is drawn in a range image, continuous 3D points can be extracted. Next, a position for the next point is checked after a seed-point selection. In this step, the position is checked to find whether a possible next point exists or not within a candidate area for point tracing. The candidate area is determined using a vector from the previous point. When a point exists within the candidate area, it is connected to the previous point. Otherwise, the point is assumed to be an outlier, and the position of the point is rectified to a suitable position using the vector from the previous point. These steps are then iterated until the geometry is closed. Finally, 3D points are connected to represent a smooth 3D polygon.

### 3. EXPERIMENT

We used the Trimble Indoor Mobile Mapping System (TIMMS) integrated with an inertial measurement unit (IMU), a wheel encoder, a LiDAR system (TX5, Trimble), and an omnidirectional camera (Ladybug 5, Point Grey) (see Figure 6). We acquired a 880-million color point cloud with TIMMS (see Figure 7) in our university.

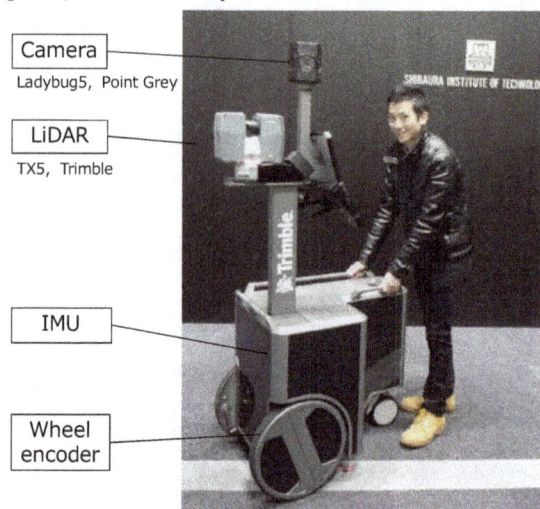

Figure 6. TIMMS

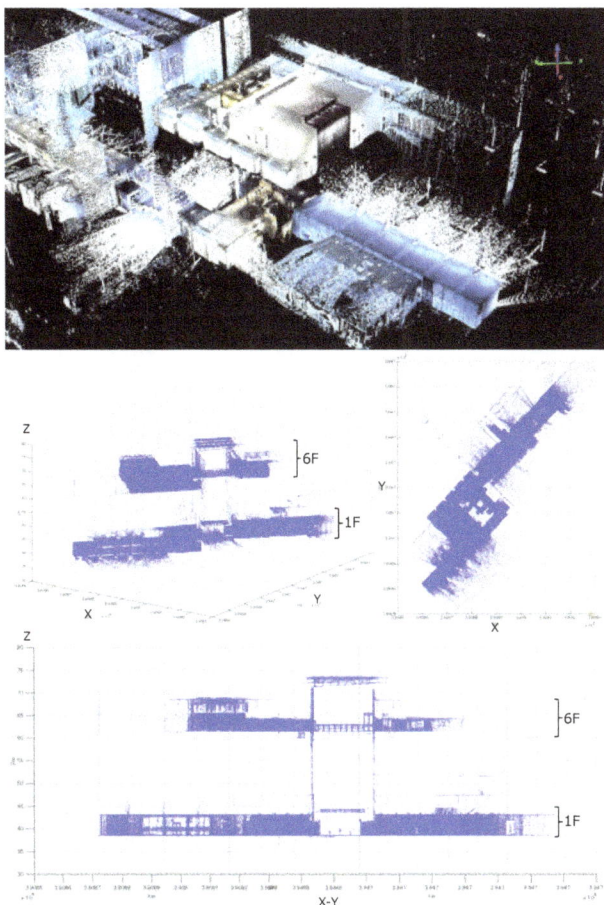

Figure 9. Rendered point cloud

## 4. RESULTS

In our experiment, 72 points were extracted as viewpoint candidates for point-based rendering, as shown in Figure 10. The point cloud taken from TIMMS was rendered from these viewpoints.

Figure 7. Acquired point cloud

An entrance foyer consisting of a large room (21.6 m × 21.6 m width) in our university were selected as our study area (see Figure 8). The study area consisted of flat and cylindrical walls, square and cylindrical pillars, a grilled ceiling, doors with glass, and windows. These objects were representative flat and cylindrical surfaces.

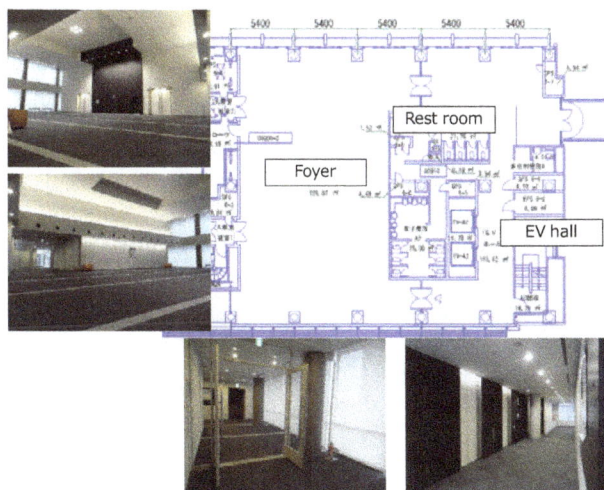

[1] Orthobinary image

[2] Viewpoint candidate network

[3] Viewpoint candidates

[4] Viewpoint candidates (filtered)

Figure 10. Viewpoint candidates

Figure 11 shows results after point-based rendering and point clustering from a viewpoint. Figure 11 includes a depth image, a depth image, normal vectors, and initial labeled surfaces (overlay of depth edge and labeled surfaces).

Each vertical axis shows height direction and each horizontal axis shows direction. Intensity values in the depth image indicate the depth from the viewpoint. Moreover, intensity values in the normal vectors and labeled surfaces indicate the horizontal direction of the point cloud. In addition, color values in the initial surfaces indicate labels of surfaces. In this experiment, spatial resolution was set as 0.2° in the horizontal direction and 2 cm in the height direction.

Figure 12 shows a rendered point cloud from a viewpoint in 3D space. The left image shows the input point cloud and the right image shows a result after polygon extraction. Processing time

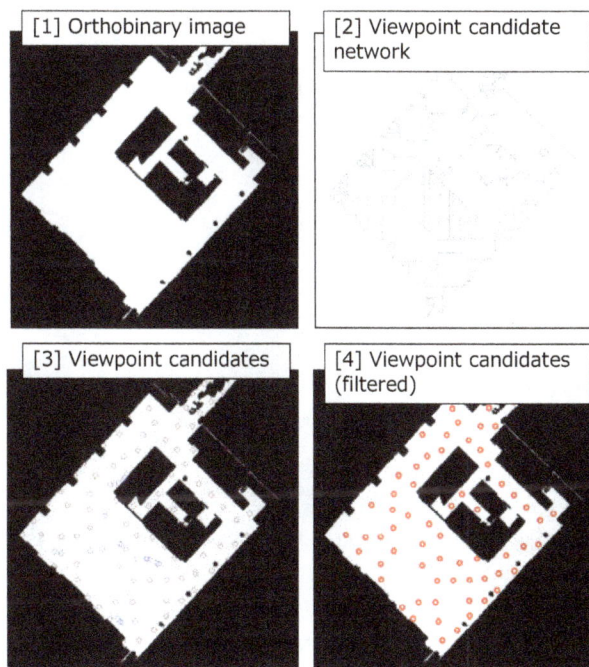

Figure 8. Study area

In the experiment, we used a 450-million color point cloud from our dataset. Figure 9 shows a part of point cloud rendering results from a viewpoint in the foyer.

for the panoramic image conversion and polygon extraction was several minutes in total for each viewpoint using an Intel core i7 2.80 GHz processor with MATLAB (single thread).

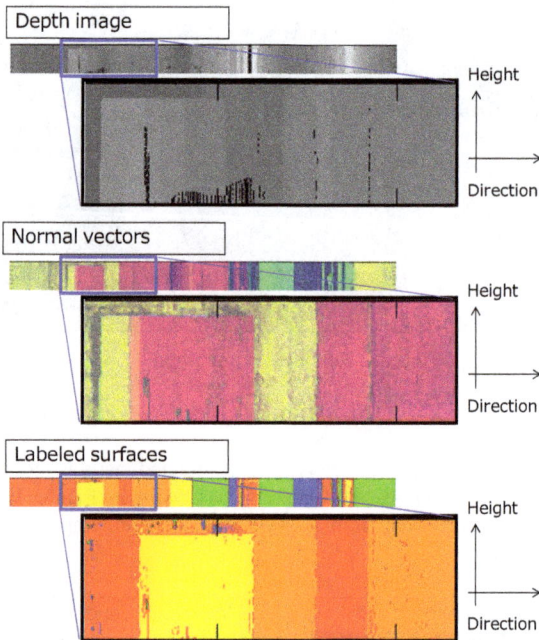

Figure 11. Results after point-based rendering and point clustering

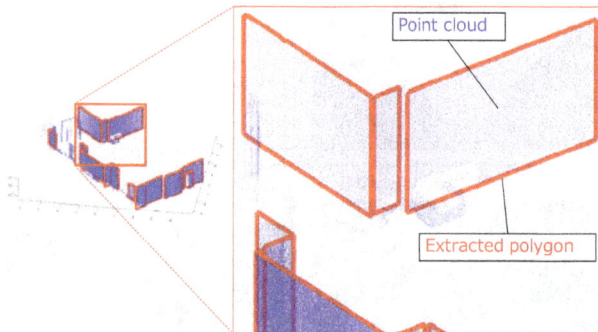

Figure 12. Point cloud and polygon extraction result

## 5. DISCUSSION

Parts of the results of polygon extraction from the point cloud are shown in Figure 13. This figure includes examples of general building features, such as a flat wall and a cylindrical wall. Each row shows a result of point-cloud visualization and extracted polygon (boundaries). We have confirmed that point-cloud interpolation in a range image achieved spike noise filtering and geometry smoothing. Moreover, we have confirmed that noise such as the pedestrian was also successfully filtered from the point cloud.

Figure 14 shows integrated results for polygon extraction from 72 viewpoints. Our approach extracted 980 polygons from the point cloud fully automatically. As shown in Figure 14, some polygons that were extracted were failures. Our investigation showed that these failures were caused by LiDAR measurement noise, such as light reflection errors and moving object measurement. Although noise was almost eliminated, the remained noise in the range image affected the point-cloud interpolation.

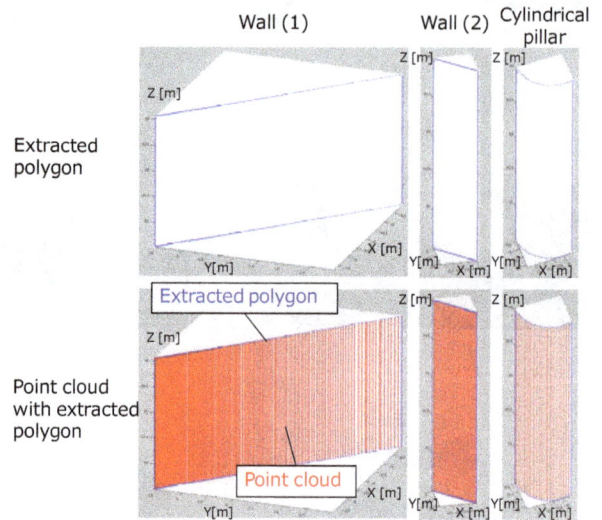

Figure 13. Parts of results of polygon extraction from point cloud

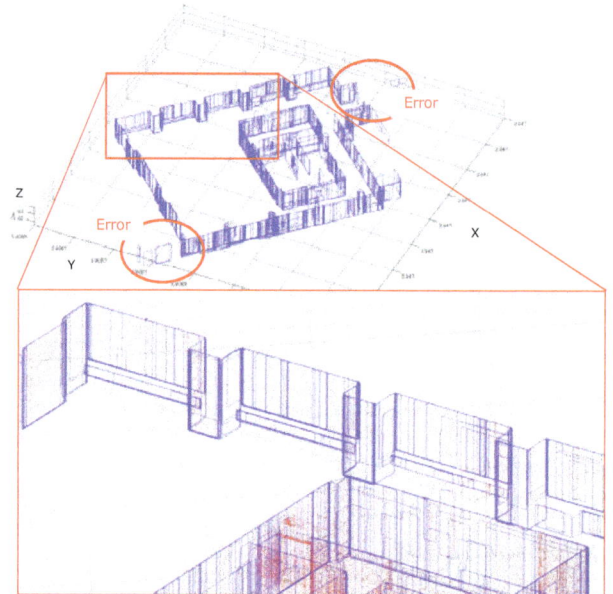

Figure 14. Integrated results in polygon extraction from 72 viewpoints

Moreover, the extraction of some features from the point cloud also failed, as shown in Figure 15. These failures were also affected by occluded areas and LiDAR measurement noise.

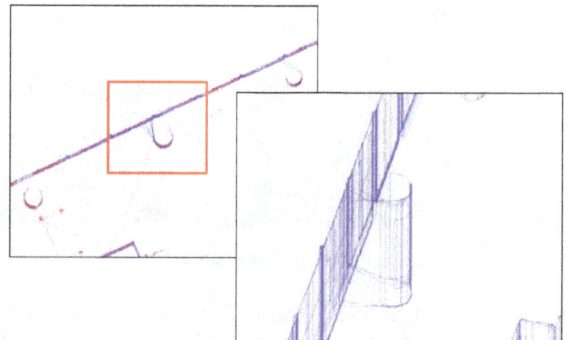

Figure 15. Failures in polygon extraction

## 6. SUMMARY

We have proposed a method for panoramic rendering-based polygon extraction from indoor mobile LiDAR data. Our aim was to improve region-based point cloud cluster modeling after point-cloud registration. First, we have proposed a point-cloud clustering methodology for polygon extraction on a panoramic range image generated with point-based rendering from a massive point cloud. Our proposed methodology consisted of the viewpoint decision for point-based rendering, the point-based rendering, the normal vector clustering for surface estimation, the point-cloud interpolation with a rectangular template, and point tracing. Next, we described an experiment that was conducted to verify our methodology with an indoor mobile mapping system (TIMMS) in an indoor environment that included flat and cylindrical surfaces. In this experiment, we extracted wall -surfaces using a rendered point cloud from 72 viewpoints over a wide indoor area. Finally, we confirmed that our proposed methodology could achieve polygon extraction through point-cloud clustering from a complex indoor environment.

## REFERENCES

Boyko, A., Funkhouser, T., 2011, Extracting roads from dense point clouds in large scale urban environment, *ISPRS Journal of Photogrammetry and Remote Sensing*, 66 (2011) S2–S12.

Jiang, X., Bunke, H., 1999, Edge Detection in Range Images Based on Scan Line Approximation, *Computer Vision and Image Understanding*, Vol.73, No.2, pp.183-199.

Vosselman, G., G. H. Gorte, B., Sithole, G., Rabbani, T., 2004, Recognising structure in laser scanning point clouds. *In: ISPRS 2004 : proceedings of the ISPRS working group VIII/2 : laser scanning for forest and landscape assessment*, pp. 33-38.

Tsai, A., Hsu, C., Hong, I., Liu, W., 2010, Plane and Boundary Extraction from LiDAR data using Clustering and Convex Hull Projection, *International Archives of the Photogrammetry, Remote Sensing and Spatial Information Sciences*, Vol. XXXVIII - Part 3A, pp.175-179.

Pu, S., Vosselman, G., 2009, Knowledge based reconstruction of building models from terrestrial laser scanning data, *ISPRS Journal of Photogrammetry and Remote Sensing*, Volume 64, Issue 6, pp. 575–584.

Zhou, Q., Neumann, U. 2008, Fast and Extensible Building Modeling from Airborne LiDAR Data, *ACM SIGSPATIAL International Conference on Advances in Geographic Information Systems (ACM GIS)*, pp.8.

Denis, E., Burck, R., Baillard, C. 2010, Towards road modelling from terrestrial laser points, *IAPRS*, Vol. XXXVIII, Part 3A, pp.293-298.

Schnabel, R., Wahl, R., Klein, R., 2007, Efficient RANSAC for Point-Cloud Shape Detection, *Computer Graphics Forum*, 26(2), 214-226.

Linsen, L., M"uller, K., Rosenthal, P., 2007, Splat-based Ray Tracing of Point Clouds, *Journal of WSCG*, Vol.15, Issue: 1-3, pp.51-58.

Nakagawa, M., 2013, Point Cloud Clustering for 3D Modeling Assistance Using a Panoramic Layered Range Image, *Journal of Remote Sensing Technology*, Vol.1, Iss.3, 10.pp.

Chevallier, N., Maillot, Y., 2011, Boundary of a non-uniform point cloud for reconstruction, *SoCG '11 Proceedings of the twenty-seventh annual symposium on Computational geometry*, pp.510-518.

Fabio, R., 2003, From point cloud to surface : the modeling and visualization problem, *International Archives of the Photogrammetry, Remote Sensing and Spatial Information Sciences*, Vol. XXXIV-5/W10.

Böhm, J., Pateraki, M., 2006, From Point Samples to Surfaces, *On Meshing and Alternatives ISPRS Image Engineering and Vision Metrology*, XXXVI, pp.50 – 55.

Golovinskiy, A., Funkhouser, T., 2009, Min-Cut Based Segmentation of Point Clouds, *IEEE Workshop on Search in 3D and Video (S3DV) at ICCV*, 6.pp.

Shapovalov, R., Velizhev, A., 2011, Cutting-Plane Training of Non-associative Markov Network for 3D Point Cloud Segmentation, *3D Imaging, Modeling, Processing, Visualization and Transmission (3DIMPVT)*, 8.pp.

Biosca, M., Luis Lerma, J., 2008, Unsupervised robust planar segmentation of terrestrial laser scanner point clouds based on fuzzy clustering methods, *ISPRS Journal of Photogrammetry and Remote Sensing*, Volume 63, Issue 1, pp.84–98.

Kitamura, K., D'Apuzzo, N., Kochi, N., Kaneko, S., 2010, Automated extraction of break lines in tls data of real environment, *International Archives of Photogrammetry and Remote Sensing*, 38(5), 331-336.

# 17

# RECONSTRUCTION OF INDOOR MODELS USING POINT CLOUDS GENERATED FROM SINGLE-LENS REFLEX CAMERAS AND DEPTH IMAGES

Fuan Tsai[a*], Tzy-Shyuan Wu[b], I-Chieh Lee[a], Huan Chang[b] and Addison Y. S. Su[c]

[a]Center for Space and Remote Sensing Research
[b]Department of Civil Engineering
[c]Research Center for Advanced Science and Technology
National Central University, Zhong-li, Taoyuan 320 Taiwan
ftsai@csrsr.ncu.edu.tw

**KEY WORDS:** Indoor modeling, RGB-D, Kinect, SFM reconstruction, Point clouds

**ABSTRACT:**

This paper presents a data acquisition system consisting of multiple RGB-D sensors and digital single-lens reflex (DSLR) cameras. A systematic data processing procedure for integrating these two kinds of devices to generate three-dimensional point clouds of indoor environments is also developed and described. In the developed system, DSLR cameras are used to bridge the Kinects and provide a more accurate ray intersection condition, which takes advantage of the higher resolution and image quality of the DSLR cameras. Structure from Motion (SFM) reconstruction is used to link and merge multiple Kinect point clouds and dense point clouds (from DSLR color images) to generate initial integrated point clouds. Then, bundle adjustment is used to resolve the exterior orientation (EO) of all images. Those exterior orientations are used as the initial values to combine these point clouds at each frame into the same coordinate system using Helmert (seven-parameter) transformation. Experimental results demonstrate that the design of the data acquisition system and the data processing procedure can generate dense and fully colored point clouds of indoor environments successfully even in featureless areas. The accuracy of the generated point clouds were evaluated by comparing the widths and heights of identified objects as well as coordinates of pre-set independent check points against in situ measurements. Based on the generated point clouds, complete and accurate three-dimensional models of indoor environments can be constructed effectively.

## 1. INTRODUCTION

Three-dimensional modeling of indoor environments is an emerging topic in the researches and applications of building modeling, indoor navigation, location-based services and related fields in recent years. One of the common objectives in three-dimensional indoor mapping is to create a digital representation of the environment. Once the rich and high-precision digital model is constructed, the complete and accurate information of an indoor environment is preserved and can be used for a variety of applications.

A popular and conventional approach of creating three-dimensional indoor models is to generate three-dimensional point clouds from multiple digital images. In this case, multiple images must be connected with each other based on identified features and three-dimensional point clouds can be created by intersecting feature points on multiple images. However, the major disadvantage of this image-based approach is the lack of points extracted in featureless areas and therefore, it may be less effective in areas such as plain walls or long, simple corridors from which few features will be identified and extracted for model reconstruction. As technology advances, new equipments and softwares are developed and have great potentials to achieve better and more effective reconstruction of three-dimensional indoor models. Among the newly developed instruments, RGB-D cameras (such as Microsoft Kinect) have attracted attentions from researchers in the fields of computer vision and photogrammetry. This type of camera can capture both RGB images and per-pixel depth information simultaneously. Although the effective range of data acquisition using Kinect sensors is short (1 to 4 meters), they can be utilized as a new tool for indoor mapping and model reconstruction (Han et al., 2013; Yue et al., 2014). However, the most popular three-dimensional reconstruction method for Kinect data is using

Kinect Fusion, which is based based on camera tracking. It depends on depth variation in the scene. Scenes must have sufficient depth variation in view to be able to track successfully, so Kinect Fusion may fail to construct the models in places where depth changes less significantly (Newcombe et al., 2011). Never the less, the integration of RGB-D based sensors and digital cameras may fill up the voids in featureless areas and create uniformly distributed points cloud of indoor environments. Accordingly, this research constructed a mobile data acquisition system consisting of multiple Kinects and digital single-lens reflex (DSLR) cameras and developed a systematic data processing procedure for integrating these two kinds of devices to generate three-dimensional point clouds of indoor environments.

## 2. MATERIAL AND METHOD

The developed data acquisition system consists of up to four RGB-D (Kinect) cameras and four DSLR cameras mounted on a steel rack, which in term can be installed on a pull-cart or similar platform. Figure 1 displays an example of the developed mobile data acquisition system and sample images from the RGB-D and DSLR cameras. The RGB-D cameras used in this study are Microsoft Kinect, which was initially used as an input device by Microsoft for Xbox game console. Microsoft Kinect can provide color and depth images synchronously. Kinect uses the technique of Light-coding. The sensor launches a laser speckle and captures the coding light back from the scene to calculate depth values. Comparing with other types of RGB-D cameras, Kinect has much lower cost than traditional ones and is more widely used in recent year (Han et al., 2013). Table 1 shows some basic parameters of Kinect.

There are some drawbacks when using Kinect for 3D mapping from its limitation. First of all, it can only be used in indoor envi-

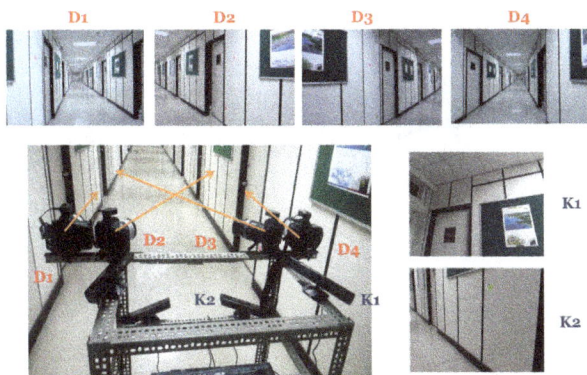

Figure 1: The developed mobile data acquisition system with multiple RGB-D and DSLR cameras.

Table 1: Basic parameters of Kinect

| Range | 0.8-4 m | |
|---|---|---|
| Image Size | Depth | 640x480 |
| | Color | 1280x960 |
| Frequency | 12 fps | |
| Accuracy | Spatial | 2-20 mm |
| | Distance | 1-70 mm |

ronment. Secondly, it can only provide limited distance in depth information (08-4 m). Therefore, it may not be appropriate to acquire data in a large, open space environment. The spatial and depth resolutions are millimeter to centimeter depending on the range. The spatial (x, y) resolution is about 2 to 20 millimeters, and distance (z) resolution is about 1 to 70 millimeters. In addition, the random error of depth measurements increases with increasing distance from the sensor, and reaches about 4 centimeter at the maximum range (Khoshelham and Elberink, 2012).

Kinect provides RGB images and per-pixel depth values. According to the information it captures, a point cloud of each frame can be generated. On the other hand, a DSLR camera captures multiple high-resolution images of the indoor environment. A visual structure from motion system (VisualSFM) is used to link the relationship of all the images captured by both Kinect and DSLR camera. The general procedure of VisualSFM is listed in Algorithm 1. This process generates a sparse (merged) point cloud in a photogrammetry way. Also, feature points of each frame were extracted to be used as tie-points. DSLR camera captures high-resolution photographs, which provide detail information of the environment. However, if the scenes are featureless or the information is not adequate, it is difficult to perform feature extraction and feature matching. This may result in lacks of point clouds in the featureless places. To overcome this disadvantage, 3D point clouds of Kinect can be used to complete the model. According to the extracted feature points, Kinect point clouds of each frame can be transformed into the same coordinate system with dense point clouds through Helmert (7-parameter) transformation as illustrated in Fig. 2. Finally, colored point clouds of indoor environments are generated. The quality of the combined point cloud can be evaluated by comparing against the coordinates of pre-set ground control points.

## 3. EXPERIMENTAL RESULTS

The developed mobile RGB-D and DSLR data acquisition system was deployed to reconstruct three-dimensional models of a long corridor in a campus building. The results are compared with pre-set control and check points to evaluate the accuracy

**Algorithm 1** General procedure of VisualSFM

1. Feature extraction (default: SIFT)

2. Feature matching

3. Sparse reconstruction

4. Bundle adjustment

5. Dense matching

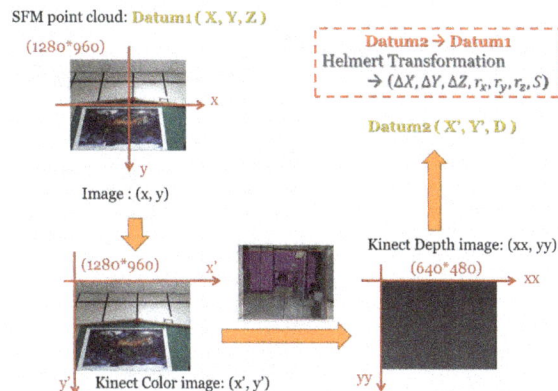

Figure 2: Determining Helmert 7-parameters from feature points

of the generated point clouds. The first experiment utilized two Kinect sensors to capture two sides of the walls at a speed of 12 frames per second. A total of 250 high resolution digital images were also captured by DSLR cameras at the same time. The high resolution photographs covered both straight and transverse sides. In front of the wall, there is about 60% overlap between adjacent images. The images used in the GCP-based SFM reconstruction included 250 high resolution photographs captured by DSLR camera and 24 RGB images captured by Kinect. The second experiment utilized four Kinect RGB-D sensors and four DSLR cameras in the same environment. In this test, 118 photographs per DSLR camera and 100 color images per Kinect were used in the SFM reconstruction.

Figure 3 displays the result of dense point clouds reconstructed from the DSLR photographs and color images of Kinect sensors. As mentioned previously, despite the high resolution of DSLR images, there are few feature points in flat, feature-less regions, thus resulting in many holes in the reconstructed point clouds. These voids can be augmented by merging the point clouds derived from Kinect depth images into the SFM generated point cloud using the Helmert transformation in order to produce a more complete model. The effect is more obvious in the second experiment as shown in Fig. 4, which displays the original (SFM reconstructed) dense matching point cloud and the merged point cloud result (combination of DSLR, Kinect color images and depth images). Figure 5 shows an inside view of the reconstructed hallway point cloud. The example in Fig. 4 and 5 clearly demonstrates that merging the data of depth images into the SFM dense matching result significantly reduces the voids in the reconstructed point cloud model, especially in the feature-less regions.

The reconstructed point cloud models were compared with pre-set ground control points and field-measurement performed with total station equipment for accuracy assessment. Table 2 lists the accuracy assessment result of the the dense matching point cloud in X, Y, Z directions and the distance. From the table, it can be seen that the average errors in the three axises are less than the 1

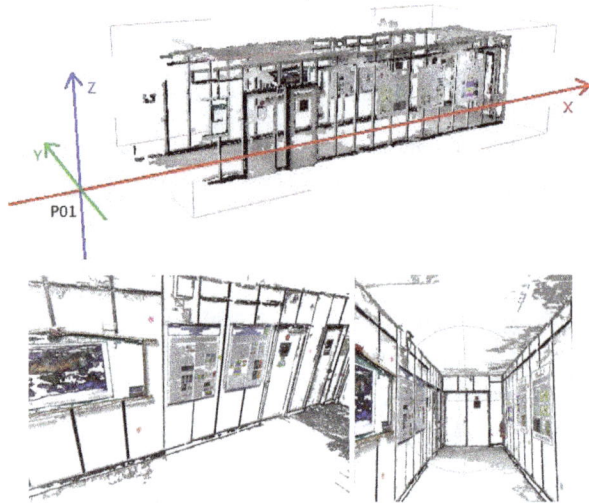

Figure 3: Point cloud generated from DSLR and Kinect color images

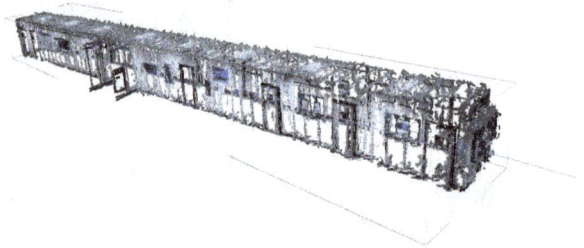

(a) dense matching point cloud from SFM

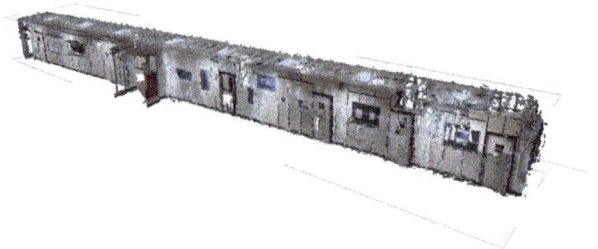

(b) merged point cloud

Figure 4: Comparison of original dense matching and merged point clouds

cm where the average distance measurement error is about 0.86 cm. Table 3 further evaluate the accuracies of the point clouds from two Kinect sensors (K1 and K2) as indicated in Fig. 1. Similarly, the average measurement errors in the three axises are also less than 1 cm. However, the standard deviation values are higher than Table 2 and the distance errors are 1.62 cm and 1.61 cm, respectively. This is reasonable because the depth sensor in Kinect is less accurate than the GRB camera, thus resulting in larger uncertainty in the measurement. Nevertheless, the results are still very accurate and the accuracy should be adequate for indoor mapping applications.

Table 2: Accuracy assessment of dense matching point cloud (unit: cm)

|  | dX | dY | dZ | Distance |
|---|---|---|---|---|
| MEAN | -0.05 | 0.61 | 0.45 | 0.86 |
| STDDEV | 0.60 | 0.70 | 0.21 | |

Table 3: Accuracy assessment of merged point cloud (unit: cm)

|  |  | dX | dY | dZ | Distance |
|---|---|---|---|---|---|
| K1 | MEAN | 0.23 | -0.27 | -0.20 | 1.62 |
|  | STDDEV | 1.01 | 0.81 | 1.08 | |
| K2 | MEAN | -0.44 | 0.08 | -0.12 | 1.61 |
|  | STDDEV | 0.95 | 1.06 | 1.04 | |

Figure 5: Inside view of the constructed point cloud model

## 4. CONCLUSIONS

This research proposes an approach to integrate the point clouds generated from DSLR and RGB-D (Kinect) cameras. In spite of short range and relatively low resolution color images, Kinect can provide real distance and scale information. The proposed method employs Structure From Motion (SFM) algorithm to integrate the DSLR and Kinect color images to generate a dense matching point cloud. However, the reconstruction may fail in feature-less regions, resulting in voids or holes in the generated point clouds. These voids are augmented by merging the point clouds from depth images into the reconstructed model using 7-parameter transformation. Using the proposed method, dense and fully colored point clouds of indoor environments can be generated effectively and accurately even in featureless areas. Experiment results indicate that the distance error of the constructed merge point clouds is less than 2 centimeters using the proposed processing procedure. This process will be an effective approach to build an indoor mapping model.

## ACKNOWLEDGEMENT

This study was supported, in part, by the Ministry of Interior and the Ministry of Science and Technology of Taiwan (ROC) under project numbers SYC1040120 and MOST-103-2221-E-008-076-MY2, respectively.

## References

Han, J., Shao, L., Xu, D. and Shotton, J., 2013. Enhanced computer vision with microsoft kinect sensor: A review. IEEE Transactions on Cybernetics 43(5), pp. 1318–1334.

Khoshelham, K. and Elberink, S., 2012. Accuracy and resolution of kinect depth data for indoor mapping applications. Sensors 12(2), pp. 1437–1454.

Newcombe, R. A., Izadi, S., Hilliges, O., Molyneaux, D., Kim, D., Davison, A. J., Kohi, P., Shotton, J., Hodges, S. and Fitzgibbon, A., 2011. KinectFusion: Real-time dense surface mapping and tracking. In: 10th IEEE Symposium on Mixed and Augmented Reality (ISMAR2011), pp. 127–136.

Yue, H., Chen, W., Wu, X. and Liu, J., 2014. Fast 3D modeling in complex environments using a single kinect sensor. Optics and Lasers in Engineering 53, pp. 104–111.

# 18

# USE OF ASSISTED PHOTOGRAMMETRY FOR INDOOR AND OUTDOOR NAVIGATION PURPOSES

D. Pagliari [a*], N.E. Cazzaniga [a], L. Pinto [a]

[a] DICA-Dept. of Civil and Environmental Engineering, Politecnico di Milano, Milan, Italy
(diana.pagliari, noemi.cazzaniga, livio.pinto)@polimi.it

**KEY WORDS:** outdoor, indoor, navigation, photogrammetry, Kinect, urban maps, GNSS, depth images

**ABSTRACT:**

Nowadays, devices and applications that require navigation solutions are continuously growing. For instance, consider the increasing demand of mapping information or the development of applications based on users' location. In some case it could be sufficient an approximate solution (e.g. at room level), but in the large amount of cases a better solution is required.
The navigation problem has been solved from a long time using Global Navigation Satellite System (GNSS). However, it can be unless in obstructed areas, such as in urban areas or inside buildings. An interesting low cost solution is photogrammetry, assisted using additional information to scale the photogrammetric problem and recovering a solution also in critical situation for image-based methods (e.g. poor textured surfaces). In this paper, the use of assisted photogrammetry has been tested for both outdoor and indoor scenarios. Outdoor navigation problem has been faced developing a positioning system with Ground Control Points extracted from urban maps as constrain and tie points automatically extracted from the images acquired during the survey. The proposed approach has been tested under different scenarios, recovering the followed trajectory with an accuracy of 0.20 m.
For indoor navigation a solution has been thought to integrate the data delivered by Microsoft Kinect, by identifying interesting features on the RGB images and re-projecting them on the point clouds generated from the delivered depth maps. Then, these points have been used to estimate the rotation matrix between subsequent point clouds and, consequently, to recover the trajectory with few centimeters of error.

## 1. INTRODUCTION

In recent years the use of devices and applications that require accurate navigation solution is continuously growing. For instance, consider the continuous demanding of 3D mapping application based on user's locations. For some of these application it is sufficient an accuracy of some meters or to know the room where the user is in. However, in the large amount of cases it is fundamental to have a more accurate solution, on the order of few decimetres or even less, as for example for the localization of fireman and paramedical personnel in case of emergencies or the positioning of sensors.

Outdoor navigation problem has been solved using GNSS (Global Navigation Satellite System) positioning in the last decades (Hofmann-Wellenhof et al., 2008). Its use has quickly widespread thank to its easiness of use, affordability of cost and capability of reach high accuracies, up to few centimetres. In order to compute a solution a GNSS receiver has to receive signal from at least four satellites, for a period long enough to reconstruct all the information encoded within the transmitted signal. This hypothesis hardly occurred in urban areas because of the presence of obstacles (such as buildings, dense foliage, tunnels, etc.) that obstructed the sky visibility. In these kinds of environments, HSGNSS (High Sensitive GNSS) receivers, having a wider level of sensitivity, could be an interesting solution. However, they could lead to large measurement error (MacCougan, 2003). Furthermore, there are scenarios like downtown areas where there are simply too few satellites in view, with a weak geometry. Concerning the number of visible satellites, for the near future, some improvements are expected using the new constellations Galileo and BeiDou. Often, GNSS antennas are coupled with INS, which are used to estimate position, velocity and orientation of a moving vehicle taking advantages of Newton's second law of motion. The potentiality

of Mobile Mapping Systems (MMS), and their ability to integrate INS/GPS (Inertial Navigation System/Global Positioning System) data, have been proved from a long time too. See for example Hassan et al., (2006). However, these solutions are very expensive when few decimetres or centimetres of accuracy are required (Al-Hamad and El-Sheimy, 2004). Tactical grade INSs can experience larger position and attitude errors in short time intervals (15 min), when they are used in stand-alone mode or there is a GNSS leakage. A more accurate solution is self-tracking total station, useful for both indoor and outdoor positioning, allowing accuracy on the order of few millimetres (Böniger and Tronche, 2010). Nevertheless, these instruments are quite expensive and require a continuous and clear visibility of the reflector, installed on the object to be tracked. Moreover, the point has to be surveyed from a stable station located at a distance smaller than the maximum instrument operational range, which is usually in the order of some hundred meters. Of course, this condition is difficult to be maintained in urban areas or inside buildings. For indoor environments some high precision solutions already exist, but the cost of the required infrastructures is still prohibitive for high accuracy applications. For instance Ultra Wide Band (UWB) is broadly used for medical applications because it allows to reach even millimetre accuracy (Mahfouz et al., 2008), while other magnetic-based systems allow to reach decimetre level positioning accuracy (Storms et al., 2010). Distances between object and source could be of 0-15 m. Radio Frequency Identification (RFID) (Zhou and Liu, 2007) or Wi-Fi (Vaupel et al., 2010) can potentially provide high accuracy resolutions too, but that accuracy is highly dependent on the number and the spacing among the installed tags. The same considerations can be extended also to the methods based on fingerprints approaches too. In this case the accuracy is strictly related to the quality of the training phase and from the number of fingerprints stored in the database.

Photogrammetry represents a low-cost solution, suited for both outdoor and indoor environments, which is potentially capable of reaching a high level of accuracy, and not yet fully exploited. The main objective of the work described in this paper is to investigate and develop a low-cost navigation solution based on an inverse photogrammetric approach. Of course, in order to obtain an image-based solution and to scale the photogrammetric problem a number of constrains are required. Moreover, the use of a pure photogrammetric solution can be impossible if only bad textured objects are framed. A possible solution for indoor positing could be to combine the complementary nature of images provided by passive and active sensors, using the Microsoft Kinect sensor for navigation purposes. The paper is organized as follows. In paragraph 2 a brief review of the use of photogrammetry for outdoor and indoor navigation is given. In section 3 the implemented solution for outdoor navigation is presented, together with a complete kinematic test. The approach developed to integrate visual and depth is introduced in paragraph 4. Finally, some conclusion and outlook for future developments are discussed.

## 2. PHOTOGRAMMETRY FOR NAVIGATION PURPOSES

The recent wide success of optical based systems is mainly due to the miniaturization and advance in technology of Charged Coupled Devices (CCDs), combined together with a huge increase in the data transmission rate and computational capabilities, as well as the development of imaging processing algorithm. For all these reasons, photogrammetry is commonly used in a variety of fields, reaching different level of accuracies, from sub-millimetre for optical metrology applications for surface inspections or reverse engineering, to tens of meters for positioning using mass-market devices for pedestrian navigation. The use of photogrammetry to define the vehicle trajectory in case of long GPS outages is discussed in Da Silva et al. (2003). They developed a low-cost MMS and used a pure photogrammetric approach to supply GPS outages. Their approach requires using some GPS positions to constrain the perspective centres of extreme stereobases. The use of image-based techniques for GPS outage bridging was also presented in Chaplin (1999) and Tao et al. (2001). In these cases the starting point was assumed to be a known position acquired with INS/GPS. A pure photogrammetric strategy is presented in Roncella et al. (2005). They proposed to automatically extract the tie points in order to compute the bundle block adjustment and recover the vehicle trajectory from the camera external orientation parameters. However, due to the absence of constrains such as Ground Control Points (GCPs) this kind of solution can be useful only for small GPS outages: along a 300 m long path they experienced a drift up to 1.5 m. According to Eugaster et al. (2012) this results can be improved adding GCPs in suitable locations along the survey. Especially concerning indoor applications, optical based systems can be classified considering how the reference information is obtained (e.g. from building models, coded targets etc.) and if they need any a-priori knowledge. A complete review of these systems can be found in Mautz (2012). For indoor positioning it is quite common to detect object on images and to match them with data stored in a previously populated database, but this approach requires a time expensive survey to collect all the information necessary to populate the database. The real challenge for indoor positioning is represented by autonomous robot navigations: in this case is fundamental to achieve a real-time

solution which is independent from any a-priori knowledge. This problem is commonly known as Simultaneous Location And Mapping (SLAM). Basically, a robot needs to know its location in the environment to navigate between places. SLAM is a typical example of chicken-and-egg problem because in order to localize a camera in the environment it is necessary to have a model of the environment itself but at the same time building the 3D model requires to know the camera poses. The new measurements are sequentially added, so the quality decreases quickly over time. A number of examples of SLAM application can be found in literature, especially in robotics community, fusing together different kind of data to enforce the computed solution. For instance, Ramos et al. (2007) combined together visual and laser data to recognize landmarks, reaching an RMS of 6.8 m. Historically, laser range systems have been mounted on robots to provide the information required for autonomous navigations. The launch on the market of Microsoft Kinect sensor opened up new opportunities for automatic robot guidance because it makes available simultaneously both passive and active imaging sensor data, maintaining low the costs.

## 3. OUTDOOR NAVIGATION WITH PHOTOGRAMMETRY IN URBAN AREAS

The proposed photogrammetric solution was initially developed in the frame of UMALS project (High Speed 3D Underground Utilities and Automatic Electrical Lying Systems), whose final goal was to perform the automatic lying of medium voltage cables. To perform this procedure it is fundamental to know with very high precision location and geometry of all the existent buried infrastructures, such as ducts, cables etc. Usually, this problem is solved performing a survey with a Ground Penetrating Radar (GPR) pulled by a vehicle or pushed by hand. This instrument has to be georeferenced using external instrumentation. To perform a correct 3D reconstruction of the location and geometry of the buried objects, the GPR has to be georeferenced with accuracy from 0.20 to 0.30 meters. As stressed before, this problem cannot be completely solved in urban areas using GNSS because of the frequent inadequacy of the sky visibility. Moreover, the GPR antenna moves very slowly (in order to meet its sample requirements), so a navigation solution using INS/GNSS will not be feasible because it will be quickly lead to unacceptable drifts. Furthermore, the GPR acquisition is performed acquiring the data along parallel strips, so the residence time in obstructed areas can continue for long periods. The proposed approach investigated the use of photogrammetry, using GCPs extracted from urban maps as constrain, as a possible solution that can overcome GNSS signal leakage in urban areas. The idea of using data extracted from urban maps to improve the navigation solution was already proposed in Crosilla and Visintini (1988), although in a different way. The method discussed in this paper requires one or more digital cameras and a GNSS antenna installed on the same vehicle carrying the GPR (see Figure 1). The inverse photogrammetric problem is solved with a bundle adjustment, using GCPs obtained from urban maps and tie points extracted from the acquired images (e.g. the building located at the roadside are generally characterized by well-textured surfaces). Because both the navigation sensors (digital cameras and GNSS antenna) and the GPR are all rigidly fixed to the vehicle, it is possible to recover the trajectory followed by the georadar. The position of the GNSS antenna phase center and the rigid transformation (rototranslation) from the camera system to a vehicle-fixed reference system are estimated during a geometric calibration phase.

Figure 1. Scheme of the vehicle equipped with two cameras and two GPS antennas used for the kinematic tests

### 3.1 Preliminary tests

A series of simulation and preliminary tests was realized to evaluate the feasibility of the proposed approach. In Barzaghi et al. (2009) a number of simulation were presented, demonstrating the potential of the proposed photogrammetric solution. Then, a series of tests was conducted to evaluate if Structure from Motion (SfM) techniques can be useful to extract tie points even in challenging urban environment in order to automatically orient even long image sequences, using a quite low number of GCPs. During these tests software EyeDEA (Roncella et al., 2011a, Roncella et al., 2011b) was used to automatically extract homologous points.

A complete preliminary test was presented in Cazzaniga et al. (2012). The test has been realized in conditions very close to the operational ones. The vehicle was equipped with two Nikon D70s cameras (with fixed focal length equal to 20 mm) and a GNSS antenna (to simulate the presence of the GPR and to have a reference trajectory to be used to evaluate the accuracy of the photogrammetric block). The survey has been realized in via Golgi (Milan, Italy) for a total length of 350 m. Eleven GCPs, extracted from urban map, have been used to georeference the photogrammetric block. It was composed by 220 images and the camera projection center have been determined with a bundle block adjustment. The precision were about 0.10 m in all directions. Then, the camera projection center has been transferred in correspondence of the GPS antenna (that simulates the presence of the GPR). The residuals were not smaller than 0.186 m, underlying the presence of a systematic error. This was probably due to the fact that the calibration vector orientation, with respect to the body frame (fixed to the vehicle) has been estimated with a limited accuracy. Furthermore, a trend between the two solutions was clearly visible. This was mainly due to two factors: a drift in the estimation of camera stations and a misalignment between the estimated reference frames. Nevertheless, the photogrammetric solution error was lower than the required tolerance for a 150 m long path. Outside it, the larger residual were mainly due to a degraded GPS accuracy.

### 3.2 Cremona city block test

During the preliminary test the proposed photogrammetric method has proven to be reliable and the results are in agreement with the accuracy needed to georeference the GPR. However, it clearly emerged that the photogrammetric solution is highly dependent on the quality of the GCPs used to georeference the block. The urban maps are characterized by different accuracy in different town areas; moreover, some outliers could be due to restitution errors or changes in urban environment. Thus, it is essential to integrate in the solution also some GNSS pseudo-observation, allowing outlier identification and rejection. These points could be easily acquired ad the beginning or at the end of the strip or in correspondence of open spaces, such as squares or intersections.

Starting from all this consideration a second kinematic test was realized in a residential area of Cremona city (see Figure 2).

The area was selected considering that only low buildings are present there, allowing acquiring a GNSS reference solution during the entire survey. The images have been acquired along close trajectories, to have an auto-consistent photogrammetric solution and to reduce possible drift, experienced during some of the preliminary test discussed so far. The vehicle was equipped with two Nikon D70s digital cameras (with fixed focal length equal to 20 mm) and 2 GPS antennas. The data acquired with one of the antennas was used to evaluate the effect of inserting some GPS pseudo-observations within the photogrammetric bundle block adjustment, while the data of the second antenna was used as a reference solution, simulating the presence of the GPR. The use of two cameras, rigidly fixed to the vehicle, allowed also evaluating the effect of introducing the further constrain of the relative orientation between the two cameras.

Figure 2. The residential area selected for the kinematic test. The selected GCPs are represented in red, while the GPS pseudo-observations in yellow.

During the survey a total of 600 images have been acquired, with a shooting time equal to one second. The system was geometrically calibrated using a building façade. On this calibration polygon 7 GCPs were previously measured using classical topographic instrumentation. During geometric calibration of the vehicle, images and satellite positions were simultaneously acquired. The components of the lever-arms were computed by comparing the camera projection centres (recovered from the bundle block adjustment) and the GPS positions. The calibration vectors components have been computed with a precision of few centimetres, and were used to insert some pseudo-observation within the bundle block adjustment (Forlani et al., 2005).

A series of different solution, assuming different constrain configurations for the photogrammetric block have been realized. In the first scenario, a pure cartographic constrain has been considered using 52 GCPs extracted from Cremona urban map (1:1000 scale). These points were selected in correspondence of building corners, shelters or pitches. These last points are very important because their height coordinate is very useful to improve the block geometry and stability. The accuracy of the GCPs used was set to 0.20 m for horizontal coordinates and 0.30 m for height, which is typical for 1:1000 scale cartography.

Since it can be reasonably assumed that in some areas, where the sky is more open, it is possible to acquire some GPS positions, we inserted few of them as pseudo-observations within the bundle block adjustment, to better constrain the block itself (GPS solution were obtained via phase double differences, achieving an accuracy of few centimetres). Furthermore, the Relative Orientation (RO) between the two cameras has been considered too. For each one of the evaluated scenario a bundle block adjustment was performed to estimate camera poses, which in turns correspond to the vehicle trajectory. The photogrammetric problem was firstly solved with the commercial software PhotoModeler® and it was refined with the scientific software Calge (Forlani, 1986), which allows introducing GPS pseudo-observations as well as the relative orientation between cameras. The automatic tie points extraction has been realized with the software EyeDEA. The extracted homologous points were filtered, maintaining an optimal distribution in the image space and preserving their multiplicity too. At the end the block was composed by more than 60,000 image observations, which correspond to about 17,600 object points (with an average multiplicity equal to 3). On average, there were 106 points per image and the image coverage was about 79%. The average number of rays per homologous points was equal to 3 (with a maximum of 22), while the average intersection angle between homologous rays is equal to 17°.

The RMSe computed for the residuals between the estimated position with the photogrammetric approach and the GPS ones, interpolated at shooting time, are reported in Table 1.

|  | N[m] | E[m] | h[m] |
|---|---|---|---|
| 52 cartographic GCP | 0.180 | 0.311 | 0.260 |
| 52 cartographic GCP+ RO | 0.195 | 0.261 | 0.248 |
| 52 cartographic GCP+ 6GPS | 0.148 | 0.140 | 1.128 |
| 52 cartographic GCP+6GPS+RO | 0.114 | 0.092 | 0.153 |

Table 1 - RMSe from the differences between GPS and photogrammetric positions (moved using the calibration vector)

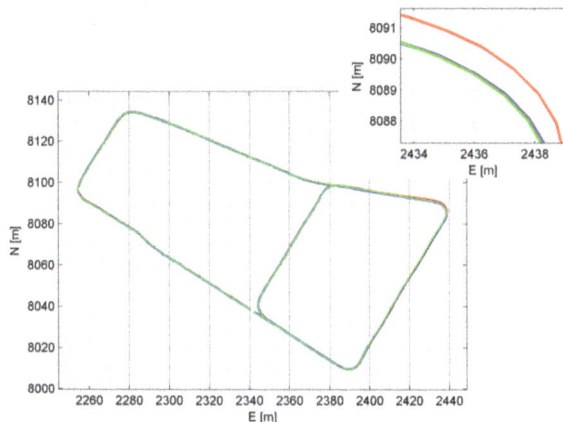

Figure 3. Trajectories estimated considering different photogrammetric block constrains and GPS reference trajectory (blue represents the GPS reference trajectory, red the photogrammetric trajectory recovered using a pure cartographic constrain and green the integrated photogrammetric-GPS solution)

From the results reported in Table 1, it is quite evident that the GPR could be better georeferenced inserting some GPS pseudo-observation in the photogrammetric solution, as could be expected. Moreover, the solution improved by adding the further constrain of the relative orientation between the two cameras. A graphic representation of the estimated trajectories, in dependence of three different constrains of the photogrammetric block and the reference GPS trajectory are represented in Figure 3.

## 4. INDOOR NAVIGATION USING MICROSOFT KINECT SENSOR

The navigation problem for indoor environments has been faced investigating the results obtainable using a low-cost RGB-D depth camera: the Microsoft Kinect sensor. The Kinect was launched in 2010 by Microsoft Corporation as a remote controller for its Xbox360 console. Unlike other human control devices (such as Nintendo Wii remote control or Sony PlayStation Move) it allows users to play and completely control the console without having to hold any device, but only with the use of voice and gesture. This is possible because during the game the players are continuously tracked and his avatar is moving according to their gestures. Kinect 1.0 is composed by a RGB camera, an IR camera, an IR projector (that projects a random pattern), a microphone, a tilt motor and a 3-axes accelerometer. On summer 2014 a new generation of Kinect, based on time-of flight technology, has been released. This new sensor is made by an RGB camera and an IR camera too, but the depths are computed performing a phase correlation between the emitted and the reflected signals. Microsoft Kinect had immediately a large diffusion being used in number of applications different from the original idea of a 3D human interface. In fact, the launch on the market of Kinect sensor extended the use of RGB-D camera to low cost projects, solving many navigation problems by integrating visual and depth data, which can reciprocally compensate their weaknesses. According to the specific environment that has to be explored, visual or depth data can represent a good solution or not. Generally, RGB images can be profitably used if there is a strongly chromatic variation or objects with highly distinctive textures. On the contrary, in case of uniformly plastered walls it can be very difficult to extract features. Instead, point clouds can be very useful in situation with low image contrast. However, in order to align the 3D models created from the acquired depth data it is necessary to have some volume variations in the frame scene (e.g. presence of furniture, room corners etc.).

The complementary nature of the gathered data and the low-cost make Kinect sensor very appealing, especially for robot navigation purposes and 3D model reconstruction.

### 4.1 Integration of visual and depth images

In literature, a number of applications based on the use of RGB and depth data can be found. See for examples Oliver et al. (2012), Endres et al. (2012). Usually the proposed solutions are based on SLAM approaches, mainly because for robot navigation a real time solution is required. However, the accuracy of the estimated trajectory decreases quickly because the estimation of the camera poses accumulates errors over time. Moreover, the large majority of the available studies about the use of Kinect for navigation purposes are strictly connected to Computer Vision communities, so it is very difficult to find information about system calibrations or the precision of the implemented approach. The solution here presented is not meant to be in real time, however the choice was driven mainly by the time requested to compute the followed trajectory. SIFT (Lowe, 2004) is the most common used interest operator because of its good performances. In fact it is highly features distinctive and it generally outperforms other interest operators, however its matching phase requires a large amount of time, especially for long image sequences, like the ones acquired by Kinect sensor. On the other hand, KLT tracking algorithm (Lucas and Kanade, 1981 - Tomasi and Kanade, 1991) can track features very quickly, but in the original implementation the points are extracted using a corner detector. Hence, it could be hard to track a high number of good features in case of homogenous frame scene. Starting from this consideration a new software has been realized. The new implemented software uses KLT algorithm, but the points tracked are extracted with more reliable interest operators, such as SIFT. Then RBG and depth images are integrated in order to reconstruct the followed trajectory. A comprehensive schematic view of this approach is given in Figure 4.

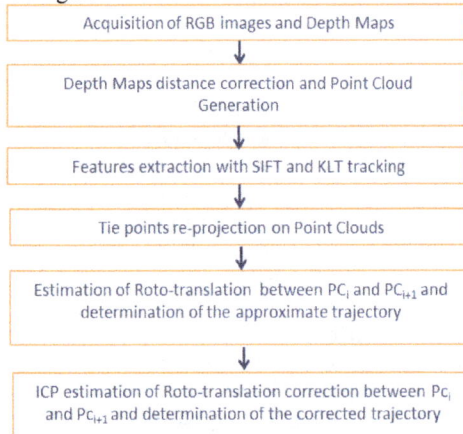

Figure 4. The proposed methodological scheme for recovering Kinect trajectory

## 4.2 Kinematic test with Kinect for XboxOne

A kinematic test was realized using the new version of Microsoft sensor. Kinect for XboxOne was placed on a cart, which has been moved in a corridor of an office building. The sensor was installed on a tripod (which was in turn fixed to the cart). On the same vehicle a reflective prism was installed too (see Figure 5). The cart was automatically tracked using a self-tracking Topcon Is203 total station, in order to have a reference trajectory for evaluating the reliability of the results. The system was geometrically calibrated acquiring simultaneously images from both the visual sensors (IR and RGB camera) and at the same time tracking the prism positions. The cart was slowly moved and, during the survey, depth and RGB images have been acquired simultaneously by Kinect 2.0. It was remotely controlled with in-house coded software based on Microsoft SDK and installed on a laptop. The data were processed using the methodological workflow presented in Figure 4.

Figure 5. The Kinect installed together with the laptop and the reflective prism on the cart

By solving the exterior orientation of the first IR frame it is possible to recover the trajectory directly in object reference system (XYZ). Once the trajectory is computed, it can be transferred to the prism reference point using the lever arm estimated during the geometric calibration.

The trajectory computed using KLT-SIFT tracking algorithm and the one refined using ICP algorithm, transferred to the reflective prism are reported in Figure 6. In the same figure is plotted also the reference trajectory, acquired with the self-tracking total station. The results show that there is a good consistency between the Kinect 2.0 solution and the reference one since the RMSs of the discrepancies in all the coordinates

are lower than 0.05 m (see Table 2). These results are in agreement with the precisions of the relative orientation parameters (between the IR and the RGB camera) and of the 3D-vector connecting the cameras and the reflective prism, reached during cart calibration phase. The ICP corrections were not significant because they were of the same order of magnitude than the error committed by Kinect 2.0 sensor when it is used as a depth measuring system.

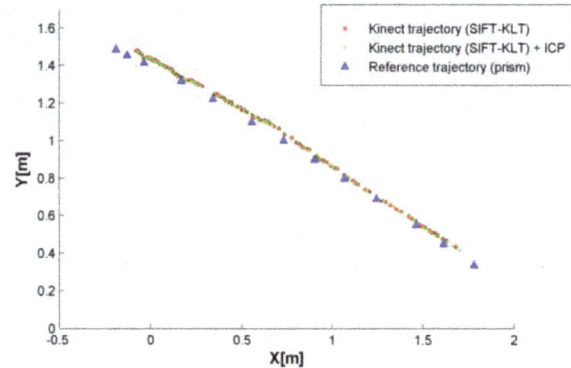

Figure 6. Trajectories estimated with the Kinect using the proposed integrated solutions

|  | X[m] | Y[m] | Z[m] |
|---|---|---|---|
| RMSe KLT-SIFT Kinect 2.0 trajectory | 0.047 | 0.035 | 0.038 |
| RMSe KLT-SIFT Kinect 2.0 trajectory +ICP correction [m] | 0.046 | 0.030 | 0.029 |

Table 2 - RMSe of GPR position between the Kinect 2.0 solution (transferred using the calibration vector) with respect to the trajectory followed by the reflective prism.

## 5. CONCLUSIONS

Both the solution here presented has been proven reliable to recover a trajectory in obstructed area, even though this solution is very dependent on the possibility of tracking at least few points throughout the entire image sequence. For outdoor navigation, photogrammetry has emerged as a reliable solution for georeferencing a slow moving vehicle with an accuracy of few decimeters while having an up-to-date and correct large scale map.

For the indoor case, the example of Kinect sensor has clearly shown that the integration of visual and depth data can be helpful to overcome possible weaknesses of a pure photogrammetric solution, in environments with poor texture. The proposed method allowed recovering the followed trajectory with an error of few centimeters, but it could be improved using RGB-D descriptors in order to have good performances in all situations.

### ACKNOWLEDGEMENTS

The authors thank 3DOM unit at Bruno Kessler Foundation for providing the software Rotoprocuster_scaling and for the help provide during the development of the indoor navigation solution.

### REFERENCES

Al-Hamad, A., El-Sheimy, N.,. 2014. Smartphones Based Mobile Mapping Systems. In: *The International Archives of Photogrammetry and Remote Sensing and Spatial Information*

*Sciences*. Vol. XL-5, Proceedings on the ISPRS Technical Commission V Symposium, Riva del Garda, Italy.

Barzaghi, R., Carrion, D., Cazzaniga, N.E., Forlani, G., 2009. Vehicle Positioning in Urban Areas Using Photogrammetry and Digital Maps, Proceedings of the ENC-GNSS09, Napoli.

Böniger, U., Troncke, J., 2010. On the Potential of Kinematic GPR survey Using Self-Tracking Total Station: Evaluating System Crosstalk and Latency. *IEEE Transaction on Geoscienze and Remote Sensing*. Vol. 28, n. 10, pp. 3792-3798.

Cazzaniga, N.E., Pagliari, D., Pinto, P., 2012. Photogrammetry for Mapping Underground Utility Lines with Ground Penetrating Radar in Urban Areas. In: *International Archives of the Photogrammetry, Remote Sensing and Spatial Information Sciences*. Volume XXXIX-B1, Melbourne, Australia

Chaplin B., 1999. Motion Estimation from Stereo Image Sequences for a Mobile Mapping System. Msc. Thesis, Department of Geomatics Engineering, University of Calgary.

Crosilla, F., Visentini, D., 1998. External Orientation of a Mobile Sensor via Dynamic Vision of Digital Map points. *Bollettino di Geodesia e Scienze Affini*. 57(1), pp.41-60.

Da Silva, J.F. de Oliviera Carmago, P., Gallis R. B. A., 2003. Development of a low-cost mobile mapping system: a South American experience. *The Photogrammetric Record*. Vol. 18 (101), pp. 5-26.

Endres, F., Hess, J., Engelhard, N., Sturn, J., Cremer, D., Burgard, W., 2012. An Evaluation of RGB-D SLAM System. *2012 IEEE International Conference on Robotics and Automation*. River Centre, Saint Paul, Minnesota.

Eugster, H., Huber, F., Nebiker, S., and Gisi, A., 2012. Integrated georeferencing of stereo image sequences captured with a stereovision mobile mapping system – approaches and practical results. In: *International Archives of Photogrammetry Remote Sensing and Spatial Information Sciences*. XXXIX-B1, pp. 309-314.

Forlani, G., 1986. Sperimentazione del nuovo programma CALGE dell'ITM. In: *Bollettino SIFET*. No. 2, pp. 63-72

Forlani G., Roncella R., Remondino F., 2005. Structure and motion reconstruction of short mobile mapping image sequences. *Proc. of the 7th Conf. On Optical 3D measurement techniques*. Vol I, pp. 265-274.

Hassan, T., Ellum, C., El-Sheimy, N., 2006. Bridging land-based mobile mapping using photogrammetric adjustments. *ISPRS Commission I Symposium*. From Sensors to Imagery.

Hoffmann-Wellenhof, B., Lichtenegger, B., Walse, E., 2008. GNSS- Global Navigation Satellite Systems. *SpringerWeinNewYork*.

Lowe, D., 2004. Distinctive Image Feature from Scale-Invariant. *International Journal of Computer Vision*. 60(2), pp. 91-110.

Lucas, B.D., Kanade, T., 1981. An Iterative Registration Techniques with an Application to Stereo Vision, *Proceedings of 7th International Joint Conference on Artificial Intelligence*. (AJCAI) 1981., pp. 674-679.

MacGougan, G. D., 2003. High Sensitive GPS Performance Analysis in Degraded Signal Environments. M.Sc Thesis. Department of Geomatics Engineering, University of Calgary.

Mautz, R. (2012). Indoor Positioning Technologies. Habilitation Thesis at ETH Zurich, 127 p, Swiss Geodetic Commission, Geodetic-Geophysical Reports of Switzerland, no. 86, ISBN 978-3-8381-3537-3.

Mahfouz, M.R., Zhang, C., Merkl, B.C., Kuhn, M.J., FAthy, A. E., 2008. Investigation of High-Accuracy Indoor 3d Positioning Using UWB Technology. *IEEE Transaction on Microwave Theory and Techniques*, Vol. 56 (6), June 2008.

Oliver, A., Kong, S., Wünsche, B., MacDonald, B., 2012. Using the Kinect as a Navigation Sensor for Mobile Robotics. *Proceedings of the 27th Conference on Image and Vision Computing*, pp. 505-514.

Ramos, F.T.; Nieto, J.; Durrant-Whyte, H.F., 2007. Recognizing and Modelling Landmarks to Close Loops in Outdoor SLAM. *2007 IEEE International Conference on Robotics and Automation*, pp.2036-204.

Roncella, R., Re, C., Forlani, G., 2011. Comparison of two Structure from Motion Strategies. In: *International Archives of Photog*rammetry, Remote Sensing and Spatial Information Science, vol. XXXVIII, 5-W16.

Roncella, R., Re, C., Forlani, G., 2011. Performance Evaluation of a Structure and Motion Strategy in Architecture and Cultural Heritage,. In: *Archives of Photogrammetry, Remote Sensing and Spatial Information Sciences*. ISPRS volume XXXVIII-5/W16.

Roncella, R., Remondino, F., Forlani, G., 2005. Photogrammetric Bridging of GPS outages in mobile mapping. *Electronic Imaging*. pp.308-319.

Storms, W., Shockley, J., Raquet, J., 2010. Magnetic Fields Navigation in Indoor Environment. *Proceeding of Ubiquitous Positioning Indoor Navigation and Location Based Service*. (UPINLBS), pp 1-10.

Tao, C. V., Chapman, M. A., Chaplin, B. A., 2001. Automated Processing of Mobile Mapping Image Sequences. *ISPRS Journal of Photogrammetry and Remote Sensing*. vol. 55, pp 330-346.

Tomasi, C., Kanade, T., 1991. Detection and Tracking of Point Feature. Carnegie Mellon University Technical Report, CMU-CS-91-132, April 1991.

Vaupel, T., Kiefer, J., Haimerl, S., Thielecke, J., 2010. Wi-Fi Positioning Systems Consideration and Devices Calibration. *2010 International Conference of Indoor Positioning and Indoor Navigation (IPIN)*.

Zhou, Y., Lui, W., 2007. Laser-Activated RFID-based Indoor Localization System for Mobile Robots. *Proceedings of the 2007 IEEE International Conference on Robotics and Automation*. Rome. Italy, 10-14 April.

# 19

# IMAGE SELECTION FOR 3D MEASUREMENT BASED ON NETWORK DESIGN

T.Fuse [a, *], R.Harada [a,]

[a] Dept. of Civil Engineering, University of Tokyo, Hongo 7-3-1, Bunkyo-ku, Tokyo, 113-8656,Japan-
fuse@civil.t.u-tokyo.ac.jp, harada@trip.t.u-tokyo.ac.jp

**Commission V, WG V/4**

**KEY WORDS:** Image Selection, Bundle Adjustment, Network Design, Visualization, Graph cut

**ABSTRACT:**

3D models have been widely used by spread of many available free-software. On the other hand, enormous images can be easily acquired, and images are utilized for creating the 3D models recently. However, the creation of 3D models by using huge amount of images takes a lot of time and effort, and then efficiency for 3D measurement are required. In the efficiency strategy, the accuracy of the measurement is also required. This paper develops an image selection method based on network design that means surveying network construction. The proposed method uses image connectivity graph. By this, the image selection problem is regarded as combinatorial optimization problem and the graph cuts technique can be applied. Additionally, in the process of 3D reconstruction, low quality images and similarity images are extracted and removed. Through the experiments, the significance of the proposed method is confirmed. Potential to efficient and accurate 3D measurement is implied.

## 1. INTRODUCTION

In late years, movements of the 3D model making and utilization are generalized by the development of the computer technology. The making of a variety of 3D models has been performed by an expert, but 3D model production software available free such as Blender or 123D is shown much on web, and the environment that anyone can easily make the 3D model is set. The constructed 3D model is used in the frequent natural disaster measures or actual situation grasp of the infrastructure other than a scene design, view simulation, a visual material for city planning and 3D model is utilized in various fields.

The general 3D model making is often carried out by photogrammetry. The three-dimensional measurement with the image presents an active state by development of Computer Vision (CV). Besides, in late years the mass image acquisition becomes possible, and an opportunity to apply photographic surveying spreads more and more. As 3D model making example from an image in large quantities, there is business to make digital 3D map of the whole world using a large quantity of satellite images and trial to make 3D model of an urban by using a large quantity of images on community websites such as Flickr (Agarwal *et al.*, 2011). Furthermore, even an individual comes to be able to easily acquire a large quantity of images as consecutive image by the spread of high-performance video cameras at a low price. It is hoped that it is not special for an individual to make 3D model by these consecutive images by using 3D model production software.

However, for the 3D model making, it is necessary to process a large quantity of images and it takes a lot of time and effort to create 3D model by using huge amount of images. Therefore efficiency of the model making is called for. Furthermore, as well as efficiency, it is important to secure the accuracy of the 3D model. In the general photogrammetry, specifications and configurations of cameras are set in advance as project planning, which is so called as network design in photogrammetry (K.B.Atkinson, 1996). The network design means surveying network construction, and mainly consists of zero-order design (ZOD: the datum problem), first-order design (FOD: the configuration problem), second-order design (SOD: the weight problem). Considering the availability of huge amount of images, the network design may be applied in order to select images among them after taking the images. The image selection based on the network design will be expected to contribute improvement of efficiency for 3D measurement and keeping of accuracy simultaneously.

A purpose of this paper is to develop a method of automatic selection of the using image from a large quantity of images based on the network design for three-dimensional measurement. Specifically, the proposed method uses image connectivity graph. The image connectivity graph represents the relationships between images by using node as an image and edges as relationships between two images. The edges in the graph have costs, which are defined by elements in the first-order design of the network design. The FOD is known as the most important aspect in the network design. The estimation accuracies of exterior orientation elements, namely variances of estimated camera positions between two images, are set as the costs. Once the image connectivity graph is constructed, the image selection problem is regarded as combinatorial optimization problem. Against the combinatorial optimization problem, the graph cuts technique (Kubo, 2000) can be applied. Here, the cost in the connectivity graph is considered as the energy, and then technique can find the optimized connectivity graph in the sense of cost minimization with smaller number of edges as possible. The remaining edges are corresponding to selected image pairs. Additionally, in the process of 3D reconstruction, low quality images and similarly images are also extracted and removed. By this, the efficiency of the image selection is improved.

## 2. AUTOMATIC IMAGE SELECTION METHOD FOR THREE DIMENSIONAL MEASUREMENT

### 2.1 Framework of method

In this section, we present perspective of the proposed method (Figure 1). Firstly, the internal orientation element of the camera for photography is found by camera calibration (Z. Zhang, 1998). Then, image matching is carried out for two pieces of images from an image in large quantities, and image coordinates of the feature points that are common to each image pairs are found. From the result of the image matching, image pairs having low accuracy of the matching are detected and removed. Furthermore, with the image coordinate of each feature point provided as a result of image matching, the three-dimensional coordinate group of the feature points and the camera external orientation element between two pieces of images are calculated. In doing so, the detection and removal of the similar images are carried out. With the three-dimensional coordinate group of the feature points and camera external orientation element between each provided image pair, the measurement accuracy of the camera external orientation element is estimated by bundle adjustment (Luhmann et al., 2006; Okaya, 2010; Iwamoto et al., 2011). Finally, based on network design, we perform graph expression of the relations between the image, and automation of the image selection is realized by a provided result and applying a combination optimization problem.

We use SURF algorithm (Bay, H. et al., 2008) for extraction of the feature points in this paper. In SURF algorithm, firstly, a color image is converted into a monochrome grayscale image, and a corner and an edge of grayscale images are detected, and feature points are extracted. The feature points here refer to the points that the pixel change (brightness change) of the image has a big. Then, quantity of feature and a feature vector are calculated and consider the points where a feature vector resembles to be the same point and perform a matching. Matching is carried out for two pieces of image pair by the method described above and reduces the number of the pair of image by comparing the number of the matching. We set a certain threshold and in certain image pairs, if the number of the matching is less than the threshold, the image pairs are removed for the reason of the matching accuracy of the pair image being low, or picture itself being low.

### 2.3 The acquisition of the three-dimensional point group and the external orientation element

Next, using the image coordinate of the feature points and camera internal orientation element in each image provided as a result of matching, three-dimensional coordinate of the feature point and the camera external orientation element are estimated. Here, after having performed the three-dimensional reconstruction between each image pair, the image pair thought to be the similar image was detected.

First of all, for similar image detection, it is necessary to find the base line length between each image pair. As for the base line length, the base line length between the adjacent frame images of consecutive images shall be defined 1. Namely, the base line length $B_{pq}$ (between the p-th frame image and the q-th frame image) is defined q-p (q>p).

Then, using the value of three-dimensional coordinate of the feature points calculated in each image pair, we calculate the distance between each image pair and the object. If a three-dimensional coordinate of $p_i$ (a certain feature point) is $(x_i, y_i, z_i)$, the distance between an image and feature point $p_i$ is found with $\sqrt{x_i^2 + y_i^2 + z_i^2}$. From the above, if the number of feature points in a certain image pair is M, the distance between the object and an image pair (the p-th frame image and the q-th frame image) is defined as follows.

$$H_{pq} = \frac{\sum_{i=1}^{M}(\sqrt{x_i^2 + y_i^2 + z_i^2})}{M} \qquad (1)$$

where     $H_{pq}$ = the distance between the object and an image pair (the p-th frame image and the q-th frame image)

         M = the number of feature points in a certain image pair

Using the base line length between each image pair and the distance between image pair and object, we detected the similar image pairs. In image pairs, the similar degree of the image becomes higher if the distance between an image pair and object is much bigger than the base line length. We set a certain threshold and detected the similar image pairs and remove one of two pieces if the value of H/B is bigger than the threshold (H means the distance between an image pair and the object, B

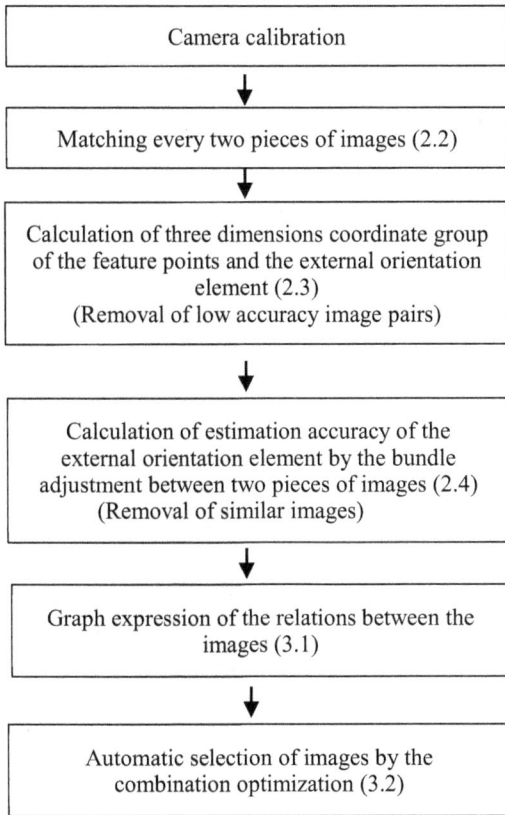

| Camera calibration |
|---|

↓

| Matching every two pieces of images (2.2) |
|---|

↓

| Calculation of three dimensions coordinate group of the feature points and the external orientation element (2.3) (Removal of low accuracy image pairs) |
|---|

↓

| Calculation of estimation accuracy of the external orientation element by the bundle adjustment between two pieces of images (2.4) (Removal of similar images) |
|---|

↓

| Graph expression of the relations between the images (3.1) |
|---|

↓

| Automatic selection of images by the combination optimization (3.2) |
|---|

Figure1. Framework of method

### 2.2 Image matching

When a three-dimensional measurement using the image of several pieces is performed, it is common to use a theory of relative orientation, and it is necessary to extract feature points with characteristic of the color and form in images and to make the point thought to be the same point between images matched. This work is called image matching, and in this paper, we performed image matching for two pieces of images from a large quantity of images.

means base line length). In this way, the removal of the similar images is carried out in a 3D reconstruction process.

## 2.4 Calculation of the external orientation element estimate accuracy by the bundle adjustment

Three-dimensional coordinate group of feature points in each image pair was found, but it is thought that these values include a photography error by all means. Here, the estimation accuracy of the external orientation element of each image pair is got by bundle adjustment for each image pair. In external orientation elements, we consider only a translation matrix expressing the position of the camera that is the most important element of the network design.

In a certain image pair, if the translation matrix that is provided as a result of 3D reconstruction is **T**, and the translation matrix that is estimated by the bundle adjustment is **T'**, the estimation accuracy is defined as follows.

$$\mathbf{T} = \begin{bmatrix} X \\ Y \\ Z \end{bmatrix} \quad \mathbf{T'} = \begin{bmatrix} X' \\ Y' \\ Z' \end{bmatrix}$$

$$\sigma_X = (X - X'), \quad \sigma_Y = (Y - Y'), \quad \sigma_Z = (Z - Z') \quad (1)$$

$$\sigma = \frac{1}{\sigma_X^2 + \sigma_Y^2 + \sigma_Z^2}$$

where   $X, Y, Z$ = components of translation matrix **T**
          $X', Y', Z'$ = components of translation matrix **T'**
          $\sigma_X, \sigma_Y, \sigma_Z$ = measurement error of each components
          $\sigma$ = the estimation accuracy of the translation matrix

## 3. AUTOMATIC IMAGE SELECTION BY THE COMBINATION OPTIMIZATION.

### 3.1 Graph expression of the relations between the images based on network design

Here, we perform the graph expression of the relation between the images based on a way of the graph theory (T.L.Basakkar, 1970). Each photographed image is set as a node of the graph, and edges mean relationships between two images. The edges in the graph have costs, which are defined by the estimation accuracy of exterior orientation elements, namely variances of estimated camera positions between two images. Camera positions mean the FOD that is thought to be the most important element in the network design. It becomes possible to express

image relation based on a network design by incorporating the estimation accuracy of the camera position that is the element of the network design for cost of the edges (Figure 2).

### 3.2 Automatic image selection by the application of the combination optimization problem

From here, we perform automatic selection of image relations by applying a combination optimization problem. Specifically, we consider applying a graph cut theory to estimate graph structure to be constructed by an edge as little as possible. For a graph expressing image relations, we establish an initial node and a terminal node and the find the set of the cut edge realizing the min-cut. As for the set of a calculated cut edge, the cost sum namely the sum of the estimation accuracy of the camera position between the images is in this way minimized. In other words, because the edge group detected is a group of the edges that the estimation accuracy of the camera position is low, it is thought that the influence that a graph gives overall estimation accuracy even if these edges are removed is small and therefore these edges are removed from a graph. The remaining edges are corresponding to selected image pairs. By reducing the number of the edges constituting a graph, the number of the image pairs to use decreases, and will be connected for the efficiency of the measurement. From the above, while securing high accuracy by consideration of the estimation accuracy of the camera position that is the element of the network design, the effective measurement by automatic selection of the image relation will be realized.

## 4 RESULTS

### 4.1 Data description

We photographed the movie around the following structures (Figure 3) using digital video camera, and converted a photographed movie into consecutive images.   The number of consecutive images got from the movie is 999. Then we applied these images to proposed method and performed an experiment to confirm the significance of the proposed method.

Figure3. Photography target structure

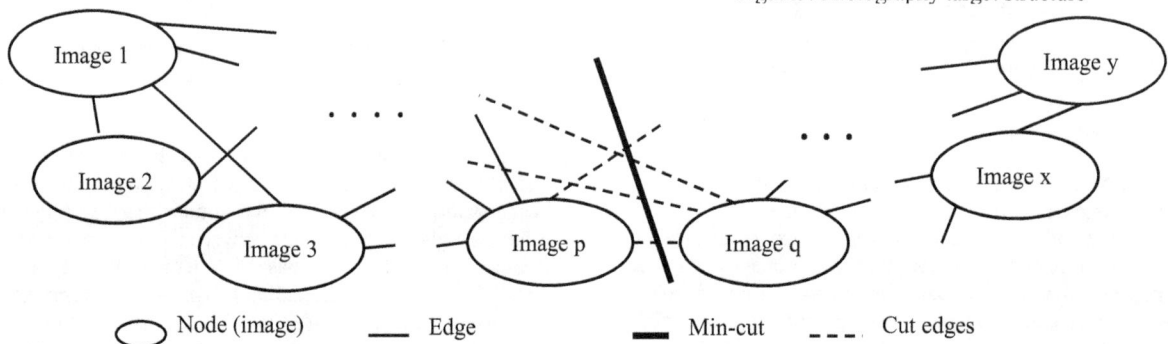

Figure 2. Graph expression of the image relation

## 4.2 Results

### 4.2.1 Image matching and removal of low quality images:
In this study, for 999 provided frames, the matching with the frame by 20 pieces before and after shall be performed, therefore the matching number of the first image pairs to perform is 19,581 pairs in total. The following figure (Figure 4) shows a distribution of the number of matching. Based on a result provided this time, we set the threshold of the number of the matching with 25 and decided to remove image pairs with number of the matching less than 25. The number of the image pairs that are removed was 1354.

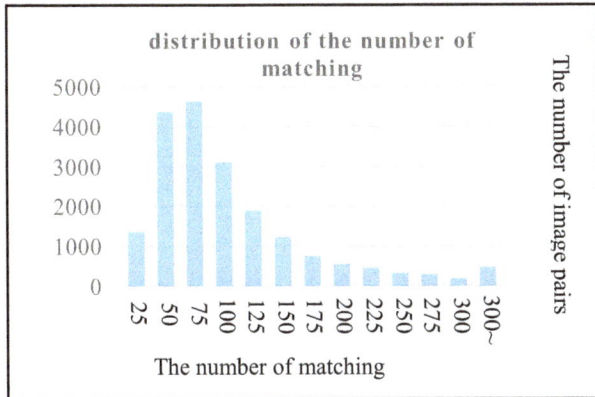

Figure 4 Distribution of the number of matching

### 4.2.2 3D reconstruction and removal of similar images:
Next, for 18,227 image pairs, we performed the three-dimensional reconstruction. We set the threshold of the number of H/B with 500, and if image pairs whose number of H/B is more than 500, we removed one of image pairs. As a result, 70 sets of image pairs were detected as similar images, and 70 pieces of images were removed. The number of the image pairs was in this way decreased from 18,227 pairs to 15,892 pairs.

### 4.2.3 Graph cut:
For 15,892 sets of image pairs, we constructed the image connectivity graph that represents the relationships between images by using node as an image and edges as relationships between two images. The number of nodes of graph was 929, and the number of edges of graph was 15,892. In this research, we made five partial graphs other than a whole graph comprised of all images. We divided all images into five group, and each image group constitutes a partial graph. For each these 6 graphs in total, we performed a graph cut. As a result, the total number of cut edges was 1,161, and the number of remaining edges was 14,731. Table 1 shows a change of the number of the edges (the number of the image pairs).

### 4.2.4 Verification of accuracy of the proposed method:
Finally, we performed verification of accuracy of the proposed method. The verification of accuracy is performed by comparing estimation accuracy of the three-dimensional coordinate of the feature point and camera external orientation element by using all image relations and these got by only using image relations chosen by proposed method. Accuracy comparison is performed by applying bundle adjustment to the selected images (after applying the proposed method) and all images (before applying the proposed method). Table 2, and Table 3 show a result of accuracy comparison of the three-dimensional coordinate group of the feature points and external orientation element.

| | Initial value | → | The low accuracy image removal (2.2) | → | The similar image removal (2.3) | → | Graph cut (3.2) |
|---|---|---|---|---|---|---|---|
| The number of the image pairs | 19,581 | → | 18,227 | → | 15,892 | → | 14,731 |

Table 1. A change of the number of the use image pairs in each stage

| | X | | Y | | Z | |
|---|---|---|---|---|---|---|
| | $\overline{X}$ | $\sigma_X$ | $\overline{Y}$ | $\sigma_Y$ | $\overline{Z}$ | $\sigma_Z$ |
| (A) | 36.648 | 1.584 | -37.019 | 1.239 | 51.935 | 1.945 |
| (B) | 35.793 | 1.226 | -35.849 | 1.324 | 50.856 | 1.803 |

(A)... Three dimensional measurement using all image relations
(B)... Three dimensional measurement using only chosen image relations
$\overline{X}, \overline{Y}, \overline{Z}$ = The average of each direction of the characteristic point three dimensions coordinate value
$\sigma_x, \sigma_{Y,} \sigma_z$ = Standard deviation of the measurement error of each direction of the characteristic point three dimensions coordinate

Table 2. A result of accuracy comparison of three-dimensional coordinate of feature point

| | X | | Y | | Z | |
|---|---|---|---|---|---|---|
| | $\overline{T_X}$ | $\sigma_{TX}$ | $\overline{T_Y}$ | $\sigma_{TY}$ | $\overline{T_Z}$ | $\sigma_{TZ}$ |
| (A) | 146.054 | 6.014 | -137.498 | 6.515 | 214.513 | 9.021 |
| (B) | 146.054 | 6.268 | -137.498 | 6.79 | 214.513 | 9.402 |

(A)… Three dimensional measurement using all image relations
(B)… Three dimensional measurement using only chosen image relations
$\overline{T_X},\overline{T_Y},\overline{T_Z}$.. The average of each value of the translation matrix of each frame
$\sigma_{TX},\sigma_{TY},\sigma_{TZ}$.. Standard deviation of the measurement error of the value of each direction of the translation matrix

Table 3. A result of accuracy comparison of camera position

## 5. DISCUSSION

Table 1 shows that by the proposed method the number of using image pairs decreased to approximately 75% in comparison with before application of method, therefore improvement of the effectiveness of 3D measurement was confirmed. In addition, Table 2 and Table 3 show that the accuracy of 3D measurement using only image pairs selected automatically by proposed method is almost the same level as that using all image pairs. From the above, an effective three-dimensional measurement was realized while keeping accuracy by an application of the proposed method.

## 6. CONCLUSION

In this paper, we examined the method to enable a more efficient 3D measurement while keeping accuracy using a large quantity of images, and developed method to select using images automatically from a large quantity of images group by considering the network design. Furthermore, we applied proposed method for real data (photographed images) and inspected effectiveness and the accuracy, and examined application possibility of the proposed method. Actually, through the experiments verifying the effectiveness and accuracy of the proposed method, the significance of the proposed method is confirmed, and it is expected that proposed method will lead to development of the future photogrammetry.

On the other hand, there are many future works to apply the proposed method practically. Firstly, it is hoped that the proposed method apply to not only sequential images but also various kinds of images such as shared images on the internet. In such cases, interior orientation elements should be introduced because these images are photographed by different cameras. Additionally, definition of cost function for the edges will be investigated. In this method, only estimation accuracy of the external orientation element between two images is set as variable for cost function. By adding the estimation accuracy of the three-dimensional coordinate of the feature points to cost function as variable, it is expected that we can express the relations between images in greater detail and can obtain result having higher accuracy. Furthermore, discussion about relationships between efficiency and accuracy will be required, because if we apply the graph cuts repeatedly, it is thought that the effectiveness improves, but the accuracy decreases. As a result, applicability of photogrammetry will be more increased.

## 7. REFERENCE

Bay, H., Ess, A., Tuytelaars, T., Gool, L.V., 2008, "Speeded-Up Robust Features (SURF)" *Computer Vision and Image Understanding*, Vol.110, No.3, pp.346-359

Iwamoto, Sugaya, Kanatani, 2011. "Bundle Adjustment for 3-D Reconstruction: Implementation and Evaluation" *IPSJ SIG Technical Report,* pp.1-8

K.B.Atkinson, 1996. *Close Range Photogrammetry and Machine Vision*, Whittles Publishing, Latheronwheel, UK

Kubo, 2000, *Kumiawasesaitekika To Algorizumu,* Kyouritsu Publishing, Tokyo, pp.85-94

Luhmann, T., Robson, S., Kyle, S. and Harley, I., 2006. Close Range Photogrammetry. Whittles, Scotland, UK.

Okaya, 2010. "Bundle adjustment" *Computer Vision saisentan gaido,* Vol.3, pp.1-31

Sameer Agarwal, Yasutaka Furukawa, Noah Snavely, Ian Simon, Brian Curless, Steven M. Seitz and Richard Szeliski, 2011. "Building Rome in a Day" *Communications of the ACM*, Vol.54, No. 10, pp.105-112,

T.L.Basakkar, 1970, *Finite Graphs and Networks An introduction with Applications*, Baifuusya, Tokyo,

Z.Zhang, 1998. "A Flexible New Technique for Camera Calibration" *IEEE Translatrions on Pattern Analysis and Machine Intelligence,* 22(11), pp.165-186

# 20

# DATA FUSION OF LIDAR INTO A REGION GROWING STEREO ALGORITHM

J. Veitch-Michaelis [a], J-P. Muller [a], J. Storey [b] D. Walton [a], M. Foster [b]

[a] Imaging Group, Mullard Space Science Laboratory (University College London, Holmbury St Mary, RH5 6NT, UK - (j.veitchmichaelis.12, j.muller, d.walton)@ucl.ac.uk
[b] IS Instruments Ltd, 220 Vale Road, Tonbridge, TN9 1SP, UK (jstorey, mfoster)@is-instruments.com

**KEY WORDS:** Stereo vision, LIDAR, Multisensor data fusion, calibration, 3D reconstruction

**ABSTRACT:**

Stereo vision and LIDAR continue to dominate standoff 3D measurement techniques in photogrammetry although the two techniques are normally used in competition. Stereo matching algorithms generate dense 3D data, but perform poorly on low-texture image features. LIDAR measurements are accurate, but imaging requires scanning and produces sparse point clouds. Clearly the two techniques are complementary, but recent attempts to improve stereo matching performance on low-texture surfaces using data fusion have focused on the use of time-of-flight cameras, with comparatively little work involving LIDAR.

A low-level data fusion method is shown, involving a scanning LIDAR system and a stereo camera pair. By directly imaging the LIDAR laser spot during a scan, unique stereo correspondences are obtained. These correspondences are used to seed a region-growing stereo matcher until the whole image is matched. The iterative nature of the acquisition process minimises the number of LIDAR points needed. This method also enables simple calibration of stereo cameras without the need for targets and trivial co-registration between the stereo and LIDAR point clouds. Examples of this data fusion technique are provided for a variety of scenes.

## 1. INTRODUCTION

Stereovision and LIDAR remain the two most commonly used methods for 3D reconstruction. Stereo camera systems can provide dense 3D information, but image matching is a computationally complex problem and reconstruction coverage and accuracy suffer if the image does not have sufficient texture. LIDAR derived ranges are accurate and are not significantly dependent on surface texture, albeit they are dependent on avoiding specular reflection angles for certain types of shiny surfaces, but in most systems the laser must be scanned to build up a 3D image.

This paper proposes a novel method for integrating LIDAR ranges into a region growing stereo matching algorithm, proposed initially by (Muller and Anthony, 1987). In particular, the image of the LIDAR spot on the scene is used to provide unambiguous seed points in regions where there is low texture.

### 1.1 Data fusion with stereo systems

In recent years, there has been a large amount of research into integrating additional sources of range information into stereo matching algorithms. Stereo data fusion algorithms generally attempt to solve the problem of ambiguous matches arising from homogenous or repetitive texture. Obvious candidate techniques for data fusion are those that actively sense the scene.

Most research has focused on the use of Time of Flight Cameras (ToFCs) (Lange and Seitz, 2001), (Foix et al., 2011), which are now available commercially at low cost (e.g. Microsoft Kinect for Xbox One). ToFCs offer active illumination and dense 3D imaging data in real-time. However, the usable range of a couple of metres depends on the modulation frequency of the illumination and performance outdoors, due to the use of near infrared illumination, is not guaranteed. ToFC range data tends to be relatively noisy with accuracy in the millimetre scale. The

use-cases for these systems are dominated by robotics, where real-time imaging is desirable for navigation and mapping.

Broadly, there are two classes of fusion algorithm:

*A priori* methods using ToFCs have been used with most of the common state-of-the-art stereo matching algorithms including dynamic programming (DP) (Gudmundsson et al., 2008), graph cuts (Hahne and Alexa, 2008); (Song et al, 2011), belief propagation (Jiejie Zhu et al., 2011) and semi-global matching (SGM) (Fischer et al., 2011). ToF range data is used to overcome the limitations of stereo in homogenous image regions while stereo range data is retained near depth discontinuities. First, a range interval is obtained for every pixel in the ToF system. This range interval is then mapped to each pixel in the stereo system and used as a constraint on the matching algorithm, either limiting the disparity search space or as an adjusted matching cost function.

*A posteriori* methods combine two (or more) complete range images, ideally producing an image that is better than any of the inputs alone. (Kuhnert and Stommel, 2006) merge ToF data with results from winner-take-all (WTA) and simulated annealing stereo. ToF data is retained in homogenous regions where stereo fails and is also used to detect blunders in the disparity map. (Beder et al., 2007) use an approach based on patchlets (Murray and Little, 2005), rectangular surface elements defined at every pixel in the disparity image. In both cases the fused data are an improvement over stereo or ToF alone, but results are limited by the low resolution of the ToF cameras.

### 1.2 Data fusion of stereo with LIDAR

Comparatively little work has involved LIDAR at close range. (Romero et al., 2004) use a 2D scanning LIDAR to seed the disparity map in a trinocular stereo system. Initial disparity estimates are calculated using LIDAR ranges which are then propagated through the image based on a set of rules that consider image texture and horizontal and vertical pixel correspondences. Results from a standard stereo matcher are not provided, but the LIDAR does improve matching performance in homogenous image regions.

(Badino et al., 2011) follow an approach largely similar to that used with ToFCs. A 3D scanning LIDAR is used to generate a minimum and maximum range map for the scene. This is then used to constrain WTA and DP stereo matchers. The presented scenes are all outdoors and the fused data contains significantly fewer blunders than stereo matching alone.

The technique presented in this paper departs from previous work. Using a visible LIDAR scanner, it is possible to perform data fusion at a lower level by imaging the LIDAR spot as it scans through the scene.

## 2. 3D IMAGING SYSTEM

The imaging system used for this research consisted of a stereo camera pair and a single point LIDAR mounted on a gimbal platform, see Figure 1.

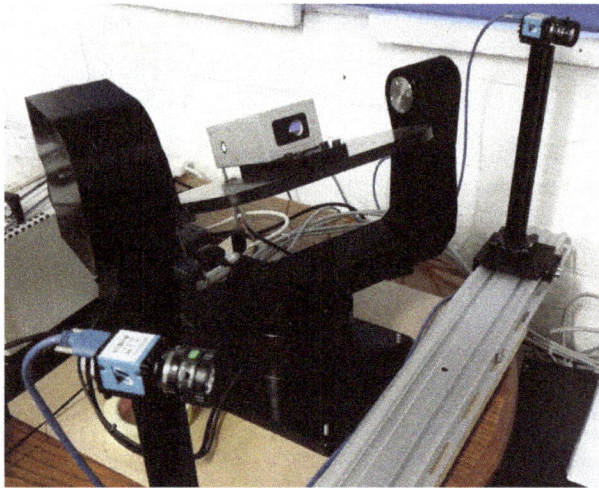

Figure 1 Custom 3D imaging system including scanning LIDAR and stereo camera pair.

### 2.1 Stereo system

The stereo system comprised two Imaging Source DMK23UM01 monochrome cameras with a resolution of 1280x960px, fitted with 8mm focal length lenses set to f/4. A custom stereo bar was built using a rail (Thorlabs XT95SP) with each camera mounted on a manual rotation stage (Thorlabs RP01) and carriage (Thorlabs XT95P11).

### 2.2 LIDAR and mount

The LIDAR used was a Dimetix FLS-C10 with a specified accuracy of ±1mm, repeatable to ±0.3mm on natural surfaces up to a distance of 65m. This sensor uses a visible (650nm) laser beam and can operate at up to 20Hz in its highest resolution mode. The LIDAR is mounted on a Newmark GM-12 gimbal mount with a specified resolution of 300μrad and a repeatability of 20μrad.

### 2.3 System geometry and model

The stereo system was modelled using the conventional pinhole camera model including radial and tangential lens distortion

(Hartley and Zisserman, 200). Camera intrinsic and stereo calibration was performed using the OpenCV (Bradski, 2000) library implementation of (Zhang, 2000) employing a planar chessboard target. This calibration is then used to epipolar rectify the images prior to stereo reconstruction. Calibration results are shown in Table 1.

| Calibration Parameter | Left Camera | Right Camera |
|---|---|---|
| Focal length (mm) | 8.378 | 8.387 |
| Principal Point (mm) | (2.459, 1.803) | (2.443, 1.780) |
| Rectified Position (m) | (0.0,0.0,0.0) | (0.463, 0, 0) |
| Reprojection Err. (px) | 0.104 | 0.103 |

Table 1 Calculated stereo camera calibration parameters

The geometric model for the LIDAR allows for systematic translational and rotational offsets, similar to (Muhammad and Lacroix, 2010). Five parameters are required: two for translational offsets orthogonal to the laser beam axis, two for rotational offsets with respect to the reported position of the mount and a distance offset. This geometry is shown in Figure 2.

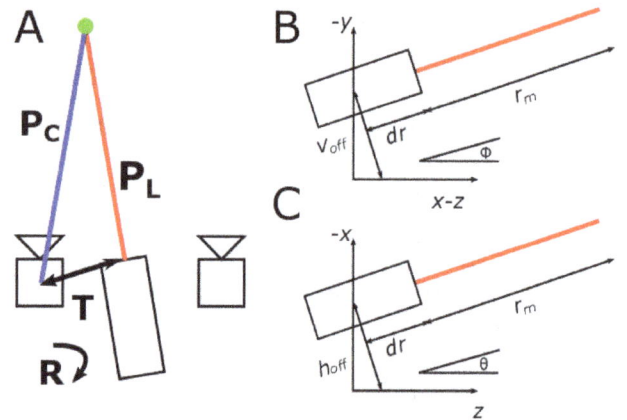

Figure 2 System geometry. A: relationship between stereo cameras and LIDAR for a single world point. B: side view of LIDAR. C: overhead view of LIDAR. Systematic angular offsets are not shown.

The model parameters are:

$r_m$ – LIDAR measured range
$\phi_m$ – Reported elevation
$\theta_m$ – Reported azimuth
$dr$ – Systematic range error
$h_{off}$ – Horizontal translation orthogonal to laser beam
$v_{off}$ – Vertical translation orthogonal to laser beam
$d_\phi$ – Systematic elevation offset
$d_\theta$ – Systematic azimuth offset
$(x_L, y_L, z_L)$ – LIDAR world coordinates

The conversion to Cartesian coordinates is as follows:

$$
\begin{aligned}
r &= r_m + dr \\
\phi &= \phi_m + d\phi \\
\theta &= \theta_m + d\theta \\
r_{xy} &= r \cos \phi - v_{off} \sin \phi \\
x_L &= r_{xy} \, r \sin \theta - h_{off} \cos \theta \\
y_L &= r \sin \phi + v_{off} \cos \phi \\
z_L &= r_{xy} \cos \theta + h_{off} \sin \theta
\end{aligned}
\tag{1}
$$

Note the use of a right hand coordinate system consistent with stereo imaging conventions – the z-axis represents depth, directed into the scene. Without loss of generality, the world origin is chosen to be the centre of projection in the left camera.

## 2.4 Cross-Calibration

The use of a visible laser allows for simple intrinsic and extrinsic calibration of the LIDAR. The LIDAR is scanned in a raster fashion across a scene. At each orientation, a stereo pair is synchronously acquired and the location of the laser spot in the epipolar-rectified images is determined using a maximum filter. A short exposure time of 1/5000s is used to ensure that the image of the spot is not saturated. Thus two point clouds are produced: one from the LIDAR ranges and one from the stereo reconstruction of the LIDAR laser spot.

Assuming a good stereo calibration, the LIDAR intrinsic and extrinsic parameters with respect to the stereo system may be determined using least squares minimisation. The error to be minimised was chosen to be the Euclidean distance between corresponding points in the stereo point cloud and the LIDAR point cloud. For a stereo reconstructed point $P_{c,i}$ and corresponding LIDAR point $P_{L,i}$, the minimisation function is:

$$\min_{R,T,\epsilon} \sum_i ||(RP_{L,i} + T) - P_{c,i}|| \qquad (2)$$

Where $R$ and $T$ are a rotation and translation matrix bringing the LIDAR point cloud onto the stereo point cloud and $\epsilon$ represents the LIDAR intrinsic parameters.

It is expected, for the system described here, that the intrinsic parameters are close to zero. Initially, $R$ and $T$ are estimated using a rigid body transformation between the two point clouds since there is a one-one correspondence between them. After transformation, any points with a large distance error (further from the mean by more than two standard deviations) are discarded. This effectively removes most spurious correspondences, for example LIDAR points which are occluded in one image. With these points removed, the rotation and translation are estimated again to get initial values for the minimisation. The LIDAR intrinsic parameters are initialised to be zero. The rotation offsets $d\theta$ and $d\phi$ were both fixed to be zero for better parameter convergence.

Figure 3 Epipolar-rectified left image of calibration scene, locations of calibration points are plotted.

The calibration scene (Figure 3) was chosen to be a wall corner to avoid coplanar calibration points. There is no particular requirement for the target to be planar, however.

| Calibration Parameter | Value |
|---|---|
| Translation, T (mm) | (-214.10, 287.55, 225.17) |
| Rotation, R (deg) | (-1.83, 0.29, 0.013) |
| Distance offset, $dr$ (mm) | 21.21 |
| Horizontal offset, $h_{off}$ (mm) | 1.05 |
| Vertical offset, $v_{off}$ (mm) | 15.02 |

Table 2 Calculated LIDAR intrinsic and extrinsic parameters.

Calibration parameters are given in Table 2. Co-registered LIDAR and stereo points are shown in Figure 4 from an overhead perspective. The LIDAR points are visibly more tightly clustered than the stereo points. The registration error between the two point clouds is shown in Figure 5.

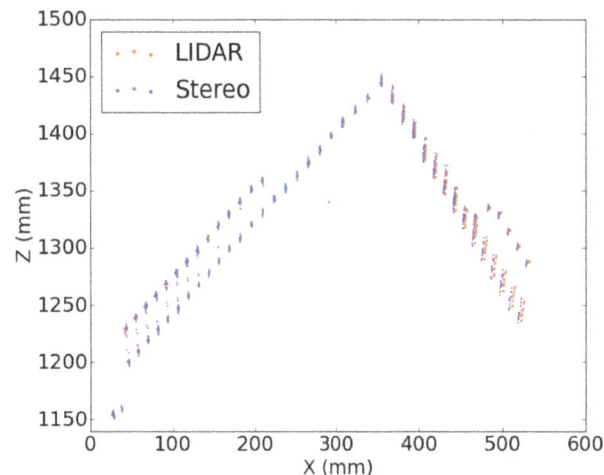

Figure 4 Overhead view of calibration points from each measurement system.

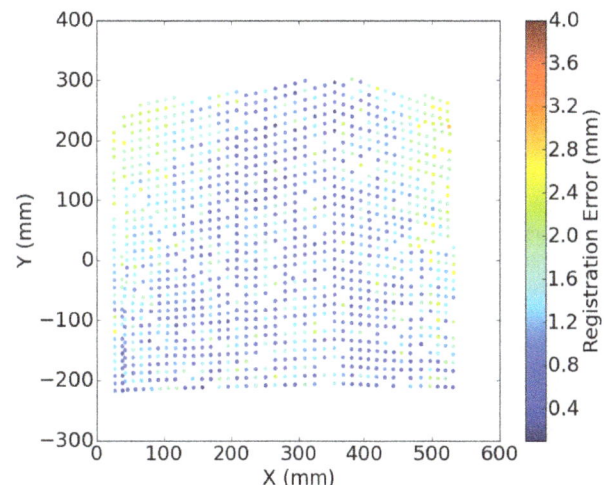

Figure 5 Registration error between LIDAR and Stereo point clouds.

The registration error was found to be (1.21±1.11mm). Some error is likely due to uncertainties in determining the laser spot location; further work will investigate ways to improve this.

### 3.   DATA FUSION ALGORITHM

The proposed algorithm operates at a lower level of fusion than prior work. It aims to directly address the issue of poor stereo matching performance in low texture regions.

### 3.1   GOTCHA stereo matcher

The stereo matcher used is based on the Gruen-Otto-Chau Adaptive Least Squares Correlation (GOTCHA) algorithm, (Otto and Chau, 1989). The actual GOTCHA used here is based on (Shin and Muller, 2012) which is a 5th generation version of the original. GOTCHA is a region growing stereo matcher taking as input a stereo pair and a list of initial, proposed correspondences or tiepoints. These tiepoints are either selected manually or are automatically generated using a feature detector such as SIFT (Lowe, 2004).

This method has proved to be accurate and robust, for example when applied to spacecraft (Day and Muller, 1989); (Thornhill et al., 1993); close-range industrial  (Muller and Anthony, 1987); (Anthony, Deacon and Muller 1988); close-range medical (Deacon et al., 1991) and Martian rover imagery (Shin and Muller, 2012).

The algorithm uses Adaptive Least Squares Correlation (ALSC) (Gruen, 1985) to refine and determine correspondences to sub-pixel accuracy, providing a disparity estimate and a confidence score. If a tiepoint is successfully matched, its neighbouring pixels are added to a priority queue, sorted by match confidence. ALSC is performed on the neighbours of the highest confidence tiepoint and any matches are added to the tiepoint queue. The process iterates until the queue is empty. Thus the disparity is grown from the initial seed points, preferentially matching from the regions with highest confidence.

### 3.2   Proposed algorithm

Normally the initial seed points are generated using SIFT keypoint matching. This approach excels in regions of high image texture, but results are poor in low texture regions where there tend to be multiple keypoints with similar descriptors.

Since the LIDAR spot is visible in both images, it can provide unique stereo correspondences. Instead of using a feature detector like SIFT, the laser spot locations are used directly as unambiguous seed points for GOTCHA. Imaging matching is performed using GOTCHA and gaps in the disparity map may be filled using LIDAR data.

This approach has several advantages over previous work. Compared to methods that use ToFCs, the LIDAR has higher accuracy and can be aimed with finer resolution. By using a visible LIDAR, cross calibration between the two coordinate systems becomes straightforward; this is particularly useful for gap filling, as the two point clouds are easily co-registered. Compared to other data fusion techniques, this method requires relatively few points from the LIDAR.

### 4.   RESULTS

A demonstration scene of a white-painted laboratory brick wall was chosen to be representative of the challenges involved with imaging indoor scenes. The left camera view is shown in Figure 6. The four kiln bricks were included to provide regions of high texture, in contrast to the wooden panel.

Figure 6 Example scene to be reconstructed. Left stereo image, epipolar rectified.

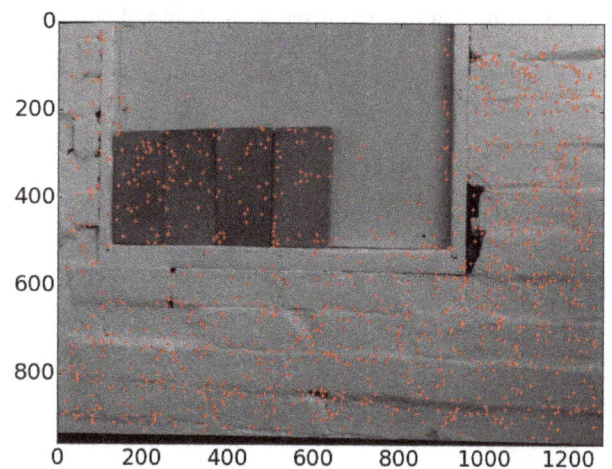

Figure 7 Matched SIFT keypoints, left stereo image, epipolar rectified.

Figure 7 shows the matched SIFT keypoints for the stereo pair. As expected, there are very few keypoints located on the panel implying a very low texture region. Figure 8 shows the result from GOTCHA using these SIFT keypoints. Match performance is good on the kiln bricks, reasonable on the painted brick wall and poor on the flat wooden panel.

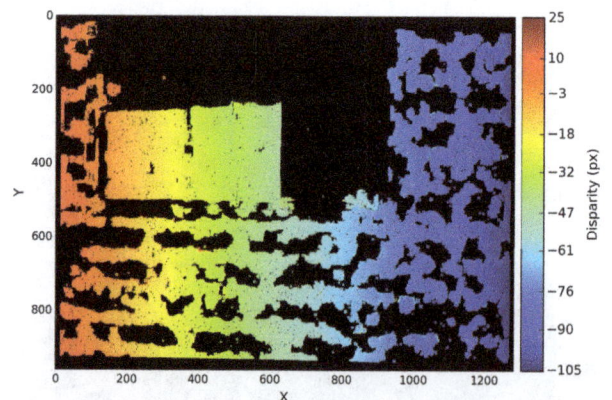

Figure 8 Disparity map using SIFT keypoints as seeds for GOTCHA stereo matching.

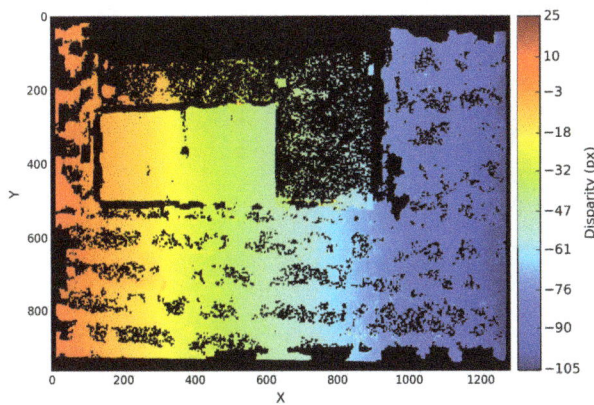

Figure 9 Disparity map using 30000 random LIDAR points as seeds for GOTCHA stereo matching.

The match results are clearly correlated with the location of the initial tiepoints. In particular the panel provides little texture for the matcher to grow into.

Figure 9 shows the same scene, but with 30000 randomly selected LIDAR points used as seeds. There is a 26.5% increase in the number of matched pixels and improvement is visible throughout the image. Performance is still degraded in low texture regions, such as the panel and between bricks, where there are isolated matched pixels from which the disparity could not be grown. These regions would be suitable for gap filling by scanning the LIDAR.

## 5. CONCLUSIONS AND FUTURE WORK

A novel data fusion algorithm has been presented, incorporating data from a visible scanning LIDAR into a region growing stereo matcher to produce denser disparity maps without loss of accuracy. This is achieved by using the image of the LIDAR laser spot to generate unambiguous seed points for the GOTCHA stereo matcher. Imaging the LIDAR spot also allows for straightforward target-less calibration of the LIDAR system and registration between the stereo and LIDAR point clouds.

Future work will look into improving cross calibration, in particular the possibility of calibrating camera intrinsic parameters using the LIDAR alone. Additionally, smarter methods of scanning the scene, based on pre-selecting areas of low image texture could focus the number of LIDAR points required for dense reconstruction on these areas where they will have the most impact.

## 6. REFERENCES

Anthony, A.G., Deacon A.T., Muller J.-P., (1988) A close range vision cell for direct input to CAD systems. IAPR Workshop on Computer Vision, October, pp. 12-14

Badino, H., Huber, D., Kanade, T., 2011. Integrating LIDAR into Stereo for Fast and Improved Disparity Computation, in:. *Presented at the 3D Imaging, Modeling, Processing, Visualization and Transmission (3DIMPVT)*, 2011 International Conference on, IEEE, pp. 405–412.

Beder, C., Bartczak, B., Koch, R., 2007. A Combined Approach for Estimating Patchlets from PMD Depth Images and Stereo Intensity Images, in: Hamprecht, F., Schnörr, C., Jähne, B. (Eds.), *Lecture Notes in Computer Science*. Springer Berlin Heidelberg, pp. 11–20.

Bradski, G., 2000. The opencv library. *Doctor Dobbs Journal*

25, 120–126. http://www.opencv.org (17 Apr 2015)

Day, T., Muller, J.-P., 1989. Digital elevation model production by stereo-matching spot image-pairs: a comparison of algorithms. *Image and Vision Computing* 7, 95–101.

Deacon, A., Anthony, A., Bhatia, S., & Muller, J. (1991). Evaluation of a CCD-Based Facial Measurement System. *Medical Informatics*, *16*(2), 213–228

Fischer, J., Arbeiter, G., Verl, A., 2011. Combination of Time-of-Flight depth and stereo using semiglobal optimization, in: *IEEE International Conference on Robotics and Automation (ICRA)* Shanghai, China, pp. 3548–3553.

Foix, S., Alenya, G., Torras, C., 2011. Lock-in Time-of-Flight (ToF) Cameras: A Survey. *IEEE Sensors J.* 11, 1917–1926

Gruen, A., 1985. Adaptive least squares correlation: a powerful image matching technique. *South African Journal of Photogrammetry* 14, 175–187.

Gudmundsson, S.A., Aanaes, H., Larsen, R., 2008. Fusion of stereo vision and Time-Of-Flight imaging for improved 3D estimation. *IJISTA* 5, 425.

Hahne, U., Alexa, M., 2008. Combining Time-Of-Flight depth and stereo images without accurate extrinsic calibration. *IJISTA* 5, 325.

Hartley, R. I. and Zisserman, A., 2004. *Multiple View Geometry in Computer Vision 2nd Ed.* Cambridge University Press.

Jiejie Zhu, Liang Wang, Ruigang Yang, Davis, J.E., Zhigeng Pan, 2011. Reliability Fusion of Time-of-Flight Depth and Stereo Geometry for High Quality Depth Maps. *IEEE Trans. Pattern Anal. Machine Intell.* 33, 1400–1414.

Kuhnert, K., Stommel, M., 2006. Fusion of Stereo-Camera and PMD-Camera Data for Real-Time Suited Precise 3D Environment Reconstruction, in:. *IEEE/RSJ International Conference on Intelligent Robots and Systems*, Beijing, China, pp. 4780–4785.

Lange, R., Seitz, P., 2001. Solid-state time-of-flight range camera. *IEEE J. Quantum Electron.* 37, 390–397.

Lowe, D.G., 2004. Distinctive image features from scale-invariant keypoints. *International Journal of Computer Vision* 60, 91–110.

Muhammad, N., Lacroix, S., 2010. Calibration of a rotating multi-beam lidar, *IEEE/RSJ International Conference on Intelligent Robots and Systems*, pp. 5648–5653.

Murray, D., Little, J.J., 2005. Patchlets: Representing Stereo Vision Data with Surface Elements. *Seventh IEEE Workshops on Application of Computer Vision.* Colorado, USA, Vol 1. , 192–199.

Muller, J-P., Anthony, A., 1987. Synergistic ranging systems for remote inspection of industrial objects *Proc. 2nd Industrial & Engineering Survey Conference,* held at University College London, 2-5 September 1987, pp270-291

Otto, G.P., Chau, T., 1989. "Region-growing" algorithm for matching of terrain images. *Image and Vision Computing* 7, 83–94.

Romero, L., Núñez, A., Bravo, S., Gamboa, L., 2004. Fusing a Laser Range Finder and a Stereo Vision System to Detect Obstacles in 3D, in: Lemaître, C., Reyes, C., González, J. (Eds.), *Lecture Notes in Computer Science*. Springer Berlin Heidelberg, pp. 555–561.

Shin, D., Muller, J.-P., 2012. Progressively weighted affine adaptive correlation matching for quasi-dense 3D reconstruction. *Pattern Recognition* 45, 3795–3809.

Song, Y., Glasbey, C., Heijden, G.A.M., Polder, G., Dieleman, J.A., 2011. Combining Stereo and Time-of-Flight Images with Application to Automatic Plant Phenotyping, in: Heyden, A., Kahl, F. (Eds.), *Lecture Notes in Computer Science*. Springer Berlin Heidelberg, pp. 467–478.

Thornhill, G., Rothery, D., Murray, J., Cook, A., Day, T.,
    Muller, J., & Iliffe, J., 1993. Topography of Apollinaris-
    Patera and Maadim-Vallis -Automated Extraction Of
    Digital Elevation Models. *Journal of Geophysical
    Research-Planets*, *98*(E12), 23581–23587
Zhang, Z., 2000. A flexible new technique for camera
    calibration. *IEEE Trans. Pattern Anal. Machine Intell.* 22,
    1330–1334.

# 21

# CROP GROUND COVER FRACTION AND CANOPY CHLOROPHYLL CONTENT MAPPING USING RAPIDEYE IMAGERY

E. Zillmann [a], M. Schönert[a], H. Lilienthal[b], B. Siegmann[c], T. Jarmer[c], P. Rosso[a], H. Weichelt [a]

[a] BlackBridge, Dept. of Application Research, 10719 Berlin, Germany - (erik.zillmann@blackbridge.com)
[b] Julius-Kühn-Institute (JKI), Federal Research Centre for Cultivated Plants, 38116 Braunschweig, Germany – (holger.lilienthal@jki.bund.de)
[c]Institute for Geoinformatics and Remote Sensing, University of Osnabrueck, 49076 Osnabrueck, Germany – (tjarmer@igf.uni-osnabrueck.de)

**KEY WORDS:** Ground Cover, Canopy Chlorophyll Content, RapidEye, Spatial Variability, Precision Agriculture

**ABSTRACT:**

Remote sensing is a suitable tool for estimating the spatial variability of crop canopy characteristics, such as canopy chlorophyll content (CCC) and green ground cover (GGC%), which are often used for crop productivity analysis and site-specific crop management. Empirical relationships exist between different vegetation indices (VI) and CCC and GGC% that allow spatial estimation of canopy characteristics from remote sensing imagery. However, the use of VIs is not suitable for an operational production of CCC and GGC% maps due to the limited transferability of derived empirical relationships to other regions. Thus, the operational value of crop status maps derived from remotely sensed data would be much higher if there was no need for re-parametrization of the approach for different situations.

This paper reports on the suitability of high-resolution RapidEye data for estimating crop development status of winter wheat over the growing season, and demonstrates two different approaches for mapping CCC and GGC%, which do not rely on empirical relationships. The final CCC map represents relative differences in CCC, which can be quickly calibrated to field specific conditions using SPAD chlorophyll meter readings at a few points. The prediction model is capable of predicting SPAD readings with an average accuracy of 77%. The GGC% map provides absolute values at any point in the field. A high $R^2$ value of 80% was obtained for the relationship between estimated and observed GGC%. The mean absolute error for each of the two acquisition dates was 5.3% and 8.7%, respectively.

## 1. INTRODUCTION

Remote sensing is a suitable tool for estimating the spatial variability of crop canopy characteristics such as green ground cover (GGC%) and canopy chlorophyll content (CCC). Both variables are often used for crop productivity analysis and site-specific crop management. Spatially high-resolution crop growth status information can provide farmers with relevant information e.g. for site-specific application of fertilizer (Scharf and Lory, 2002, Emerine, 2006), growth regulator (Maas et al. 2004), irrigation requirements (Hunsaker et al., 2005, Er-Raki, 2010), and crop productivity analysis (Schulthess et al., 2013). Since field management decisions are often time-critical, an almost real time production and provision of spatially high-resolution CCC and GGC% maps is desired.

Leaf chlorophyll absorption in the visible part of the electromagnetic spectrum provides the basis for using remotely sensed reflectance as a tool for the determination of crop development status. Often spectral vegetation indices (VI) are used to derive crop status information. Several studies have proven the existence of empirical relationships between different VIs and both CCC and GGC%. Even though the normalized difference vegetation index (NDVI) (Rouse et al. 1973) is the most commonly used VI, it has the limitation that it tends to saturate when LAI exceeds 2, and it is also strongly influenced by soil background conditions (Baret et al., 1991).

Several other VIs have been proposed for estimating CCC of various crops (Daughtry et al., 2000, Haboudane et al., 2002). In particular, the red-edge region of the spectrum showed strong potential for estimating canopy chlorophyll content. The main advantage of red-edge based indices is their reduced saturation effect due to a lower absorption by the chlorophyll in the red-edge spectral region compared to the red spectrum (Gitelson and Merzlyak, 1996). Thus, red-edge based indices are still sensitive to chlorophyll absorption at higher crop canopy densities. Since CCC varies widely over the growing season and among crops, any VI requires a large dynamic range for chlorophyll estimation. Eitel et al., (2007) have proven the general suitability of the RapidEye red-edge band for CCC estimation in winter wheat.

Accordingly, BlackBridge (as the owner and distributor of RapidEye data) has been using the Normalized Difference Red-Edge Index (NDRE) (Barnes et al., 2000), to produce relative chlorophyll maps. Although these maps are known to reflect the variability of CCC within individual fields, the variable nature of the NDRE-Chlorophyll relationship in different situations, has prevented the establishment of a universal relationship for estimating actual CCC values on the field. This limited transferability of empirical relationships hinders its incorporation into operational production processes.

One option to overcome this limitation is to explore the possibility of establishing a relationship NDRE-CCC on a case by case basis, but with a procedure that is simple enough to be applied by farmers with little effort and previous knowledge. With this goal in mind, in this study a test was performed to determine the accuracy of NDRE maps as predictors of values of CCC when using a set of a few measurements on the ground.

Often VIs used for fractional GGC% estimation rely on the soil line in the red and near-infrared (NIR) spectral feature space. Richardson and Wiegand (1977) introduced the bare soil line concept to improve the discrimination between bare soil and sparse vegetation cover. In particular the weighted difference vegetation index (WDVI) (Clevers, 1988) and the perpendicular vegetation index (PVI) (Jackson et al., 1980) were successfully tested to estimate GGC% from multi-spectral imagery (Bouman et al., 1992, Schulthess et al., 2013, Maas et al., 2008, Rajan et al., 2009).

Even though the basic concept of the PVI is to reduce the influence of bare soil reflection, this index cannot be considered fully insensitive to soil brightness. Huete et al. (1985) found out that for same amounts of vegetation cover PVI shows lower values on dark soils than on bright soils. Moreover, the initial assumption of an existing "global" soil line encompassing a wide range of soil conditions has been disproved. Soil type specific conditions cause variations in the slope and intercept of the soil line, and consequently influence the value of the particular VI (Huete et al., 1985). Therefore, regional specific soil lines are necessary to enable accurate GGC% estimation by utilizing soil line based VIs.

Maas et al. (2008) developed a non-empirical and self-calibrating approach for estimating GGC% based on the bare soil line and full canopy point (FCP) reflectance. The FCP is defined as the canopy reflectance in the NIR and the red spectral band at 100% ground coverage when seen directly from above. With parameters derived from the scatterplot of the NIR vs. the red band values of a particular multi-spectral image, they achieved a GGC% estimation error below 6% on average.

The operationalization of this approach for the GGC% map production relies on the automated determination of an adequate soil line and FCP in each particular image, which is highly error-prone without a proper image screening beforehand. The accidental inclusion of urban areas, lakes, clouds and cloud shadows prevents the accurate identification of the soil line (Maas et al., 2008, Xu et al., 2013). Therefore, one of the goals of the study reported here is to find a procedure to improve the automatic extraction of a soil line representative of the area under study.

The identification of the FCP requires the existence of full canopy within the image, which is not guaranteed for acquisitions early and late in the season. Furthermore, leaves transmit and reflect light in the NIR spectrum and absorb only a small fraction. As a result, the NIR reflection for a pixel of full canopy can continue to increase with increasing leaf density. Thus, it is very likely that automatically extracted NIR values are above the normal values of full canopy. To overcome this challenge, the use of an empirical FCP is recommended (Maas et al., 2008).

The aim of this work is to demonstrate the feasibility of the automated generation of relative canopy chlorophyll maps (CCC) and absolute GGC% maps for individual fields based on RapidEye imagery, with the least amount of manual intervention. The resulting maps were compared to corresponding ground truth measurements of chlorophyll content and GGC% to assess the accuracy. The produced maps provide information about the spatial variability of crop growth that has potential use in precision agriculture as a means for directed field scouting and variable rate management.

## 2. MATERIALS AND METHODS

### 2.1 Study Area

The study area is located in the federal state of Saxony-Anhalt, Germany (11°54′E, 51°47′N) in an intensively used agricultural landscape. The region is characterized by Chernozem in conjunction with Cambisols and Luvisols as the predominant soil types of the Loess covered Tertiary plain. The test site is characterized by highly variable spatial soil properties. Within the study area, one winter wheat (*Triticum aestivum L*) field

with a size of 90 ha was selected for the assessment of wheat CCC and GGC%. The field showed two areas with no vegetation as a result of waterlogging in early spring 2011.

### 2.2 Field Measurements and Data Extraction

Field data were collected close to image acquisition on the 8th of May and 22nd of June 2011. The first campaign was conducted within one day of image acquisition to avoid any distortions of the results due to high daily growth rates at this stage of crop development. The sample locations were defined aiming at covering the entire crop variability within the field as described in Siegmann et al. (2013). A total number of 24 and 18 sample plots were measured at winter wheat's stem elongation and early ripening stage, respectively (Figure 1).

Figure 1. Test site with the sampling locations measured on 8th of May (dark) and 22nd of June 2011 (bright). The image shows the situation on the 7th of May 2011.

#### 2.2.1 Ground Measurements of Chlorophyll

Field data collection included leaf area index (Licor LAI-2200©, Delta-T Sunscan©) and leaf chlorophyll meter readings in the upper canopy (Minolta SPAD-502©). SPAD measurements represent a unit-less relative measurement of leaf chlorophyll content and have been proven to be positively correlated to chlorophyll content of wheat (Reeves et al., 1993) and other crops (Zhu et al., 2012).

Since satellite images represent the spectral reflectance from 3-dimensional crop canopies, the SPAD readings at leaf level were also transformed to a 3-dimensional CCC. The $CCC_{SPAD}$ was derived by calculating the product of the corresponding leaf SPAD reading and leaf area index (Gitelson et al., 2005). Samples showed a considerable $CCC_{SPAD}$ range from 18 at minimum to 151 at maximum on the 8th of May 2011 and from 20 at minimum to 240 at maximum on the 22nd of June. The average $CCC_{SPAD}$ from the two dates ranged from 58 to 119.

#### 2.2.2 Ground Measurements of Ground Cover

Photographs of the wheat canopy were taken with a standard digital camera looking downward from a distance of approximately 1.5 m to allow the estimation of green crop ground cover. Photographs were subject to supervised classification aiming at the objective determination of reference GGC%. Trimble eCognition Developer 8© software was used to perform an object-based supervised classification of green vegetation and non-green vegetation. The image was cropped to include only the central portion for the GGC% determination to minimize the effects of optical distortions on the plant canopy present near the edges of the image. GGC% was calculated as the area of green vegetation of the resulting polygon shape file divided by the total area of the photograph.

The mean reference GGC% of the field was 71% on the 8th of May and 65% on the 22nd of June 2011. The variability of GGC% observed was considerably higher in June ranging from 23% at minimum and 95% at maximum compared to 42% at minimum and 96% at maximum in May.

### 2.3 Remote Sensing Data Processing

#### 2.3.1 Satellite Imagery

The RapidEye (RE) satellite system is a constellation of five identical earth observation satellites with the capability to provide large area, multi-spectral images with frequent revisits in high resolution (6.5 m at nadir). In addition to the blue (440–510 nm), green (520–590 nm), red (630–685 nm) and NIR (760–850 nm) bands, RapidEye has a red-edge band (690–730 nm), especially suitable for vegetation analysis. The RapidEye level 3A standard product covers an area of 25x25 km, and is radiometrically calibrated to radiance values (Anderson et al. 2013), ortho-rectified, and resampled to 5 m spatial resolution. All the images used were calibrated to top of atmosphere reflectance. The two images used (Tile ID 3363006) for crop status mapping were acquired on the 7th of May and 27th of June 2011.

#### 2.3.2 Chlorophyll Mapping

Since the relationship between canopy chlorophyll and the spectral VI used may vary between crop types or different areas, it is more appropriate to restrict the comparisons to individual crop fields. For this reason, the Chlorophyll Map focuses on differences within single fields, thus providing a relative chlorophyll level scale.

The NDRE was calculated for the entire satellite image (1), as:

$$NDRE = (\rho NIR - \rho REdge) / (\rho NIR + \rho REdge) \qquad (1)$$

where $\rho NIR$ and $\rho REdge$ are the reflectance values of the near infrared and red-edge spectral region.

The NDRE layer obtained was clipped to the test field area and all included pixels were encoded as a relative chlorophyll level index (RCLI) into a 0 – 100 grey value scale.

The relative chlorophyll values were calibrated to the field specific conditions by obtaining three ground measurements of $CCC_{SPAD}$ for each of the three chlorophyll level (low, moderate and high) areas previously delineated. A linear regression analysis between the $CCC_{SPAD}$ values and the corresponding relative chlorophyll map values allowed for generating a linear transfer function to be applied to the relative chlorophyll map in order to estimate the spatial distribution of CCC.

#### 2.3.3 Green Ground Cover Mapping

Green ground cover (GGC) is defined as the fraction of an area covered with green plant canopy. GGC percent maps are generated based on a modification of the original approach developed by Maas et al. (2008). The required bare soil line slope and intercept were obtained by calculating the arithmetic mean from automatically generated soil lines of a set of multi-temporal images using the slightly modified procedure from Fox et al. (2004). Images before and after the main vegetation period were used to guarantee a sufficient number of pixels representing bare soil. The empirical FCP reflectance was determined by averaging the reflectance values from multiple locations within the field known to be more than 90% covered with vegetation at the time of image acquisition.

The GGC% (2) was calculated from the ratio of the PVI (3) value to the corresponding full-canopy PVI ($PVI_{FC}$) value (4) as:

$$GGC\% = PVI / PVI_{FC} \qquad (2)$$
where
$$PVI = (\rho NIR - a * (\rho Red)) - b) / (1 + a^2)^{0.5} \qquad (3)$$
and
$$PVI_{FC} = (\rho NIR_{FC} - a * (\rho Red_{FC})) - b) / (1 + a^2)^{0.5} \qquad (4)$$

in which a and b are the slope and the intercept of the bare soil line respectively; and $\rho NIR$ and $\rho red$ are the reflectance values of the corresponding spectral band.

The final GGC% map expresses the percentage of ground covered by the crop green foliage (0%, no green vegetation, and 100%, ground entirely covered with green vegetation).

### 2.4 Accuracy Assessment

The sampling points were buffered with a radius of 10 m to extract the average estimated CCC ($CCC_{est}$) and GGC% values which were then stored in a shape file for subsequent analysis. Linear regression analysis between $CCC_{est}$ and ground measured $CCC_{SPAD}$, as well as between the estimated and observed GGC% was performed to assess the estimation accuracy, respectively.

### 3. RESULTS AND DISCUSSION

Correlation analysis between the $CCC_{SPAD}$ data obtained during the field sampling campaign and six spectral VIs derived from multi-spectral RapidEye imagery revealed best correlations for those indices incorporating the red-edge reflection (Table 1). The results revealed strong linear correlation between $CCC_{SPAD}$ and NDRE and $CI_{red-edge}$ for two different development stages of winter wheat.

Table 1. Correlation coefficient (r) between $CCC_{SPAD}$ of winter wheat and selected vegetation indices (n=24 in May; n=18 in June; level of significance, $p = 0.01$).

| Date | NDRE | MCARI | MTVI2 | $CI_{red-edge}$ | OSAVI | NDVI |
|------|------|-------|-------|-----------------|-------|------|
| May 2011 | 0.90 | 0.63 | 0.89 | 0.90 | 0.89 | 0.86 |
| June 2011 | 0.87 | 0.82 | 0.87 | 0.87 | 0.87 | 0.86 |

MCARI (Daughtry et al., 2000), MTVI2 (Haboudane et al., 2004), CIred-edge (Gitelson et al., 2003), OSAVI Rondeaux et al., 1996)

Figure 2 shows the generated RCLI map for the winter wheat field on 7th of May 2011. The dimensionless chlorophyll levels are represented in eight classes of colour tones representing relative differences in chlorophyll content within the field.

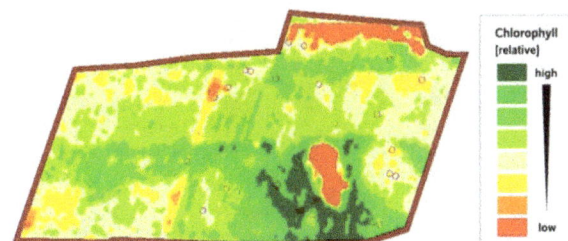

Figure 2. Spatial distribution of classes representing relative chlorophyll level differences in winter wheat on 7th of May 2011.

This relative map provides accurate information about the spatial variability of CCC within the field and facilitates directed field scouting to obtain SPAD and LAI measurements in specific areas of the field. These measurements allow for the calibration the RCLI map to field condition specific CCC$_{SPAD}$ values and enables the farmer to make in-season N fertilization rate decisions and applications.

Three CCC$_{SPAD}$ measurements corresponding to each of the three classes representing high, moderate, and low relative chlorophyll levels were selected and used to calibrate the relative chlorophyll values to CCC$_{SPAD}$ values. The relationship between relative chlorophyll levels and CCC$_{SPAD}$ for both dates is shown in Figure 3-A and Figure 4-A, respectively.

The derived linear transfer function was then used to estimate the spatial variability of CCC based on relative chlorophyll values derived from remotely sensed data. Results of estimating CCC$_{est}$ in the described way are plotted in Figure 3-B and Figure 4-B. The least-squares linear regression line fit to the points of CCC$_{est}$ vs. ground-based CCC$_{SPAD}$ observations tend to overestimate low CCC$_{SPAD}$ values and underestimate higher CCC$_{SPAD}$ values. Based on the prediction, the two outliers (Figure 4-B) should belong to the high CCC group of points but the observed values are significantly lower than that. From an inspection of the sample locations in the satellite images, no particular situation was found to explain these unexpected values. One certain possibility would be a field measurement error.

However, the resulting coefficients of determination ($R^2$) indicate that more than 75% of the total variance among the points can be explained by the models. This strong correlation shows that the procedure applied is able to provide information on the spatial variability of CCC.

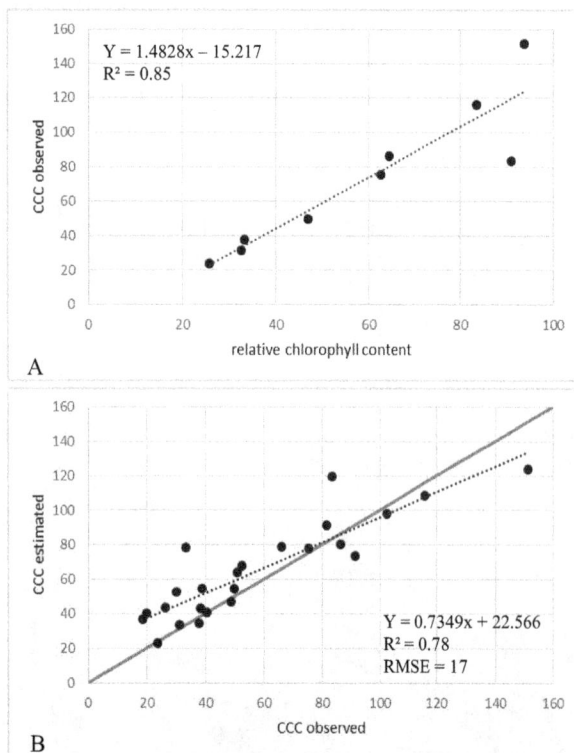

A

B

Figure 4. A: Calibration model for the June data set to turn relative chlorophyll level values into CCC$_{SPAD}$. B: Estimated CCC$_{est}$ plotted vs. the corresponding observed CCC$_{SPAD}$ measurements on June 27th 2011. The solid diagonal line represents the 1:1 line.

To better evaluate the fit of the model, RMSE, mean absolute error (MAE), and average accuracy (MAE as a percentage of observed mean plus 100) were calculated. On the 8th of May, the prediction accuracy yielded an RMSE of 17 and an MAE of 13 suggesting that on average the estimates of CCC$_{est}$ are within 13 SPAD units of their true values. On the 22nd of June the RMSE of 36 and MAE of 25 were slightly higher mainly due to the larger range of values. However, both models were capable of predicting CCC$_{SPAD}$ with an average accuracy of 78% and 77%.

The final accuracy of the CCC$_{est}$ strongly depends on the field measurements used for establishing the calibration model. In case of inadequate measurements, the CCC estimation results will be less accurate. Nevertheless, the results demonstrated that RapidEye imagery is suitable to provide sufficiently accurate relative chlorophyll level maps, which can potentially support site-specific management decisions.

The automatically derived soil line slope (a) and intercept (b) for multi-temporal RapidEye images of the project area are summarized in Table 2. Both parameters are variable over time. Since the area of study is the same in all images, the only remaining reasons for the observed variability could be haze, clouds, cloud shadows, or different non-photosynthetic crop residuals left on the fields. Additionally, an influence of different sensor viewing angles cannot be excluded.

The GGC% was estimated based on the two RapidEye images acquired on the 7th of May and 27th of June 2011 using equation (2) with the derived average slope of 1.05291 and average intercept of -0.04858 (Table 2) together with the reflectance values for the empirical FCP of winter wheat in the Red = 0.04 and NIR = 0.4 spectral band.

A

B

Figure 3. A: Calibration model to turn relative chlorophyll level values into CCC$_{SPAD}$ for the May data set. B: Estimated CCC$_{est}$ plotted vs. the corresponding observed CCC$_{SPAD}$ measurements on May 7th 2011. The solid diagonal line represents the 1:1 line.

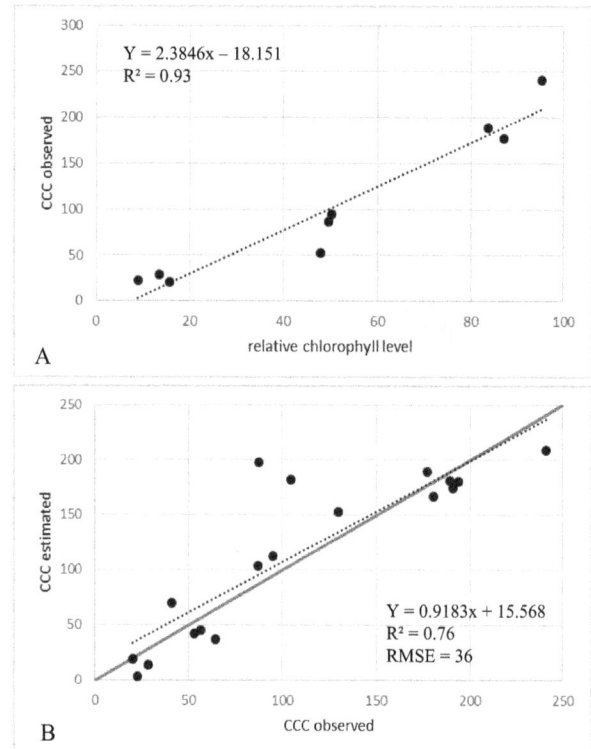

Table 2. Bare soil line characteristics of multi-temporal images and the mean values used for estimating GGC% from RapidEye images.

| Imaging date | Slope (a) | Intercept (b) |
|---|---|---|
| March 1st ,2011 | 1.1400 | -0.07887 |
| September 2nd , 2011 | 1.0463 | -0.05811 |
| September 24th, 2011 | 1.0627 | -0.04996 |
| October 2nd , 2011 | 0.95041 | -0.03103 |
| March 26th , 2012 | 0.98531 | -0.02531 |
| March 4th , 2013 | 1.02156 | -0.03717 |
| March 10th , 2014 | 1.17059 | -0.06735 |
| March 29th , 2014 | 1.12263 | -0.05741 |
| October 1st , 2014 | 0.97676 | -0.03198 |
| **Mean** | **1.05291** | **-0.04858** |
| STDEV | 0.07770 | 0.01836 |

The results revealed a strong correlation of r = 0.89 between estimated and observed GGC% for the sampling locations within the test field. Regression analysis indicated that 80% of the total variance among the points could be explained by the model (Figure 5) for the pooled data set with an RMSE of 9. The average estimated GGC% was 68%, which is about 2% higher than the average observed GGC%. The approach tends to overestimate lower GGC% and underestimate higher GGC% values. The derived MAE for the pooled data set was 6.8% (Table 3). This suggests that estimates of GGC% using the proposed approach based on an averaged regional soil line and an empirical FCP were on average within 7% of their true values.

Since field management decisions have to be made based on the actual crop status in the field, the GGC% estimation accuracy at each particular date is more important than the overall accuracy across the season. The observation date specific accuracy measures are reported in Table 3. For both dates, the calculated MAE values suggest that the GGC% estimations were on average within 6% and 9% of their true values.

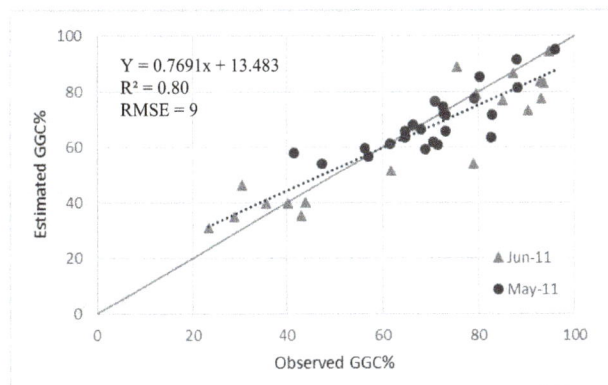

Figure 5. Scatter plot of GGC% estimated from RapidEye imagery vs. corresponding in-situ measurements of GGC% for pooled data of the May and June campaign.

Table 3: Accuracy measures for GGC% estimations based on RapidEye imagery using a regional specific average soil line and an empirical full canopy point reflectance.

| | R² | MAE | RMSE |
|---|---|---|---|
| May 7th, 2011 | 0.61 | 5.3 | 7.3 |
| June 27th , 2011 | 0.85 | 8.7 | 10.9 |
| Pooled data | 0.80 | 6.8 | 9.0 |

Even though the use of an averaged soil line and empirical FCP deprives, to a certain extent, the original GGC% estimation procedure of the self-calibration capability as its main advantage, the procedure is still able to accurately estimate GGC%.

Figure 6 shows the GGC% maps produced for the test field based on RapidEye images acquired on the 7th of May and 27th of June 2011. In early May, the GGC% shows considerable spatial variation in the winter wheat canopy, which might be suitable for variable rate management. The areas with the lowest GGC% values correspond to the areas affected by waterlogging in early spring. In June, the wheat canopy shows more large-scale variation in GGC% compared to May. The larger areas with lower GGC% represent areas affected by drought-stress-induced abnormally premature ripening due to extreme weather conditions in early summer 2011.

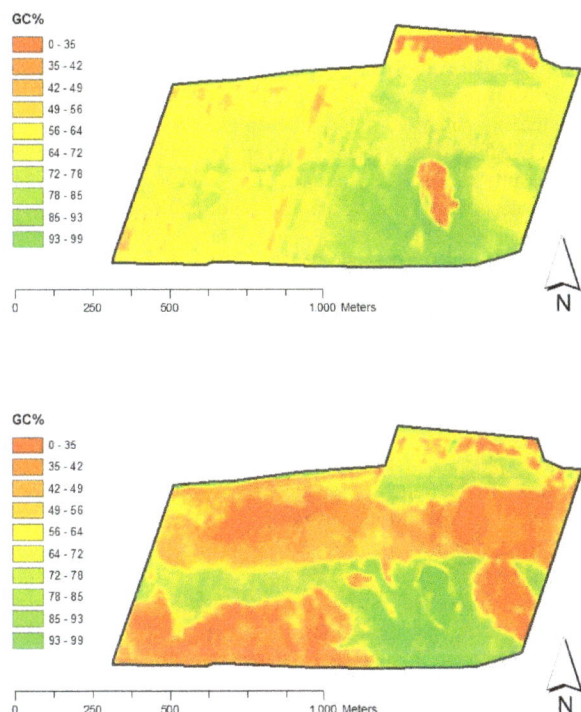

Figure 6. GGC% maps for winter wheat derived from RapidEye imagery acquired on 7th of May 2011 (top) and on 27th of June 2011 (bottom).

The successful use of a pre-defined, region-specific soil line and FCP parameters instead of determining them directly from the image can probably be attributed to the stability of the sensor calibration over time. Nevertheless, the influence of haze or other atmospheric effects on the GGC% estimation accuracy have to be investigated further. In this context, focus has to be placed especially on the transferability of the empirical full canopy point reflectance across regions, seasons, and crop types. Since it is known that VIs show different values for crops with planophile and erectrophile leaf canopy architecture, the proposed utilization of an empirical FCP has to be tested on crops other than winter wheat to help reduce estimation errors.

## 4. CONCLUSION

Results demonstrated that high-resolution RapidEye imagery is suitable for providing accurate spatial information on CCC and GGC% during the growing season, and is potentially useful for site-specific crop management.

Relationships between CCC$_{SPAD}$ and red-edge based VIs showed highest correlations for different development stages of winter wheat. Since such empirical relationships are crop type or even cultivar specific, it is almost impossible to achieve reliable and accurate estimations solely from reflectance measurements. A method has been presented that is capable of providing spatially accurate and fast relative chlorophyll level maps, which allow for directed field scouting. The cultivar specific calibration can be realized afterwards at low cost by obtaining a small number of in-situ measurements in different areas representing high, moderate, and low relative chlorophyll content using a portable chlorophyll-meter. This approach is capable of accurately estimating CCC$_{SPAD}$ with an average accuracy of 77%. Such chlorophyll maps provide valuable complementary information for the derivation of site-specific N-fertilizer recommendations. Furthermore, in conjunction with soil fertility and yield maps, chlorophyll level maps can aid in the delineation of site-specific management zones.

The results of this study showed that GGC% of winter wheat can be accurately estimated from RapidEye images based on a region-specific bare soil line and an empirical reflectance of crop full canopy. Effective GGC% prediction of winter wheat across the growing season yielded a coefficient of determination of $R^2 = 0.8$. On average, the estimated GGC% values were within 7% of their true values.

The great operational value of the proposed procedure is that it is does not rely on any empirical relationship. According to Rajan et al (2009) the major advantage of such GGC% maps is that they provide an absolute measure of crop canopy density at any point in the field, allowing for direct comparison of GGC% derived from multi-temporal images. Therefore, GGC% maps enable the monitoring of seasonal crop growth dynamics within individual fields.

RapidEye's short revisit cycle and the red-edge channel make the satellite constellation unique for agricultural monitoring. Its capability for simultaneous acquisitions of crop status information for large areas can greatly reduce the workload for conducting field surveys and the crop analysis necessary to obtain relevant input for precision agriculture.

## 5. ACKNOWLEDGEMENT

The authors wish to express appreciation to the Julius-Kühn-Institute (Federal Research Centre for Cultivated Plants in Germany) and the Institute for Geoinformatics and Remote Sensing of the University of Osnabrueck for kindly supplying the field data used in this study, based on the "Hyland" project supported under code 50 EE 1014.

## REFERENCES

Anderson, C., Thiele, M., Brunn, A., 2013. Calibration and validation of the RapidEye constellation. In: *Anais XVI Simpósio Brasileiro de Sensoriamento Remoto* - SBSR, Foz do Iguaçu, PR, Brasil, 13 a 18 de abril de 2013.

Baret, F., Guyot, G., 1991. Potentials and limits of vegetation indices for LAI and APAR assessment. *Rem. Sens. of Environ.* 36:161.

Barnes, E.M., Clarke, T.R., Richards, E.S., Colaizzi, P.D., Haberland, J.,. Moran, M.S., 2000. Coincident detection of crop water stress, nitrogen status and canopy density using ground-based multispectral data. In: *Proc. 5th Int. Conference on Precision Agriculture*. Bloomington, MN, USA.

Bouman, B.A.M., van Kasteren, H.W.J., Uenk, D., 1992. Standard relations to estimate ground cover and LAI of agricultural crops from reflectance measurements. Eur. J. Agron. 1, pp. 249–262.

Clevers, J.G.P.W., 1988. The derivation of a simplified reflectance model for the estimation of leaf area index. *Rem. Sens. of Environ.*, 25, pp.53-69.

Daughtry, C.S., Walthall, C.L., Kim, M.S., Brown De Colstoun, E., McMurtrey, J. E., 2000. Estimating corn leaf chlorophyll concentration from leaf and canopy reflectance. *Rem. Sens. of Environ.*, 74, pp. 229–239.

Eitel, J.U.H., Long, D.S., Gessler, P.E.; Smith, A.M.S., 2007. Using in-situ measurements to evaluate the new RapidEye satellite series for prediction of wheat nitrogen status. *Int. Journal of Remote Sensing*, 28, pp. 1-8.

Emerine, D.M., 2006. Variable and reduced rate nitrogen application based on multispectral aerial imagery and directed field sampling. In: D. Richter (Ed.), *Proc. of the 2006 beltwide cotton conference*. San Antonio, TX, 3–6 Jan 2006. Memphis, TN, pp. 2128–2131.

Er-Raki, S., Chehbouni, A., Duchemin, B, 2010. Combining Satellite Remote Sensing Data with the FAO-56 Dual Approach for Water Use Mapping. In: *Irrigated Wheat Fields of a Semi-Arid Region*. Remote Sens., 2, pp. 375-387.

Fox, G.A., Sabbagh, G.J., Searcy, S.W., Yang, C., 2004. An automated soil line identification routine for remotely sensed images. *Soil Sci. Soc. Am. J.*, 68, pp. 1326–1331.

Gitelson, A.A., Merzlyak, M.N., 1996. Signature analysis of leaf reflectance spectra: Algorithm development for remote sensing of chlorophyll. J. of Plant Phys. 148(3-4), pp. 494-500.

Gitelson, A.A., Merzlyak, M.N., 2003. Relationships between leaf chlorophyll content and spectral reflectance and algorithms for non-destructive chlorophyll assessment in higher plant leaves. *J. of Plant Phys.*, 160(3), pp. 271-282.

Gitelson AA, Vina A., Rundquist, D.C., Ciganda, V., Arkebauer T.J., 2005. Remote estimation of canopy chlorophyll content in crops. *Geophys Res Lett*; 32.

Haboudane, D., Miller, J.R., Tremblay, N., Zarco-Tejadad, P.J., Dextrazec, L., 2002. Integrated narrow-band vegetation indices for prediction of crop chlorophyll content for application to precision agriculture. *Rem. Sens. of Environ.*, 81, pp. 416–426.

Haboudane, D., Miller, J.R., Pattey, E., Zarco-Tejada, P.J., Strachan, I.B., 2004. Hyperspectral vegetation indices and novel algorithms for predicting green LAI of crop canopies: Modeling and validation in the context of precision agriculture. *Rem. Sens. of Environ.*, 90, pp. 337-352.

Huete, A.R., 1985. Spectral response of a plant canopy with different soil backgrounds. *Rem. Sens. of Environ.* 17, pp. 37-53.

Hunsaker, D.J., Barnes, E.M., Clarke, T.R., Fitzgerald, G.J., Pinter, P.J., 2005. Cotton irrigation scheduling using remotely sensed and FAO-56 basal crop coefficients. *American Society of Agricultural Engineers*, St.Joseph, MI.

Jackson, R.D., Pinter, P.J., Paul, J., Reginato, R.J., Robert, J., Idso, S.B., 1980. Hand-held radiometry. *Agricultural Reviews and Manuals ARM-W-19*. Oakland, California: U.S.

Maas, S.J., Brightbill, J., Hooton, J., 2004. Remote sensing for precision agriculture in the Texas High Plains. In: P. Dugger, D. Richter (Eds.), *Proc. of the 2004 beltwide cotton conferences* San Antonio, TX, 5–9 Jan 2004. Memphis, TN, pp. 184–187.

Maas, S.J., Rajan, N., 2008. Estimating ground cover of field crops using medium-resolution multispectral satellite imagery. *Agron. J.* 100, pp. 320–327.

Rajan, N., Maas, S.J., 2009. Mapping crop ground cover using airborne multispectral digital imagery. *Precision Agriculture* 10(4), pp. 304-318.

Reeves, D.W., Mask, P.L., Wood, C.W., Delano, D.P., 1993. Determination of Wheat Nitrogen Status with Hand-held Chlorophyll Meter: Influence of Management Practices, *J. Plant Nutrition*, 16, pp. 781–796.

Richardson, A.J., Wiegand, C.L., 1977. Distinguishing vegetation from soil background information. *Photogrammetric Engineering and Remote Sensing,* 43(12), pp. 1541-1552.

Rondeaux, G., Steven, M., Baret, F., 1996. Optimization of Soil adjusted vegetation indices. *Rem. Sens. of Environ.*, 55, pp. 95-107.

Rouse, J.W., Haas, R.H., Schell, J.A., Deering, D.W., 1973. Monitoring vegetation systems in the Great Plains with ERTS, Third ERTS Symposium, NASA SP-351 I, pp. 309-317.

Scharf, P.C., Lory, J.A., 2002. Calibrating corn color from aerial photographs to predict sidedress nitrogen need. Contrib. from the Missouri Agric. Exp. Stn. J. Ser. No.13086. *Agron. J.* 94, pp. 397–404.

Schulthess, U., Timsina, J., Herrera, J. M., McDonald, A., 2013. Mapping field-scale yield gaps for maize: An example from Bangladesh, *Field Crops Research*, 143(1) pp. 151-156.

Siegmann, B., Jarmer, T., Lilienthal, H., Richter, N., Selige, T., Höfle, B., 2013. Comparison of narrow band vegetation indices and empirical models from hyperspectral remote sensing data for the assessment of wheat nitrogen concentration. In: *Proc. 8th EARSeL SIG IS workshop*, Nantes, France.

Xu D, Guo X., 2013. A Study of Soil Line Simulation from Landsat Images in Mixed Grassland. *Remote Sens.*, 5(9), pp. 4533-4550.

Zhu, J., Tremblay, N., Liang, Y., 2012. Comparing SPAD and atLEAF values for chlorophyll assessment in crop species, *Canadian J. of Soil Science*, 92, pp. 645-648.

# 22

# A DENOISING OF BIOMEDICAL IMAGES

Thanh D.N.H[a], Dvoenko S.D.[a]

[a] Tula State University, Institute of Applied Mathematics and Computer Sciences, Lenin ave. 92 Tula city, Russian Federation - (dnhthanh@hueic.edu.vn, dsd@tsu.tula.ru)

**Commission VI, WG VI/4**

**KEY WORDS:** Total variation, ROF model, Gaussian noise, Poisson noise, Mixed Poisson-Gaussian noise, Image processing, Biomedical image, Euler-Lagrange equation.

**ABSTRACT:**

Today imaging science has an important development and has many applications in different fields of life. The researched object of imaging science is digital image that can be created by many digital devices. Biomedical image is one of types of digital images. One of the limits of using digital devices to create digital images is noise. Noise reduces the image quality. It appears in almost types of images, including biomedical images too. The type of noise in this case can be considered as combination of Gaussian and Poisson noises. In this paper we propose method to remove noise by using total variation. Our method is developed with the goal to combine two famous models: ROF for removing Gaussian noise and modified ROF for removing Poisson noise. As a result, our proposed method can be also applied to remove Gaussian or Poisson noise separately. The proposed method can be applied in two cases: with given parameters (generated noise for artificial images) or automatically evaluated parameters (unknown noise for real images).

## 1    INTRODUCTION

One of the important types of digital technique that has many applications in many fields of life is digital image. It is a type of a signal that is obtained from a real analogous signal by discretization and quantization. There are many devices can create digital images, such as digital camera, X-ray scanner, and so on. Ordinarily these devices can give unexpected effects. One of them is noise. Noise reduces image quality and efficiency of image processing.

The problem of noise removal from digital images is very actual today. In order to remove noise more effectively, we need to classify it. There are many types of noise, for example, Gaussian noise (almost for digital image by using digital camera), Poisson noise (for X-ray image), speckle noise (for ultrasonogram), and so on.

There are many developed strong approaches to solve noise removal problem. The approach that uses total variation (Chan, 2005; Burger, 2008; Chambolle, 2009; Xu, 2014; Rankovic, 2012; Lysaker, 2006; Li, 2006; Zhu, 2012; Tran, 2012; Getreuer, 2012; Caselles, 2011; Rudin, 1992; Chen, 2013) is well-known and very promising.

Rudin (1992) is pioneer to apply this concept. He proposed the total variation to solve many problems in image processing. Especially, he built a model to remove noise on digital images. This model is named as ROF (Rudin, 1992; Chen, 2013).

However, ROF model is usually used to remove only Gaussian noise. Of course it can also remove other types of noise, but not very effectively. Another popular noise in medical images is Poisson noise. For example, this noise appears in medical X-ray images. ROF model cannot treat this noise effectively. Therefore, Le T. (2007) developed so called modified ROF model.

Both of Gaussian and Poisson noises is popular, but their combination is also important (Luisier, 2011). This combination of noises usually appears in biomedical images, for example, in electronic microscopy images (Jezierska, 2011; Jezierska 2012).

As we talk above, ROF and modified ROF models ineffectively treat this combination. ROF model gives priority to Gaussian noise, but modified ROF model gives it to Poisson noise.

In order to treat this combination of noise, we will combine ROF model (for Gaussian noise) and modified ROF model (for Poisson noise). Our model will treat this combination by considering proportion of noise between them.

In experiments, we used a real image and add noise into them. We performed denoising of digital images by proposed method and other methods, such as ROF model, median filter (Wang, 2012) and Wiener filter (Abe, 2012). In order to evaluate an image quality after denoising, we used well-known criteria MSE (Mean Square Error), PSNR (Peak Signal-to-Noise Ratio) and SSIM (Structure SIMilarity) (Wang, 2004; Wang, 2006). We give priority to PSNR, because it is most popular and used to evaluate the quality of restored signal in signal processing in general, and in image processing, especially.

## 2    DENOISING MODEL FOR MIXED POISSON-GAUSSIAN NOISE

Let in $R^2$ space a bounded domain $\Omega \subset R^2$ be given. Let us call functions $u(x,y) \in R^2$ and $v(x,y) \in R^2$, respectively, ideal (without noise) and observed images (noisy), where $(x,y) \in \Omega$.

If the function $u$ is smooth, then its total variation is defined by

$$V_T[u] = \int_\Omega |\nabla u| \, dxdy \,,$$

where $\nabla u = (u_x, u_y)$ is a gradient (nabla operator), $u_x = \partial u / \partial x$, $u_y = \partial u / \partial y$, $|\nabla u| = \sqrt{u_x^2 + u_y^2}$. In this paper, we only consider function $u$ that always has limited total variation $V_T[u] < \infty$.

## 2.1 Denoising Model

According to results (Chang, 2005; Burger, 2008; Rudin, 1992; Chen, 2013; Scherzer, 2009), image smoothness is characterised by the total variation. The total variation of noisy image is always greater than the total variation of smoothed image.

When Rudin solved the problem $V_T[u] \to \min$, he used this characteristic and assumed, that Gaussian noise variance is fixed by the additional constraint

$$\int_\Omega (u-v)^2 \, dxdy = const .$$

He proposed the ROF model to remove Gaussian noise from an image

$$u^* = \arg\min_u \left( \int_\Omega |\nabla u| \, dxdy + \frac{\lambda}{2} \int_\Omega (u-v)^2 \, dxdy \right) ,$$

where $\lambda > 0$ is a Lagrange multiplier.

Le T. (2007) proposed another model to remove Poisson noise based on ROF model:

$$u^* = \arg\min_u \left( \int_\Omega |\nabla u| \, dxdy + \beta \int_\Omega (u-v\ln(u)) \, dxdy \right) ,$$

where $\beta$ is a regularization coefficient. We call it a modified ROF model for Poisson noise.

In order to develop the denoising model for mixed noise, we also solve the problem based on the smooth characteristic of the total variation

$$V_T[u] \to \min .$$

And we also define a constrained condition. We assume that with given image $u$, the mixed noise in image is fixed too (Poisson noise is unchangeable, and Gaussian noise only depends on noise variance):

$$\int_\Omega \ln(p(v\,|\,u)) \, dxdy = const , \quad (1)$$

where $p(v\,|\,u)$ is a conditional probability.

Let us consider Gaussian noise. Its probability density function (p.d.f.) is

$$p_1(v\,|\,u) = \exp\left( -\frac{(v-u)^2}{2\sigma^2} \right) / (\sigma\sqrt{2\pi}) .$$

For Poisson noise the p.d.f. is

$$p_2(v\,|\,u) = \frac{\exp(-u)u^v}{v!} .$$

We have to note that intensity levels of image colours are integer (for example, the intensity interval for an 8-bit grayscale image is from 0 to 255), so we regard $u$ as an integer value, but this will ultimately be unnecessary (Le 2007).

In order to treat combination of Gaussian and Poisson noises, we assume the following linear combination

$$\ln(p(v\,|\,u)) = \lambda_1 \ln(p_1(v\,|\,u)) + \lambda_2 \ln(p_2(v\,|\,u)) ,$$

where $\lambda_1 > 0$, $\lambda_2 > 0$, $\lambda_1 + \lambda_2 = 1$.

According to (1), we obtain the denoising problem with constrained condition as following:

$$\begin{cases} u^* = \arg\min_u \int_\Omega |\nabla u| \, dxdy \\ \int_\Omega \left( \frac{\lambda_1}{2\sigma^2}(v-u)^2 + \lambda_2(u-v\ln(u)) \right) dxdy = \kappa , \end{cases}$$

where $\kappa$ is a constant value.

We can transform this constrained optimization problem to the unconstrained optimization problem by using Lagrange functional

$$L(u,\tau) = \int_\Omega |\nabla u| \, dxdy + \tau \left( \frac{\lambda_1}{2\sigma^2} \int_\Omega (v-u)^2 \, dxdy + \right.$$
$$\left. \lambda_2 \int_\Omega (u-v\ln(u)) \, dxdy - \kappa \right)$$

to find

$$(u^*,\tau^*) = \arg\min_{u,\tau} L(u,\tau) , \quad (2)$$

where $\tau > 0$ is a Lagrange multiplier.

This is our proposed model to remove mixed Poisson-Gaussian noise from digital image. We have to notice that, if $\lambda_1 = 0$ and $\beta = \lambda_2\tau$, we obtain modified ROF model for removing Poisson noise. If $\lambda_2 = 0$ and $\lambda = \lambda_1/(2\sigma^2)$, then we obtain ROF model for removing Gaussian noise. In the case of $\lambda_1 > 0, \lambda_2 > 0$ we get the model for removing mixed Poisson-Gaussian noise.

## 2.2 Model Discretization

In order to solve the problem (2), we can use the Lagrange multipliers method (Zeidler, 1985; Rubinov, 2003; Gill, 1974).

However, in this paper, we will solve it by using the Euler-Lagrange equation (Zeidler, 1985).

Let function $f(x,y)$ be defined in limited domain $\Omega \subset R^2$ and be the second-order continuous differentiable one by $x$ and $y$ for $(x,y) \in \Omega$.

We consider the special convex functional $F(x,y,f,f_x,f_y)$, where $f_x = \partial f / \partial x$, $f_y = \partial f / \partial y$.

The solution of the optimization problem

$$\int_\Omega F(x,y,f,f_x,f_y) \, dxdy \to \min$$

satisfies the following Euler-Lagrange equation

$$F_f(x,y,f,f_x,f_y) - \frac{\partial}{\partial x} F_{f_x}(x,y,f,f_x,f_y) - \frac{\partial}{\partial y} F_{f_y}(x,y,f,f_x,f_y) = 0 ,$$

where

$$F_f = \partial F / \partial f , \quad F_{f_x} = \partial F / \partial f_x , \quad F_{f_y} = \partial F / \partial f_y .$$

We use the result above to solve the problem (2). The solution of the problem (2) is given by the following Euler-Lagrange equation:

$$-\frac{\lambda_1}{\sigma^2}(v-u) + \lambda_2\left(1 - \frac{v}{u}\right) - $$
$$\mu\frac{\partial}{\partial x}\left( \frac{u_x}{\sqrt{u_x^2 + u_y^2}} \right) - \mu\frac{\partial}{\partial y}\left( \frac{u_y}{\sqrt{u_x^2 + u_y^2}} \right) = 0, . \quad (3)$$

where $\mu = 1/\tau$. We can reduce (3) to

$$\frac{\lambda_1}{\sigma^2}(v-u) - \lambda_2\left(1 - \frac{v}{u}\right) + \quad (4)$$

$$\mu \frac{u_{xx}u_y^2 - 2u_xu_yu_{xy} + u_x^2u_{yy}}{(u_x^2 + u_y^2)^{3/2}} = 0,$$

where

$$u_{xx} = \frac{\partial^2 u}{\partial x^2}, \ u_{yy} = \frac{\partial^2 u}{\partial y^2},$$

$$u_{xy} = \frac{\partial}{\partial x}\left(\frac{\partial u}{\partial y}\right) = \frac{\partial}{\partial y}\left(\frac{\partial u}{\partial x}\right) = u_{yx}.$$

In order to discretize the equation (4), we add an artificial time parameter and consider the function $u = u(x,y,t)$. Then the equation (4) relates to the following diffusion equation

$$u_t = \frac{\lambda_1}{\sigma^2}(v - u) - \lambda_2(1 - \frac{v}{u}) +$$

$$\mu \frac{u_{xx}u_y^2 - 2u_xu_yu_{xy} + u_x^2u_{yy}}{(u_x^2 + u_y^2)^{3/2}}, \tag{5}$$

where $u_t = \partial u / \partial t$.

We can write the discretized form of the equation (5) as following:

$$u_{ij}^{k+1} = u_{ij}^k + \xi\left(\frac{\lambda_1}{\sigma^2}(v_{ij} - u_{ij}^k) - \right.$$

$$\left. \lambda_2(1 - \frac{v_{ij}}{u_{ij}^k}) + \mu\varphi_{ij}^k\right), \tag{6}$$

where

$$\varphi_{ij}^k = \frac{\nabla_{xx}(u_{ij}^k)(\nabla_y(u_{ij}^k))^2}{((\nabla_x(u_{ij}^k))^2 + (\nabla_y(u_{ij}^k))^2)^{3/2}} +$$

$$\frac{-2\nabla_x(u_{ij}^k)\nabla_y(u_{ij}^k)\nabla_{xy}(u_{ij}^k) + (\nabla_x(u_{ij}^k))^2\nabla_{yy}(u_{ij}^k)}{((\nabla_x(u_{ij}^k))^2 + (\nabla_y(u_{ij}^k))^2)^{3/2}},$$

$$\nabla_x(u_{ij}^k) = \frac{u_{i+1,j}^k - u_{i-1,j}^k}{2\Delta x}, \ \nabla_y(u_{ij}^k) = \frac{u_{i,j+1}^k - u_{i,j-1}^k}{2\Delta y},$$

$$\nabla_{xx}(u_{ij}^k) = \frac{u_{i+1,j}^k - 2u_{ij}^k + u_{i-1,j}^k}{(\Delta x)^2},$$

$$\nabla_{yy}(u_{ij}^k) = \frac{u_{i,j+1}^k - 2u_{ij}^k + u_{i,j-1}^k}{(\Delta y)^2},$$

$$\nabla_{xy}(u_{ij}^k) = \frac{u_{i+1,j+1}^k - u_{i+1,j-1}^k - u_{i-1,j+1}^k + u_{i-1,j-1}^k}{4\Delta x\Delta y},$$

$$u_{0j}^k = u_{1j}^k; \ u_{N_1+1,j}^k = u_{N_1,j}^k; \ u_{i0}^k = u_{i1}^k; \ u_{i,N_2+1}^k = u_{i,N_2}^k;$$

$$i = 1,...,N_1; j = 1,...,N_2;$$

$$k = 0,1,...,K; \Delta x = \Delta y = 1; \ 0 < \xi < 1.$$

Here $K$ is enough great number. In this paper, we use $K = 500$.

## 2.3    Finding Optimal Parameters

We can use the procedure (6) to perform image denoising. In this procedure, values of parameters $\lambda_1, \lambda_2, \mu, \sigma$ need to be given. In some cases, we have to define these parameters to perform image denoising automatically. Then parameters $\lambda_1, \lambda_2, \mu$ in process (6) need to be changed into $\lambda_1^k, \lambda_2^k, \mu^k$ for each step $k$. So we obtain new procedure that allows us to

calculate values of these parameters automatically in iteration steps.

### 2.3.1    Optimal Parameters $\lambda_1$ and $\lambda_2$

Let $(u, \tau)$ be a solution of the problem (2). Then we get the condition

$$\frac{\partial L(u, \tau)}{\partial u} = 0.$$

This condition gives us the optimal parameters $\lambda_1, \lambda_2$:

$$\lambda_1 = \frac{\int\limits_{\Omega}(1 - \frac{v}{u})dxdy}{\frac{1}{\sigma^2}\int\limits_{\Omega}(v - u)dxdy + \int\limits_{\Omega}(1 - \frac{v}{u})dxdy},$$

$$\lambda_2 = 1 - \lambda_1.$$

Its discretized form is

$$\lambda_1^k = \frac{\sum\limits_{i=1}^{N_1}\sum\limits_{j=1}^{N_2}(1 - \frac{v_{ij}}{u_{ij}^k})}{\sum\limits_{i=1}^{N_1}\sum\limits_{j=1}^{N_2}(\frac{v_{ij} - u_{ij}^k}{\sigma^2} + 1 - \frac{v_{ij}}{u_{ij}^k})},$$

$$\lambda_2^k = 1 - \lambda_1^k,$$

where $k = 0,1,...,K$.

### 2.3.2    Optimal Parameter $\mu$

In order to find an optimal parameter $\mu$, we multiply (3) by $(u - v)$ and integrate by parts over $\Omega$. Finally, we obtain the formula to find the optimal parameter $\mu$:

$$\mu = \frac{\int\limits_{\Omega}(-\frac{\lambda_1}{\sigma^2}(u - v)^2 - \lambda_2\frac{(u - v)^2}{u})dxdy}{\int\limits_{\Omega}(\sqrt{u_x^2 + u_y^2} - \frac{u_xv_x + u_yv_y}{\sqrt{u_x^2 + u_y^2}})dxdy}.$$

Its discretized form is

$$\mu^k = \frac{\sum\limits_{i=1}^{N_1}\sum\limits_{j=1}^{N_2}(-\frac{\lambda_1^k}{\sigma^2}(u_{ij}^k - v_{ij})^2 - \lambda_2^k\frac{(u_{ij}^k - v_{ij})^2}{u_{ij}^k})}{\sum\limits_{i=1}^{N_1}\sum\limits_{j=1}^{N_2}\eta_{ij}^k},$$

where

$$\eta_{ij}^k = \sqrt{(\nabla_x(u_{ij}^k))^2 + (\nabla_y(u_{ij}^k))^2} -$$

$$\frac{\nabla_x(u_{ij}^k)\nabla_x(v_{ij}) + \nabla_y(u_{ij}^k)\nabla_y(v_{ij})}{\sqrt{(\nabla_x(u_{ij}^k))^2 + (\nabla_y(u_{ij}^k))^2}},$$

$$\nabla_x(u_{ij}^k) = \frac{u_{i+1,j}^k - u_{i-1,j}^k}{2\Delta x}, \ \nabla_y(u_{ij}^k) = \frac{u_{i,j+1}^k - u_{i,j-1}^k}{2\Delta y},$$

$$\nabla_x(v_{ij}^k) = \frac{v_{i+1,j}^k - v_{i-1,j}^k}{2\Delta x}v, \ \nabla_y(v_{ij}^k) = \frac{v_{i,j+1}^k - v_{i,j-1}^k}{2\Delta y},$$

$$u_{0j}^k = u_{1j}^k; \ u_{N_1+1,j}^k = u_{N_1,j}^k; \ u_{i0}^k = u_{i1}^k; \ u_{i,N_2+1}^k = u_{i,N_2}^k;$$

$$v_{0j} = v_{1j}; \ v_{N_1+1,j} = v_{N_1,j}; \ v_{i0} = v_{i1}; \ v_{i,N_2+1} = v_{i,N_2};$$

$$i = 1,...,N_1; \; j = 1,...,N_2; \; k = 0,1,...,K; \; \Delta x = \Delta y = 1 .$$

### 2.3.3  Optimal Parameter $\sigma$

In order to evaluate this parameter $\sigma$, we use the result of Immerker (1996):

$$\sigma = \frac{\sqrt{\pi/2}}{6(N_1-2)(N_2-2)} \sum_{i=1}^{N_1}\sum_{j=1}^{N_2} |u_{ij} * \Lambda| , \text{ where}$$

$$\Lambda = \begin{pmatrix} 1 & -2 & 1 \\ -2 & 4 & -2 \\ 1 & -2 & 1 \end{pmatrix} \text{ is the mask of an image.}$$

Operator $*$ is a convolution operator, where

$$u_{ij} * \Lambda = u_{i-1,j-1}\Lambda_{33} + u_{i,j-1}\Lambda_{32} + u_{i+1,j-1}\Lambda_{31} + u_{i-1,j}\Lambda_{23} +$$

$$u_{ij}\Lambda_{22} + u_{i+1,j}\Lambda_{21} + u_{i-1,j+1}\Lambda_{13} + u_{i,j+1}\Lambda_{12} + u_{i+1,j+1}\Lambda_{11},$$

$$i = 1,...,N_1; j = 1,...,N_2;$$

$$u_{ij} = 0 , \text{ if } i = 0, \text{ or } j = 0, \text{ or } i = N_1 + 1 ,$$

$$\text{or } j = N_2 + 1 .$$

We have to notice, that the parameter $\sigma$ is just evaluated at first time of the iteration process.

### 2.4  Image Quality Evaluation

In order to evaluate image quality after denoising, we use criteria MSE (Mean Square Error), PSNR (Peak Signal-to-Noise Ratio) and SSIM (Structure SIMilarity) (Wang 2004, 2006):

$$Q_{MSE} = \frac{1}{N_1 N_2} \sum_{i=1}^{N_1}\sum_{j=1}^{N_2} (u_{ij} - v_{ij})^2 ,$$

$$Q_{PSNR} = 10\lg\left(\frac{N_1 N_2 L^2}{\sum_{i=1}^{N_1}\sum_{j=1}^{N_2}(u_{ij}-v_{ij})^2}\right),$$

$$Q_{SSIM} = \frac{(2\bar{u}\bar{v} + C_1)(2\sigma_{uv} + C_2)}{(\bar{u}^2 + \bar{v}^2 + C_1)(\sigma_u^2 + \sigma_v^2 + C_2)} ,$$

where

$$\bar{u} = \frac{1}{N_1 N_2}\sum_{i=1}^{N_1}\sum_{j=1}^{N_2} u_{ij} , \; \bar{v} = \frac{1}{N_1 N_2}\sum_{i=1}^{N_1}\sum_{j=1}^{N_2} v_{ij} .$$

$$\sigma_u^2 = \frac{1}{N_1 N_2 - 1}\sum_{i=1}^{N_1}\sum_{j=1}^{N_2} (u_{ij} - \bar{u})^2 ,$$

$$\sigma_v^2 = \frac{1}{N_1 N_2 - 1}\sum_{i=1}^{N_1}\sum_{j=1}^{N_2} (v_{ij} - \bar{v})^2 ,$$

$$\sigma_{uv} = \frac{1}{N_1 N_2 - 1}\sum_{i=1}^{N_1}\sum_{j=1}^{N_2} (u_{ij} - \bar{u})(v_{ij} - \bar{v}) ,$$

$$C_1 = (K_1 L)^2, C_2 = (K_2 L)^2; K_1 \ll 1; K_2 \ll 1 .$$

For example, $K_1 = K_2 = 10^{-6}$ , $L$ is image intensity, where, for example, $L = 2^8 - 1 = 255$ for an 8-bit greyscale images.

The greater value $Q_{PSNR}$, the better image quality. If $Q_{PSNR}$ is between 20 and 25, then an image quality is

acceptable, for example, for the wireless transmission (Thomos, 2006).

$Q_{SSIM}$ is used to evaluate image quality by comparing similarity of both images. Its value is between -1 and 1. The greater value $Q_{SSIM}$, the better image quality.

$Q_{MSE}$ is a criterion to evaluate the difference between two images. $Q_{MSE}$ is mean-squared error. The lower value $Q_{MSE}$, the better restoration result. The value of $Q_{MSE}$ also relates to the value of $Q_{PSNR}$.

### 2.5  Initial solution

Because the iteration process uses the initial solution to perform finding solution, so the restoration result of automatically evaluated parameters case also depends on this initial solution. This dependency affects to restoration result, but it is not too much.

The initial solution can be given by one from two methods: directly given or given through a set of initial parameters $(\lambda_1^0, \lambda_2^0, \mu^0)$.

If the initial values of parameters $\lambda_1, \lambda_2, \mu$ are given, the obtained solution is not very good. Because when we set up the initial parameters to find initial solution, the priority of processing of Gaussian and Poisson noise on initial solution is fixed and the result will depend on these initial parameters.

If the initial solution is constant value, the total variation and the differences by $x$-direction and $y$-direction will be 0. This is very bad for our iteration process.

If we make an artificial image by randomizing, the restoration result is bad. Because the random function will affect to the noise property.

The best case is the initial solution need to be enough different with noisy image but not much. In experiment, we make this solution by using average neighbour pixels (closed similar with noise assessment method of Immerker).

### 2.6  Experiments

We use an example to test our model in the case of processing a real image. In this case, we use an image of human skull with the size 300x300 pixel (Figure 1a). We zoom, crop and show the part of the original image under processing (Figure 1b – 1f).

First, we create the noisy image by adding Gaussian noise (Figure 1c) and second, create noisy image by adding Poisson noise (Figure 1d).

In order to calculate proportion between intensities of Gaussian and Poisson noises, we calculate the variance of Poisson noise. The value of variance of Gaussian noise is calculated via Poisson noise variance. Let the variance of Gaussian noise be four times greater than the variance of Poisson noise.

First, let us consider Poisson noise. Its distribution is $p_2(v|u)$, value of the variance of Poisson noise is $\sigma_2 = \sqrt{u_{ij}}$, respectively, with $u_{ij}$ at every pixel $(i,j)$ of image, where $i = 1,...,N_1; j = 1,...,N_2$. We denote this Poisson noisy image as $v^{(2)}$. Obviously, intensity value of $v^{(2)}$ ought to be between 0 and 255. If the intensity value of some pixels are out of this interval, they need to be reset to intensity value of respective pixel of the original image $u$, that means $v^{(2)}{}_{ij} = u_{ij}$ .

In this case, number of them is 5 (0.0056%). The variance of Poisson noise can be calculated as average value $\sigma_2 = 10.0603$.

Now, we consider Gaussian noise. Its variance need to be 40.2412 (because we explained above, variance of Gaussian noise is four times over variance of Poisson noise). We denote this Gaussian noisy image as $v^{(1)}$. As above case, intensity value of $v^{(1)}$ also need to be between 0 and 255. In this case, there are 5780 pixels out of this interval, respectively 6.42% of all image pixels.

We create resulting noisy image (Figure 1e) by combining first noisy and second noisy images with proportion 0.5 for Gaussian noisy image $v^{(1)}$ and 0.5 for Poisson noisy image $v^{(2)}$.

This means $v = 0.5v^{(1)} + 0.5v^{(2)}$. Hence:

$$\lambda_1 / \lambda_2 = \frac{40.2412 \cdot 0.5}{10.0603 \cdot 0.5} = 4/1.$$

As a result: $\lambda_1 = 4/5 = 0.8$, $\lambda_2 = 1/5 = 0.2$.

Values of $Q_{MSE}$, $Q_{PSNR}$ and $Q_{SSIM}$ of noisy image are respectively 427.9526, 21.4168, and 0.4246.

|  | $Q_{PSNR}$ | $Q_{SSIM}$ | $Q_{MSE}$ |
|---|---|---|---|
| Noisy | 21.4168 | 0.4246 | 427.9526 |
| ROF | 26.5106 | 0.8465 | 145.2183 |
| Median | 25.6477 | 0.7871 | 177.1364 |
| Wiener | 24.2657 | 0.6596 | 243.5077 |
| Proposed method with $\lambda_1$=0.8, $\lambda_2$=0.2, $\mu = 0.0857$, $\sigma = 40.2412$. | **27.4315** | 0.8198 | **117.4713** |
| Proposed method with evaluated parameters $\lambda_1$=0.8095, $\lambda_2$=0.1905, $\mu = 0.0970$, $\sigma = 38.2310$. | 27.2567 | **0.8383** | 122.2941 |

Table 1. Quality comparison of noise removal methods for real image of human skull.

The Table 1 shows the result of denoising for real image in cases: given parameters and automatically evaluated parameters.

We have to notice, that in the case of the real image, the value of $Q_{PSNR}$ of denoising for given ideal parameters is better, than the value of $Q_{PSNR}$ of denoising for automatically evaluated parameters, but the value of $Q_{SSIM}$ is inversed.

Based on experimental results, we can see that the restoration result of automatically evaluated parameters case depends on initial solution.

We use convolution operator to make a new image. The Table 2 shows the dependency of restoration result on initial solution, where:

(a) Initial parameters $\lambda_1^0 = 0, \lambda_2^0 = 1, \mu = 1$;

(b) Initial parameters $\lambda_1^0 = \lambda_2^0 = 0.5, \mu = 1$;

(c) Initial solution $u^0 = rand(\cdot,\cdot)$ with $rand(\cdot,\cdot)$ to create randomized two-dimensional matrix with specific size;

(d) Initial solution $u^0 = v * A$ with $A = \frac{1}{9}\begin{pmatrix} 1 & 1 & 1 \\ 1 & 1 & 1 \\ 1 & 1 & 1 \end{pmatrix}$.

Table 2 shows the best denoising result for the case (d) with respect two most important criteria (PSNR and MSE).

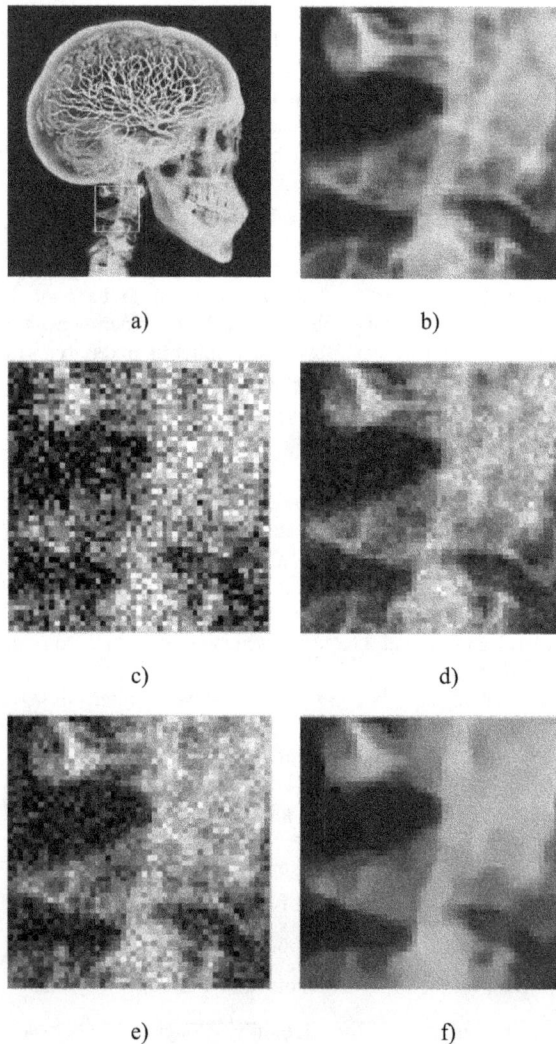

Figure 1. Denoising of real image: a) original image, b) cropped image, c) with Gaussian noise, d) with Poisson noise, e) with mixed noise, f) after denoising.

|  | (a) | (b) | (c) | (d) |
|---|---|---|---|---|
| $\lambda_1$ | 0.8095 | 0.8114 | 0.9256 | 0.8069 |
| $\lambda_2$ | 0.1905 | 0.1886 | 0.0744 | 0.1931 |
| $\mu$ | 0.0970 | 0.0985 | 0.1026 | 0.0965 |
| $\sigma$ | 38.2310 | | | |
| $Q_{PSNR}$ | 27.2567 | 27.1327 | 26.4279 | **27.2571** |
| $Q_{MSE}$ | 122.2941 | 125.8371 | 148.0081 | **121.6320** |
| $Q_{SSIM}$ | 0.8383 | 0.8381 | **0.8497** | 0.8384 |

Table 2. Dependency of restoration result on initial solution.

## 3    CONCLUSIONS

In this paper, we proposed the method that is based on variational approach to remove combination of Poisson and Gaussian noises (mixed noise).

The denoising result depends on parameters, especially on coefficients of linear combination $\lambda_1$ and $\lambda_2$. We can specify the values of parameters or these values can be automatically evaluated. In order to apply this model to real image, we need to use the proposed method with automatically evaluated parameters. In this case, the solution also depends on initial solution.

The proposed method can be applied to remove separate Gaussian or Poisson noise (respectively ROF model and modified ROF model for Poisson noise), or mixed Poisson-Gaussian noise as well.

We also can use this variational approach to remove other kinds of noise, such as noise of magnetic resonance images (MRI), ultrasonogram, etc.

## REFERENCES

Abe C., Shimamura T., 2012, 'Iterative Edge-Preserving adaptive Wiener filter for image denoising', *ICCEE*, vol. 4, no. 4, pp. 503-506.

Burger M., 2008, *Level set and PDE based reconstruction methods in imaging*, Springer.

Caselles V., Chambolle A., Novaga M., 2011, *Handbook of mathematical methods in imaging*, Springer.

Chan T.F., Shen J., 2005, *Image processing and analysis: Variational, PDE, Wavelet, and stochastic methods*, SIAM.

Chambolle A., 2009, 'An introduction to total variation for image analysis', *Theoretical foundations and numerical methods for sparse recovery*, vol. 9, pp. 263-340.

Chen K., 2013, 'Introduction to variational image processing models and application', *International journal of computer mathematics*, vol. 90, no. 1, pp. 1-8.

Getreuer P., 2012, 'Rudin-Osher-Fatemi total variation denoising using split Bregman'. *IPOL 2012*, 'http://www.ipol.im/pub/art/2012/g-tvd/'.

Gill P.E., Murray W., 1974, *Numerical methods for constrained optimization*, Academic Press Inc.

Immerker J., 1996, 'Fast noise variance estimation', *Computer vision and image understanding*, vol. 64, no.2, pp. 300-302.

Jezierska A., 2012, 'Poisson-Gaussian noise parameter estimation in fluorescence microscopy imaging', *IEEE International Symposium on Biomedical Imaging 9th*, pp. 1663-1666.

Jezierska A., 2011, 'An EM approach for Poisson-Gaussian noise modelling', *EUSIPCO 19th*, vol. 62, is. 1, pp. 13-30.

Le T., Chartrand R., Asaki T.J., 2007, 'A variational approach to reconstructing images corrupted by Poisson noise', *Journal of mathematical imaging and vision*, vol. 27, is. 3, pp. 257-263.

Li F., Shen C., Pi L., 2006, 'A new diffusion-based variational model for image denoising and segmentation', *Journal mathematical imaging and vision*, vol. 26, is. 1-2, pp. 115-125.

Luisier F., Blu T., Unser M., 2011, 'Image denoising in mixed Poisson-Gaussian noise', *IEEE transaction on Image processing*, vol. 20, no. 3, pp. 696-708.

Lysaker M., Tai X., 2006, 'Iterative image restoration combining total variation minimization and a second-order functional', *International journal of computer vision*, vol. 66, pp. 5-18.

Nick V., 2009, Getty images, 'http://well.blogs.nytimes.com/2009/09/16/what-sort-of-exercise-can-make-you-smarter/'.

Rankovic N., Tuba M., 2012, 'Improved adaptive median filter for denoising ultrasound images', *Advances in computer science*, WSEAS ECC'12, pp. 169-174.

Rubinov A., Yang X., 2003, *Applied Optimization: Lagrange-type functions in constrained non-convex optimization*, Springer.

Rudin L.I., Osher S., Fatemi E., 1992, 'Nonlinear total variation based noise removal algorithms', *Physica D.* vol. 60, pp. 259-268.

Scherzer O., 2009, Variational *methods in Imaging*, Springer.

Thomos N., Boulgouris N.V., Strintzis M.G., 2006, 'Optimized Transmission of JPEG2000 streams over Wireless channels', *IEEE transactions on image processing*, vol. 15, no.1, pp .54-67.

Tran M.P., Peteri R., Bergounioux M., 2012, 'Denoising 3D medical images using a second order variational model and wavelet shrinkage', *Image analysis and recognition*, vol. 7325, pp. 138-145.

Wang C., Li T., 2012, 'An improved adaptive median filter for Image denoising', *ICCEE*, vol. 53, no. 2.64, pp. 393-398.

Wang Z., 2004, 'Image quality assessment: From error visibility to structural similarity', *IEEE transaction on Image processing*, vol. 13, no. 4, pp. 600-612.

Wang Z., Bovik A.C., 2006, *Modern image quality assessment*, Morgan & Claypool Publisher.

Xu J., Feng X., Hao Y., 2014, 'A coupled variational model for image denoising using a duality strategy and split Bregman', *Multidimensional systems and signal processing*, vol. 25, pp. 83-94.

# 23

# CAMERA CALIBRATION IN 3D MODELLING FOR UAV APPLICATION

Hideharu Yanagi [a, c], Hirofumi Chikatsu [b]

[a] Japan Association of Surveyors, 1-33-18, Hakusan, Bunkyo-ku, Tokyo 113-0001, Japan – yanagi@jsurvey.jp
[b] Division of Architectural, Civil and Environmental Engineering, Tokyo Denki University, Hatoyama, Hiki-gun, Saitama 350-0394, JAPAN - chikatsu@g.dendai.ac.jp
[c] School of Science and Engineering, Tokyo Denki University, Hatoyama, Hiki-gun, Saitama 350-0394, JAPAN

**Commission V, WG V/4**

**KEY WORDS:** Consumer grade digital camera, Calibration, Lens distortion, 3D modelling

**ABSTRACT:**

In recent times, small types of Unmanned Aerial Vehicles (UAVs) have been receiving attention in areas such as 3D modelling, maintenance engineering, and personal interest (hobby) usage. However, the payload a small type of UAV is able to carry is limited. Given these circumstances, small consumer grade digital cameras are often used for UAV photogrammetry with small types of UAV. Though, digital photogrammetry using the consumer grade cameras is enormously expected in various application fields such as UAV photogrammetry.
There is a large body of literature on camera calibration. However, the lens distortion of small consumer grade digital cameras is still an issue from the viewpoint of accuracy aspect.
The authors have been investigating camera calibration using various cameras. However, small consumer grade digital cameras, which is called as entry-cameras, have an accuracy degradation problem. The issue was addressed by conducting calibration tests using two kinds of consumer grade digital cameras, each with a resolution of 12 mega pixels. This paper presents the results of an investigation into the cause of accuracy degradation in digital photogrammetry using entry-cameras.

## 1. INTRODUCTION

In recent times, small types of UAVs have been receiving more attention, because they are increasingly in areas such as 3D measurements, maintenance engineering, and for hobby use and so on, because of their simplicity of operation and lightness of body. However, the payload a small type of UAV is able to transport, is limited. Therefore, small consumer grade digital cameras are often used for UAV photogrammetry. On the other hand, consumer grade digital cameras have been rapidly gaining interest, with many low-priced consumer grade models available on the market. Therefore, performance evaluations for consumer grade digital cameras have been investigated from the viewpoint of digital close range photogrammetry (Fraser, 1997, Habib and Morgan, 2003).

In digital close range photogrammetry using consumer grade digital cameras, lens distortion is one of important parameters. The authors have been investigating lens distortion using various cameras (Chikatsu and Takahashi, 2009). From these results, digital close range photogrammetry using consumer grade digital camera is expected for various application fields. However, not all of these cameras are necessarily suitable for use in 3D modeling. In particular, small consumer grade digital cameras which is called as entry-cameras have an accuracy degradation problem. In other words, in some of these cameras the accuracy is not improved after camera calibration.

In order to resolve the issue, at first, the factors for accuracy degradation have been investigating using a camera of which the accuracy did not improve after camera calibration. Finally, the authors' results led them to propose a calibration method for improving the accuracy of the abovementioned camera in this study.

## 2. EXPERIMENTS

### 2.1 Cameras

In order to confirm the camera which does not show the accuracy improvement after calibration, the two consumer grade digital cameras shown below was investigated in this study (Table 1, Figure 1).

| Cameras | Nikon Coolpix S3000<br>Image Size: 4000 × 3000 pixels<br>Image sensor: 1/2.3-in. CCD<br>Focal length: 4.9mm |
| --- | --- |
| | Canon IXY 200F<br>Image Size: 4000 × 3000 pixels<br>Image sensor: 1/2.3-in. CCD<br>Focal length: 5.0mm |

Table 1. Specifications of cameras

(a) Nikon Coolpix S3000          (b) Canon IXY 200F

Figure 1. Consumer grade digital cameras

Table 1 lists the specifications of both of the selected cameras and shows that the image and sensor size of both models are the same. However, it was unclear whether both of the cameras were equipped with the same sensor, because the technical details of the image sensor had not been published. Furthermore, the technical details of the lens used in each camera were also unknown.

## 2.2 Test Target

Figure 2 shows a test target that was used in the study. The size of the test target is H: 640 mm, W: 480 mm and D: 20 mm. The black circular targets in the test target were manufactured with an accuracy of ±0.05 mm. The diameter of each circular target was 20 mm, and the targets were arranged in intervals of 40 mm. The height of the central targets (three rows in the center) was 20 mm. GCPs (ground control points) and checkpoints were arranged as shown in Figure 2. The pixel coordinates of these points were obtained as the area gravity by using image processing procedures.

White circles: GCPs, Black squares: Checkpoints

Figure 2. Test target

## 2.3 Lens distortion model

In this study, the lens distortion was evaluated by using Brown's model (Brown, 1971), as follows.

$$\bar{x} = x + \frac{x}{r}\left(K_1 r^3 + K_2 r^5 + K_3 r^7\right) + P_1\left(r^2 + 2x^2\right) + 2P_2 xy$$
$$\bar{y} = y + \frac{y}{r}\left(K_1 r^3 + K_2 r^5 + K_3 r^7\right) + P_2\left(r^2 + 2y^2\right) + 2P_1 xy \quad (1)$$

where, $r = \sqrt{x^2 + y^2}$ , $\bar{x}, \bar{y}$ are the corrected image coordinates; x, y are the measured image coordinates; $K_1$, $K_2$, and $K_3$ are coefficients of radial distortion; $P_1$ and $P_2$ are coefficients of tangential distortion.

## 3. EXPERIMENTAL RESULTS

In this study, five stereo images were taken as a base-height ratio of 0.34, and altitudes of 667 mm and 736 mm for the S3000 and IXY 200F, respectively. The focal length of both lenses was set to the largest possible wide angle setting. Camera calibrations were performed using bundle adjustment based on 13 GCPs (Figure 2).

Figure 3 shows the average of five calculations as the experimental results. In Figure 3, the proportional accuracy of the results was computed using equation (2). In addition, Figure 3 also shows the curve of the standard error, which was computed using equations (2) and (3) from the results of the camera calibration for the various cameras.

$$\text{Proportional accuracy} = \frac{\sqrt{(\sigma_X^2 + \sigma_Y^2 + \sigma_Z^2)/3}}{\sqrt{DX^2 + DY^2 + DZ^2}} \quad (2)$$

where, $DX$, $DY$, $DZ$ are the object field diameter; $\sigma_X$, $\sigma_Y$, and $\sigma_Z$ are the root mean square error for the check points.

$$\sigma_{X_0} = \sigma_{Y_0} = \frac{H}{f}\sigma_p \qquad \sigma_{Z_0} = \sqrt{2}\frac{H}{f}\frac{H}{B}\sigma_p \quad (3)$$

where, $\sigma_{X0}$, $\sigma_{Y0}$, and $\sigma_{Z0}$ are the standard errors; $H$ is the altitude; $f$ is the focal length; $B$ is the baseline; and $\sigma_P$ is the pointing accuracy (estimated to be 0.1 pixels (Schaefer and Murai, 1988)).

The calculated results obtained for the IXY 200F closely approximated the curve of the standard error in Figure 3. On the other hand, those obtained for the S3000 were positioned further from the curve of the standard error.

Figure 4 shows the results of the camera calibration for cases in which lens distortion was not considered and considered. Note: Figure 4 shows that the proportional accuracy was normalized by the standard error.

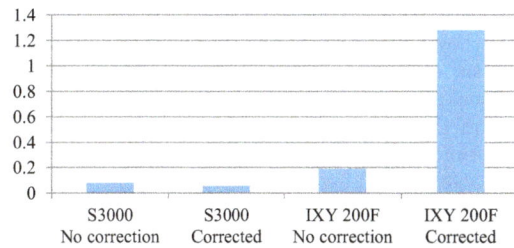

Figure 4. Results of camera calibration

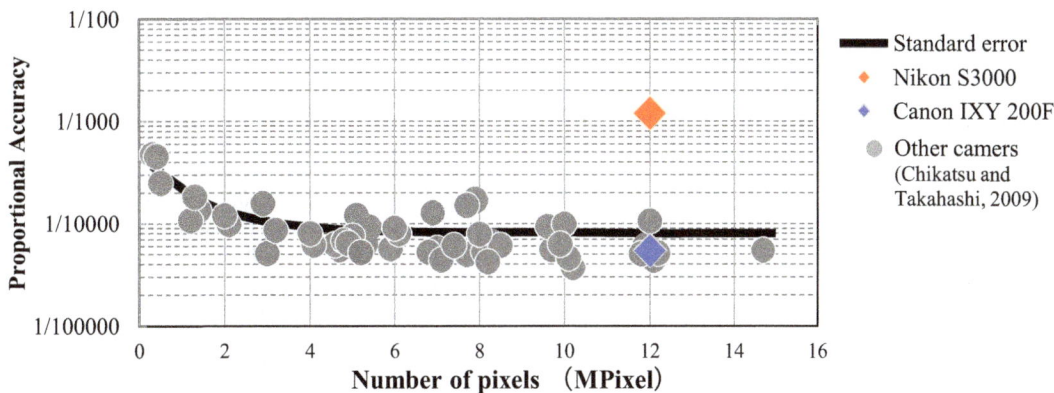

Figure 3. Experimental results

| Distortion model | Coefficients of lens distortion | | Abbreviation |
| | Radial distortion | Tangential distortion | |
| --- | --- | --- | --- |
| 3rd polynomial function | $K_1$ | - | 3R |
| 5th polynomial function | $K_1, K_2$ | - | 5R |
| 7th polynomial function | $K_1, K_2, K_3$ | - | 7R |
| 3rd polynomial function and tangential distortions | $K_1$ | $P_1, P_2$ | 3RD |
| 5th polynomial function and tangential distortions | $K_1, K_2$ | $P_1, P_2$ | 5RD |
| 7th polynomial function and tangential distortions | $K_1, K_2, K_3$ | $P_1, P_2$ | 7RD |

Table 2. Distortion model and its abbreviation (Wakutsu and Chikatsu, 2011)

From Figure 4, it is clearly understood that the accuracy of the IXY 200F in terms of lens distortion was improved after camera calibration. However, the accuracy of the S3000 was not improved, even after camera calibration.

Based on these results, the cause of accuracy degradation in the S3000 was investigated and the results are presented in the next section.

## 4. VERIFICATION

### 4.1 Distortion

In previous experiments, the S3000 was calibrated by using equations (1). However, Brown's model is able to divide into six models such as shown in Table 2 (Wakutsu and Chikatsu, 2011). Therefore, suitable distortion model for the S3000 is investigated in this session.

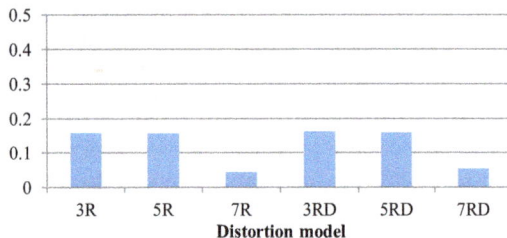

Figure 5. Results of camera calibration

Figure 5 shows the results of the camera calibration using each of the six distortion models and the results show the proportional accuracy normalized by the standard error. Figure 5 indicates that the 3rd polynomial function with tangential distortions model (3RD) shows the best accuracy. However, the accuracy show quite different low accuracy from the viewpoints of digital photogrammetric aspect.

The cause of the degradation in accuracy was investigated using the 3RD model.

### 4.2 Chromatic aberration

As one of the cause of accuracy degradation, chromatic aberration was considered. Chromatic aberrations are well known to occur in relation to the wavelength of light. The green channel in an image has been attention for correction of the chromatic aberration, and Guan, et al. were corrected chromatic aberration based on green channel (Guan, H., 2005).

Consequently, the green channel images that were extracted from the stereo image were used for the investigation in this study.

Figure 6 shows the results of the camera calibration for which the proportional accuracy was normalized by the standard error using the normal images and the green channel image. The distortion model that was used was the 3RD model.

It can be seen that the result for which the green channel images were used show almost the same accuracy as the result obtained

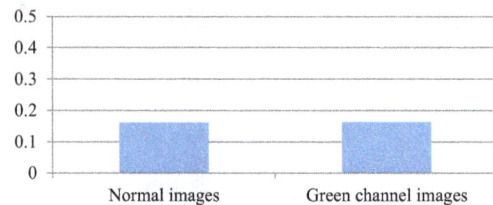

Figure 6. Results of camera calibration
using green channel images

using a normal image. It is concluded that chromatic aberration does not show a significant factor.

### 4.3 Verification Result

In this study, lens distortion models and chromatic aberration were investigated. However, neither of these approaches led to an improvement in the accuracy. At this stage, the reason as to why it has not been possible to improve the accuracy is not known, neither is it known whether the aberration of the lens is responsible.

On the other hand, there are many consumer grade digital cameras with an internal image processing function. It is unclear as to whether image distortion is automatically corrected by the internal image processing function.

Thus, it is necessary to treat the situation as a black box by inferring whether the image data are automatically adjusted. If automatic image adjustment were used, it would be difficult to obtain an original image which had not been adjusted.

Therefore, a calibration method for the camera which does not improve the accuracy even after calibration was proposed in this study.

## 5. PROPOSED CALIBRATION METHOD

### 5.1 Verification for distortion

In this study, it was taken center image to confirm the shape of the distortion for the S3000.

Figure 7 shows the result of the camera calibration based on the center of the image using the 3RD model.

Generally, it is known that lens distortion is symmetrical and many calibration models were developed based on symmetrical images. However, as seen in Figure 7, it is clearly understood that lens distortion for the S3000 is not symmetrical.

Generally, it is well known that wide-angle lenses, such as that of the S3000, show barrel distortion. The barrel distortion was automatically adjusted previously using internal image processing function. The influence of the distortion adjustment appears in the center of the right-left row in the image. Therefore, the proposed calibration method focuses on this row for adjustment to compensate for the large distortion, such as shown in the row on the right in Figure 7.

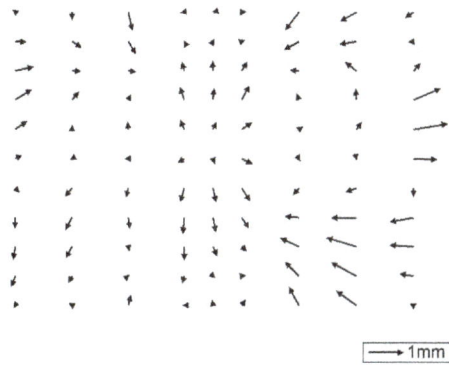

Figure 7. Result of camera calibration of center image

## 5.2 Proposed Method

In the first step of the proposed method, the block in which the largest distortion occurs (e.g., the row on the right side in Figure 7) is extracted after ordinary camera calibration. Second, a fitted curve is computed by using GCPs in the extracted block. The extracted points are adjusted along the computed fitted curve (Figure 8). And in order to adjust left-right, the block of opposite side was computed by the same procedure. Finally, the camera calibration is repeated using all the points, including the adjusted points.

It is observed that the points of the row on the right side in the center of the image (Figure 7) have the largest distortion. Therefore, the points in this row were adjusted by using the proposed method. Furthermore, the row on the left side was also adjusted in this work.

Figures 8 and 9 show the adjusted points of the side rows on the left and right. Table 3 compares the results of an ordinary camera calibration and the proposed method. These results show that the proportional accuracy was improved by the proposed method. However, the accuracy improvement was still issue.

## 6. CONCLUSION

In this study, the cause of accuracy degradation for consumer grade digital cameras, known as entry-cameras, was investigated. It is concluded that one of the reason for the accuracy could not improve even after procedure of camera calibration is not known whether the internal image processing function had been applied previously or not. The second reason is the occurrence of asymmetrical distortion.

Generally, it is well known that wide-angle lenses, such as that of the S3000, show barrel distortion. Previously, this distortion was automatically adjusted using the internal image processing function. The influence of the distortion adjustment appears in the center of the right-left row in the image.

Based on the above facts and assumption, the authors proposed a camera calibration method for the camera that does not show accuracy improvement after the camera calibration procedure. As a result of our method, it is observed that the accuracy shows some improvement and is expected to be improved even more by upgrading the proposed method.

## References

Fraser, C.S., 1997. Digital camera self-calibration. ISPRS Journal of Photogrammetry and Remote Sensing, Vol. 52, No. 4, pp. 149-159.

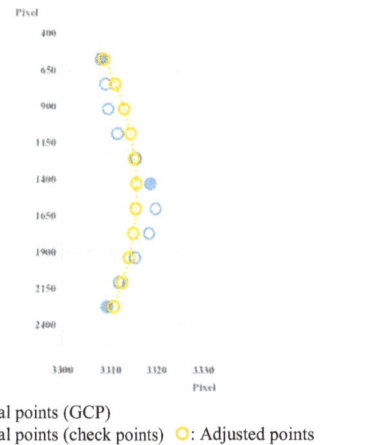

●: Original points (GCP)
○: Original points (check points)   ○: Adjusted points

Figure 8. Adjusted points

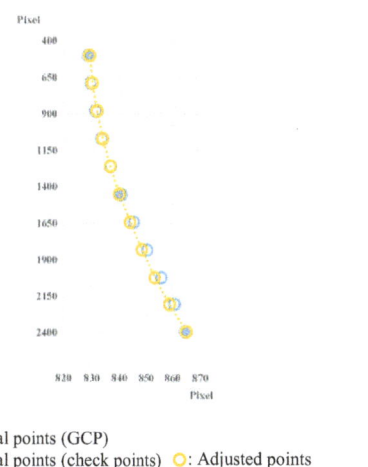

●: Original points (GCP)
○: Original points (check points)   ○: Adjusted points

Figure 9. Adjusted points (opposite of side)

|  | Normal | Proposed method |
|---|---|---|
| Proportional accuracy | 1/1906 | 1/3108 |

Table 3. Result of proposed method

Habib, A. and Morgan, M., 2003. Automatic Calibration of Low-Cost Digital Cameras. Journal of Optical Engineering, Vol. 42, No. 4, pp. 948-955.

Chikatsu, H. and Takahashi, Y., 2009. Comparative Evaluation between Consumer Grade Cameras and Mobile Phone Cameras for Close Range Photogrammetry. Videometrics, Range Imaging, and Applications X, Proc. of SPIE, 74470H.

Brown, D. C., 1971. Close-range camera calibration, Photogrammetric Engineering, Vol. 37, No. 8, pp. 855-866.

Schaefer, H. and Murai, S., 1988. Automated target detection for real-time camera calibration, Proceeding of 9th Asian Conference on Remote Sensing, pp. H.2-4–1-H.2-4-8.

Wakutsu, R. and Chikatsu, H., 2011. Practical calibration for consumer grade digital camera with integrated high zooming lens. Videometrics, Range Imaging, and Applications XI, 80850T.

Guan, H., Shiraishi, K.,Watanabe, K., Fukuoka, H. and Ohashi, K., 2005. Digital Image Correcting Method for Digital Camera. Ricoh Technical Report, Vol. 31, pp. 103-110.

# 24

# IMAGE-BASED LOCALIZATION
# FOR INDOOR ENVIRONMENT USING MOBILE PHONE

Yewei Huang, Haiting Wang, Kefei Zhan, Junqiao Zhao, Popo Gui, Tiantian Feng

**KEY WORDS:** Indoor Localization, Image Matching, SIFT, HoG, Mobile phone

**ABSTRACT:**

Real-time indoor localization based on supporting infrastructures like wireless devices and QR codes are usually costly and labor intensive to implement. In this study, we explored a cheap alternative approach based on images for indoor localization. A user can localize him/herself by just shooting a photo of the surrounding indoor environment using the mobile phone. No any other equipment is required. This is achieved by employing image-matching and searching techniques with a dataset of pre-captured indoor images. In the beginning, a database of structured images of the indoor environment is constructed by using image matching and the bundle adjustment algorithm. Then each image's relative pose (its position and orientation) is estimated and the semantic locations of images are tagged. A user's location can then be determined by comparing a photo taken by the mobile phone to the database. This is done by combining quick image searching, matching and the relative orientation. This study also try to explore image acquisition plans and the processing capacity of off-the-shell mobile phones. During the whole pipeline, a collection of indoor images with both rich and poor textures are examined. Several feature detectors are used and compared. Pre-processing of complex indoor photo is also implemented on the mobile phone. The preliminary experimental results prove the feasibility of this method. In the future, we are trying to raise the efficiency of matching between indoor images and explore the fast 4G wireless communication to ensure the speed and accuracy of the localization based on a client-server framework.

## 1. INTRODUCTION

In recent years, the increasingly matured global positioning technology has breed various location-based applications, such as the navigation system of mobile devices and even unmanned vehicles and aircrafts. However, the signals of the global navigation system usually fails to penetrate into indoor environment, which leads to the unreliability of the indoor localization. One of the solutions to this problem is by establishing an indoor referencing framework based on wireless communication, i.e. Wi-Fi, Zigbee, Bluetooth etc. [1]. Nevertheless various issues exist in current approaches. They are vulnerable to signal interference and multiple reflections, and the layout scheme of hotspots dramatically affects the accuracy of localization [1]. In addition, this technique can only be applied in coarse indoor positioning with meter level accuracy. Other localization methods based on QR code or RFID [2] are simpler to implement. However these landmarks which provide location reference can only be embedded in limited locations therefore can hardly provide continuous localization in the indoor space.

In this study, we explored a simple, yet accurate image-based approach for indoor localization. A structured image database of an indoor environment is constructed, and used as the indoor location referencing, since each image's relative pose can be estimated. A user's location is determined based on a new captured image by first quickly searching the similar images from the database and then precisely matching them and conduct relative orientation. This idea is not new and has been already explored in many fields such as robotics [3]. However, this study tries to examine the practicability and limitations of such methods for human users in specified environment such as exhibitions and museums. We also try to implement the whole pipeline on off-the-shell mobile phones and plan to explore the fast LTE wireless communication to ensure the speed and accuracy of the localization by using a client-server framework.

The remainder of this paper is organized as follows. In section 2, we discuss related work on image matching and indoor localization. In section 3, we introduce how to build a structured image database and how to localize by image searching and matching. The result of the experiment is presented in section 4. At last, we conclude the paper and describe future improvements in section 5.

## 2. RELATED WORK

Vision-based localization has drawn intensive attention because of its passive nature and is analogous to human localization [1]. Many methods have been proposed, such as the visual vocabulary tree [4, 5], which packs the SIFT features extracted from images into vectors of visual words. Although the searching is speeded up but constructing the visual words is complex and costly. [3, 6] use landmarks to implement indoor localization. The landmarks are features or group of features detected from the images. They must be stationary, distinctive in the map, repeatable and robust against noise [6]. During the searching period, features which are detected from the query image are matched to the landmarks. Comparing to the previous method, this method is more efficient, but is vulnerable to similar features presented in various locations [6]. In topology-based method [8], a topological map is built from a series of images or video sequences, and then is refined by learning vector quantization (LVQ). During the searching stage, nearest neighbor rule is used to detect the similar region in the query image. This method assumes that the navigation path is unique in the topological map. Therefore, the query image can be misclassified in some case.

There are also methods using stereo images [1, 3, 7]. The stereo images can provide depth information which is helpful for 3D reconstruction. However, they are more or less the same as monocular images in localization applications.

## 3. INDOOR LOCALIZATION BASED ON STRUCTURED IMAGE DATABASE

The whole pipeline of our image-based localization is shown in Fig. 1. It can be divided into two main stages. The first is the building of structured image database. Our goal is to estimate all the camera poses and position information of pre-captured indoor images. All these information compose a structured image database. The second stage is the localization. Once the user takes a photo, the image features are extracted on the fly. These features are then matched with the feature descriptions extracted in the former step, from which the image with the richest matches can be detected. Then the parameters of the relative orientation of the newly taken photo can be estimated. As a result, the location of the user can be identified.

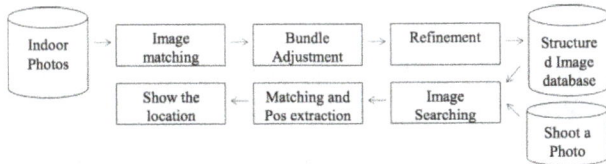

Fig. 1 The pipeline of image-based indoor localization

### 3.1 Collection of indoor environment images

It is significant to collect proper indoor images because its quality has a great effect on the construction of structured image database and the localization. Ideally, any images of an indoor environment can be incorporated in the database. However, due to the robustness problem of feature descriptors and the richness of image texture, we find the following aspects are important.

First, the collected indoor images should have at least 70% overlap and the camera should not be too close to the target. Otherwise, it is difficult to capture enough textures and would cause difficulties in image matching.

Second, we should avoid view directions containing surfaces of strong reflection, transparency, or vegetation and light sources. All these situations will introduce ambiguous to matching.

Finally, the orientation of camera should be roughly perpendicular to the target and the moving path, so that to relieve the influence of rotation and the problem of reflection for the target objects such as glasses. Fig. 2 show several indoor images used for building the structured image database.

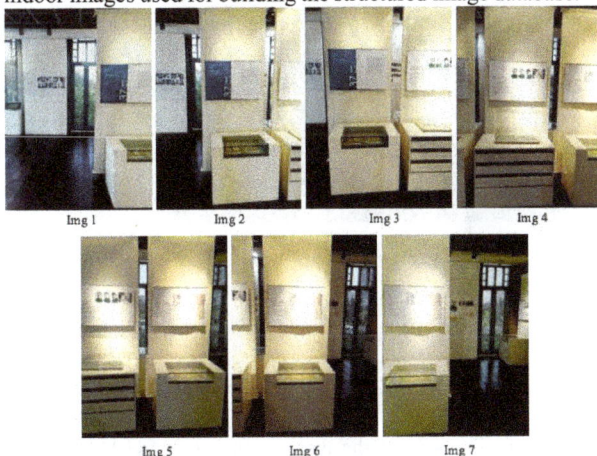

Fig. 2 Examples of the pre-captured indoor image sets

### 3.2 Construction of structured image database

In this period, the first step is to detect features from the images. We compared the SIFT descriptor and two SIFT-like descriptors: Speeded Up Robust Features (SURF) [8] and Affine-SIFT (ASIFT) [9]. These methods are known to be robust in feature extraction and description. The captured photos are resized to 800x600 to balance the computational

cost and the richness of image details. Then the images are matched to their neighbors according to the Flann-based matcher [10]. Fig. 3 show the results of matching using different feature descriptors, from which the results of ASIFT are more accurate and 649 matched points are extracted. The implementation of SIFT and SURF extract more feature points but also introduce distractions to the matching stages, since the RANSAC estimation is influenced by the outliers. As a result, the ASIFT is chosen in our method.

In the second step, all the indoor images are aligned in a queue and we match each image with its neighbors. After that the relative orientation can be conducted on each image pair, hence the relative image and camera poses are calculated. To globally optimize the poses, the bundle adjustment is used [11]. Then we get sparse point cloud and the poses of the images as well.

To construct the structured image database, the poses are described in a unified coordinate frame. We define the frame based on the pose of the first image and all the consequent images are transformed into this frame. The global coordinates are not needed at this moment because the relative location conjugated with the semantic description of the image, e.g. the 2nd floor, room 3, can already provide enough information for localization. However, if the accurate distance is demanded for routing, we have to introduce the global scale factor.

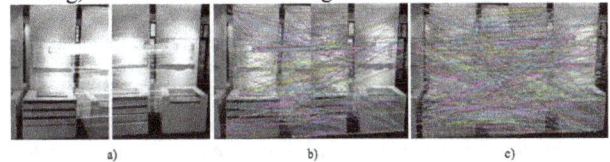

Fig. 3 The comparison of image matching based on ASIFT a), SIFT b) and SURF c)

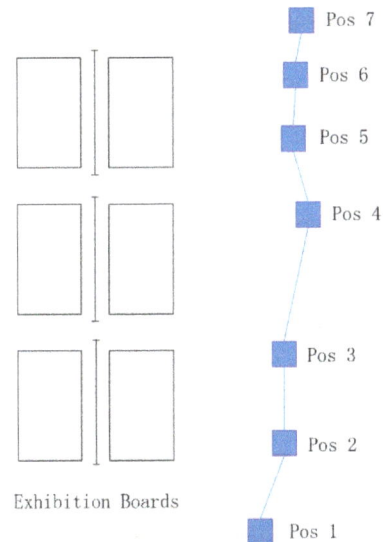

Fig. 4 The top-down view of the extracted positions of cameras from the input image set.

### 3.3 Image searching based on HoG features

In a localization scenario, we provide the possibility that a user can be aware of his/her ego location simply by just shooting one photo of the indoor scene. This is realized by the technique of fast image searching and deducing the camera's pose using image matching and orientation.

Image searching is known to be nontrivial since the description of an image can be complex which leads to high computation cost. In this study, a two stage searching mechanism is developed. Firstly, a simple statistic-based image description, i.e. the histogram, is used to filter out the non-similar images from the database. Fig. 5 is the newly captured photo by the user and Fig. 6 is the histograms of the previously shown indoor images and the query image. It can be seen that Img5, Img6 and Img7 are most unlike to be the similar scene. Before comparing, the captured image is enhanced by histogram equalization on mobile phone.

The histogram is only used to roughly rank the images by similarity. We should further use more precise features such as SIFT features to directly match the similar images. The number of matched points then can be used as an evaluation of the similarity of images [8]. However, this approach is too burdensome in practice because matching N image pairs are needed (where N is equal to the number of images in the structured image set). Building the image pyramid could hardly help in this case because SIFT features extracted from multi-scaled image set can hardly be compared directly. An alternative is to store and match SIFT features based on a binary-tree index. However, this can be overwhelming when the structure image database is built from too many photos. As a result, the Histogram of Oriented Gradients (HoG) [12] is introduced to detect images containing similar scene (as shown in Fig. 7).

At first, we decompose the query image into 3 levels using quadtree and calculation the HoG features for each image block (Fig. 8). The reason is that all these image patches may contain part of overlapping scene in the database. During localization, the 3-level HoG features of the shoot photo is compared with all the images stored in the image database using a moving window. Therefore once a similar image patch is found, we can deduce that the image may overlap part with the query image. To solve the problem of scale variance, we constructed image pyramid for all the images in the database. After the "similar image" is detected, the image matching using ASIFT is then conducted. And the pose of the query camera can be extracted by the same method as described in section 1.2. However, once the matching is failed. The next image in the ranked list is selected to match with the query image.

The full pipeline of this approach is still in construction but the preliminary results show the effectiveness of this solution. The computational cost of HoG feature extraction and patch detecting is much less then direct SIFT-based image matching. Another problem is that HoG is not rotation invariable. Nevertheless we make an assumption in our application that the photos are always taken either vertically or horizontally. A detection of the aspects of the photo would solve the most of the problem instead of rotating images in multiple degrees.

Fig. 5 The newly captured image to be used for localization

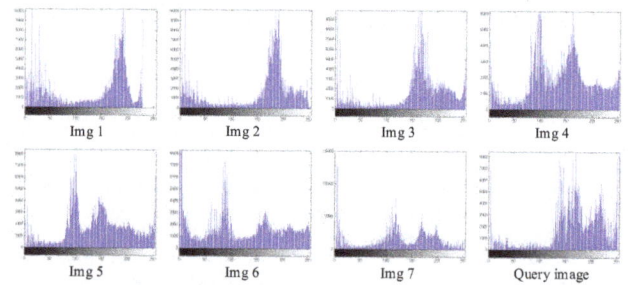

Fig. 6 The histograms of the image set

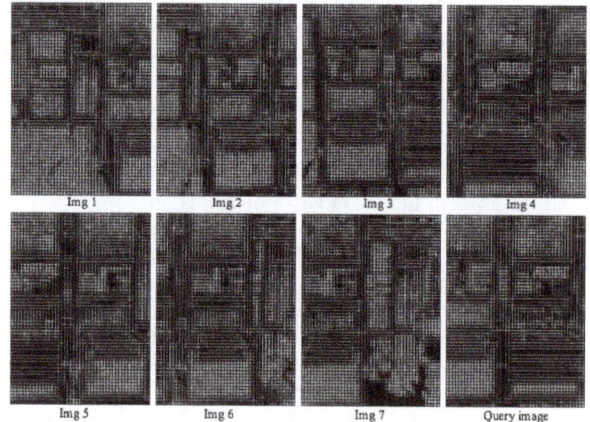

Fig. 7 The HoG feature extracted from the indoor images

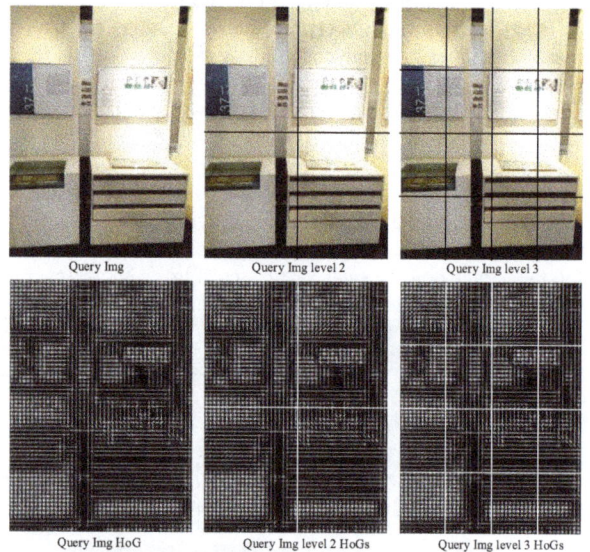

Fig. 8 The 3 level HoG features extracted from the query image

## 4. RESULTS

We take photos of the second and the third floor of the History Museum of Tongji University with iPhone 5s at a fixed focal length. The camera is control and calibrated by own developed program. The size of the every image is 3264×2448 pixels. Taking the speed of calculation into account, the images were down sampled into the size of 800 ×600 pixels.

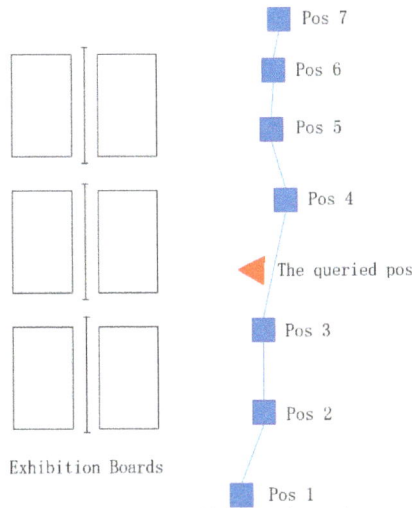

Fig. 9 The calculated camera position from the newly captured image

SIFT and SURF we deployed is based on OpenCV [13], while ASIFT is provide by the author[1]. The original version cost around 20 seconds to matching an image pair. There remains space to lessen the computing time.

The query image can be located with the proposed method and Fig. 9 shows its correct location.. The precision of the result still needs evaluation, however comparing to the wireless-based method this location yields much higher accuracy.

The localization result is shown in Fig. 10, which is an app developed on Android. The red spot displays the location of a user in the floor plan of the museum, and the semantic information is shown on the side.

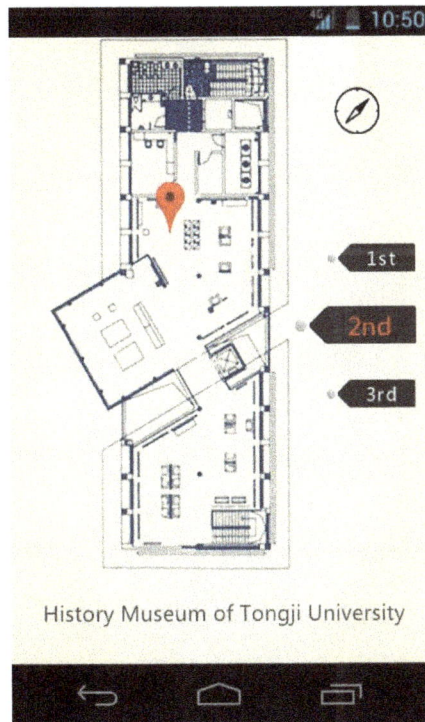

Fig. 10 The localization result shown in mobile app

## 5.  CONCLUSION

In this paper, we present an image-based indoor localization system implemented on mobile phone, which is free from specified devices thus are easy to implement and flexible to use in practice. The key adopted techniques are the image matching and searching. Although the preliminary experiments has shown its feasibility, the accuracy of the results has to be examined in detail and the performance remains space to be improved. To optimize the construction of the database, the geometric constrains of indoor image set can be incorporated. Our HoG-based image searching still needs further refinements. At present, only the photo capture and enhancement is fully implemented on mobile phone using OpenCV. The image searching and matching are migrating from desktop to the mobile platform. We planned to use the client-server framework to relieve the task load on the mobile phone by uploading the image to the server-side and return results on the mobile phone. This can be done by using the fast 4G LTE communication.

## Acknowledgement
This research is supported by the NSFC young scholar funding (No.41201379), the Program for Young Excellent Talents in Tongji University (No.2014KJ027) and the National Student Innovation Training Program (No.0250107010).

## 6.  REFERENCES

[1]Mautz, R. (2012). *Indoor positioning technologies* (Doctoral dissertation, Habilitationsschrift ETH Zürich, 2012).

[2]Alghamdi, S., Van Schyndel, R., & Alahmadi, A. (2013, April). Indoor navigational aid using active RFID and QR-code for sighted and blind people. In *2013 IEEE Eighth International Conference on Intelligent Sensors, Sensor Networks and Information Processing,* (pp. 18-22).

[3]Lategahn, H., & Stiller, C. (2012, July). City gps using stereo vision. In *2012 IEEE International Conference on Vehicular Electronics and Safety (ICVES),* (pp. 1-6).

[4]Sattler, T., Leibe, B., & Kobbelt, L. (2011, November). Fast image-based localization using direct 2D-to-3D matching. In *2011 IEEE International Conference on Computer Vision (ICCV),* (pp. 667-674). IEEE.

[5]Li, Y., Snavely, N., & Huttenlocher, D. P. (2010). Location recognition using prioritized feature matching. In *Computer Vision–ECCV 2010* (pp. 791-804). Springer Berlin Heidelberg.

[6]Sinha, D., Ahmed, M. T., & Greenspan, M. (2014, May). Image Retrieval using Landmark Indexing for Indoor Navigation. In *2014 Canadian Conference on Computer and Robot Vision (CRV),* (pp. 63-70).

[7]Lategahn, H., & Stiller, C. (2014). Vision-only localization. *Intelligent Transportation Systems IEEE Transactions on, 15, 3,* 1246 - 1257.

[8]Bay, H., Tuytelaars, T., & Van Gool, L. (2006). Surf: Speeded up robust features. In *Computer vision–ECCV 2006* (pp. 404-417). Springer Berlin Heidelberg.

[9]Morel, J. M., & Yu, G. (2009). ASIFT: A new framework for fully affine invariant image comparison. *SIAM Journal on Imaging Sciences, 2*(2), 438-469.

[10]Muja, M., & Lowe, D. G. (2009). Fast approximate nearest neighbors with automatic algorithm configuration. *Visapp International Conference on Computer Vision Theory & Applications,* 331--340.

[11]B. Triggs, P. McLauchlan, & R. Hartley, & A. Fitzgibbon (1999). "Bundle Adjustment — A Modern Synthesis".

[1] http://www.cmap.polytechnique.fr/~yu/research/ASIFT/

*ICCV '99: Proceedings of the International Workshop on Vision Algorithms.* Springer-Verlag. (pp. 298–372).

[12] Suard, F., Rakotomamonjy, A., Bensrhair, A., & Broggi, A. (2006, June). Pedestrian detection using infrared images and histograms of oriented gradients. In *Intelligent Vehicles Symposium, 2006 IEEE* (pp. 206-212).

[13] Lowe, D. G., "Distinctive Image Features from Scale-Invariant Keypoints", *International Journal of Computer Vision, 60, 2* (pp. 91-110, 2004).

# 25

# A RAPID CLOUD MASK ALGORITHM FOR SUOMI NPP VIIRS IMAGERY EDRS

M. Piper [a,*], T. Bahr [b]

[a] INSTAAR / CSDMS, University of Colorado Boulder, U.S.A. – Mark.Piper@colorado.edu
[b] Exelis VIS GmbH, 82205 Gilching, Germany – Thomas.Bahr@exelisinc.com

**KEY WORDS:** Suomi NPP, VIIRS, Cloud Mask Algorithm, Classification, Skill Scores

**ABSTRACT:**

Suomi National Polar-orbiting Partnership (NPP) is the first of a new generation of NASA's Earth-observing research satellites. The Suomi NPP Visible Infrared Imaging Radiometer Suite (VIIRS) collects visible and infrared views of Earth's dynamic surface processes. This NPP mission produces a series of Environmental Data Records (EDRs). As accurate information on cloud occurrence is of utmost importance for a wide range of remote-sensing applications and analyses, we developed a cloud mask algorithm, adapted from the Landsat 7 Automatic Cloud Cover Assessment, for use with the VIIRS Imagery EDRs. The algorithm consists of a sequence of pixel-based tests that use thresholds on VIIRS top-of-atmosphere reflectances and brightness temperatures. Our cloud mask algorithm provides a simpler, though less informative and robust, alternative to the VIIRS Cloud Mask (VCM) Intermediate Product, with the advantage in that it can be applied to a higher spatial resolution VIIRS Imagery EDR. The algorithm is compared with the VCM in three case studies.

## 1. INTRODUCTION

Suomi National Polar-orbiting Partnership (NPP) is the first of a new generation of NASA's Earth-observing research satellites that observes many facets of our changing Earth. Since 2011 it collects and distributes remotely-sensed land, ocean, and atmospheric data to the meteorological and global climate change communities. The Suomi NPP Visible Infrared Imaging Radiometer Suite (VIIRS) has a 22-band radiometer similar to the MODIS instrument. It collects visible and infrared views of Earth's dynamic surface processes, such as wildfires, land changes, and ice movement. VIIRS also measures atmospheric and oceanic properties, including clouds and sea surface temperature. This NPP mission produces a series of Environmental Data Records (EDRs).

As accurate quality information on cloud occurrence is of utmost importance for a wide range of remote-sensing applications and analyses, we developed a new cloud mask algorithm, based on the Landsat 7 automatic cloud cover assessment (Irish, 2000; Irish et al., 2006), for use with the VIIRS Imagery EDRs. Though the cloud mask algorithm produces results that are less accurate than the VIIRS Cloud Mask (VCM) Intermediate Product (JPSS, 2015), it has an advantage in that it can be quickly calculated from a VIIRS Imagery EDR and used to assess cloud cover at the EDR's 375 m nadir resolution.

## 2. CLOUD MASK ALGORITHM

The cloud mask algorithm exploits the similarity between the multispectral bands of Landsat-7 ETM+ and NPP VIIRS Imagery, as shown in Figure 1.

The cloud mask algorithm (hereafter VIBCM, for VIIRS I-Band Cloud Mask) consists of a sequence of six pixel-based tests that use thresholds on VIIRS top-of-atmosphere reflectances,

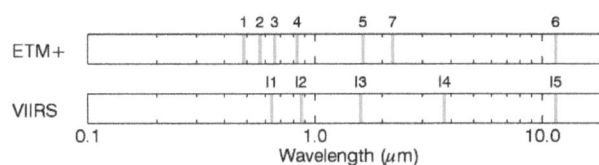

Figure 1. A comparison of the central wavelengths of the Landsat 7 ETM+ bands and the VIIRS I-Bands.

brightness temperatures, and combinations of these measurements. Each test returns a binary (pixel is clear or cloudy) result. For a pixel to be classified as cloudy, it must pass all six tests:

1. **Brightness threshold.** Pixels in I1 with a reflectance greater than 0.08 are classified as cloudy.

2. **Normalized difference snow index.** Pixels with a Normalized Difference Snow Index (NDSI) greater than 0.7 and I2 reflectance greater than 0.11 are classified as cloudy. With VIIRS, NDSI is calculated as (I1 - I3) / (I1 + I3) (JPSS, 2011).

3. **Temperature threshold.** Pixels in I5, the thermal band, with brightness temperatures less than 312 K are classified as cloudy. This threshold value is higher than the value of 300 K used in Irish (2000) because warm clouds were being excluded.

4. **Band I3-I5 composite.** In the composite defined by (max(I3) - I3)*I5, pixels less than a threshold of 410 K, determined by sensitivity analysis, are classified as cloudy. In Irish (2000), a threshold of 225 K was chosen by a similar analysis.

5. **Band I2/I1 ratio.** In this vegetation index proxy, pixels less than a threshold of 2.0 are classified as cloudy.

---

[*] Corresponding author

6.  **Band I2/I3 ratio.** Useful for identifying rocky or sandy areas, pixels in this test with ratios greater than 1.0 are classified as cloudy.

Unless otherwise noted, threshold values in the tests are taken from Irish (2000). Filters 6 and 8 from Irish (2000) use Landsat 7 ETM+ bands that don't match the VIIRS Imagery bands, and so are omitted in the VIBCM. Irish also identifies ambiguously cloudy pixels, then uses a second pass over the data to help classify these pixels. This second pass is not employed in the VIBCM.

## 3.  EXPERIMENT

The VIBCM was tested against the VCM on three daytime, ascending, VIIRS scenes, listed in Table 1.

| Location | Acquisition Start Time | Orbit | Description |
|---|---|---|---|
| Scene 1 **Hawaii** | 2015-Feb-06 23:18:58.504 | 16991 | This scene is dominated by warm ocean and warm clouds, with some cold clouds. The Hawaiian Islands, partially obscured by clouds, are visible at the bottom left. |
| Scene 2 **Eastern United States** | 2012-Jun-04 18:27:52.535 | 3127 | This scene contains mostly land surface, with cold clouds from a mid-latitude cyclone over the Northeastern U.S., and a line of thunderstorms from a trailing cold front in the Southeast. |
| Scene 3 **Northern Europe** | 2012-Nov-06 11:39:51.214 | 5322 | Cold clouds from two weather systems dominate this scene, with an area of clearing between the systems. Snow is visible on the ground in Norway. |

Table 1. The VIIRS scenes used to compare the VIBCM to the VCM.

In each scene, the VCM, at 750 m resolution, was registered to the VIBCM and interpolated to the VIIRS Imagery resolution of 375 m using nearest-neighbor sampling. We chose to use UTM, with the WGS-84 datum, as the common projection for the masks. We chose to count VCM pixels that are probably or confidently cloudy, with medium to high mask quality, as cloudy for the comparison.

To quantify the relationship between the VCM and the VIBCM, we constructed a 2 x 2 contingency table (Stanski et al., 1989) for each of the three scenes, with the VCM representing the "observed" variable and the VIBCM the "forecast" variable.

As shown in Table 2, each pixel has a binary classification – cloudy or not cloudy – for each algorithm. With the VCM as

the basis for comparison, there are two cases where the VIBCM gives the correct result: a "hit" (cell (a) in Table 2), when a pixel is classified as cloudy by both algorithms, and a "correct negative" (d) when both algorithms classify the pixel as not cloudy. The VIBCM is incorrect in the remaining two cases: a "miss" occurs when VIBCM doesn't classify a cloudy pixel, and a "false alarm" occurs when it classifies a pixel as cloudy that isn't.

|  |  | Cloud in VCM | | |
|---|---|---|---|---|
|  |  | **Yes** | **No** | Totals |
| **Cloud in VIBCM** | **Yes** | Hit (a) | False alarm (b) | a + b |
|  | **No** | Miss (c) | Correct negative (d) | c + d |
|  | Totals | a + c | b + d | Grand total n = a+b+c+d |

Table 2. A contingency table for the frequency of cloudy pixels in the VCM and the VIBCM.

Skill scores can be derived from the contingency table values (Stanski et al., 1989). The skill scores used in this experiment are listed in Table 3 and described below.

| Name | Formula |
|---|---|
| Bias | $(a + b) / (a + c)$ |
| Hit rate | $a / (a + c)$ |
| Accuracy | $(a + d) / n$ |
| False alarm rate | $b / (b + d)$ |
| Critical Success Index (CSI) | $a / (a + b + c)$ |
| Heidke Skill Score (HSS) | $2 (ad - bc) / ((a + c)(c + d) + (a + b)(b + d))$ |
| Hanssen-Kuiper Skill Score (KSS) | $a / (a + c) - b / (b + d)$ |

Table 3. Skill scores derived from the contingency table in Table 2.

1.  **Bias** compares the frequency of forecasts to the frequency of actual occurrences. Bias ranges from zero to infinity, with an unbiased score of one. Here, a bias less than one indicates fewer cloudy pixels are present in the VIBCM than in the VCM.

2.  **Hit rate** measures the proportion of observed events that were correctly forecast. The range of the hit rate is zero to a perfect score of one.

3.  **Accuracy** is the ratio of correct events (both cloud and no cloud) to the total number of events. It ranges from zero to a perfect score of one.

4.  **False alarm rate** scores false alarms given the event did not occur. It ranges from one to a perfect score of zero.

5. **Critical Success Index** accounts for false alarms and misses after removing correct negatives. It ranges from zero to a perfect score of one.

6. **Heidke Skill Score** measures the fraction of correct forecasts after removing those due to chance. This score ranges from minus infinity to a perfect score of one, with a score of zero equal to chance, and negative scores indicating skill less than chance.

7. **Hanssen-Kuiper Skill Score** separates the forecasted "Yes" cases from "No" cases. It ranges from minus one to one, with a perfect score of one, and chance equal to zero.

All code used in creating the VIBCM and in comparing it with the VCM are open source and freely available from Piper (2015), under the MIT License.

## 4. RESULTS AND DISCUSSION

To provide a qualitative comparison of the cloud masks, the three VIIRS scenes are previewed in Figures 2, 3, and 4.

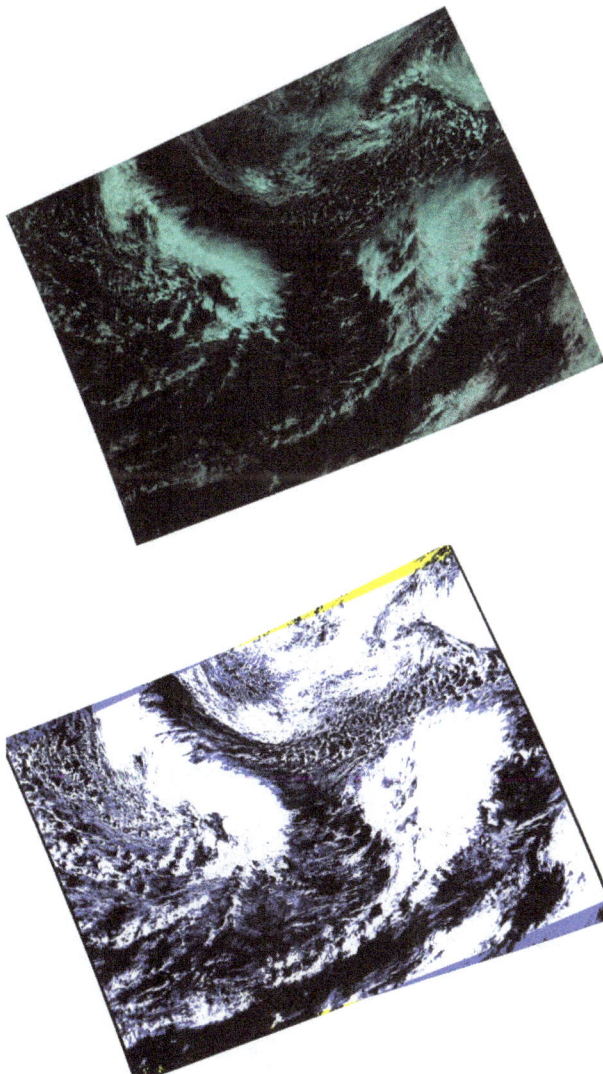

Figure 2. VIIRS I3-I2-I1 false color composite (above) and cloud mask comparison (below) images for Scene 1 (Hawaii).

In the first position of each Figure, an I3-I2-I1 false-color composite image, with a two percent linear stretch applied, is displayed to give a visual depiction of the cloud cover in the scene. In the second position, the VIBCM and VCM computed for a scene are added graphically as binary images. In the result, a white pixel is identified as cloudy by both masks, a black pixel is not a cloud in either mask, a blue pixel is a cloud only in the VCM, and a yellow pixel is a cloud only in the VIBCM. The white, black, blue, and yellow colors in these images provide a visual representation of the contingency table computed for each scene. The areas around the edges of these images, where the registered scenes do not overlap, are excluded from further calculations.

Tables 4, 5, and 6 give quantitative comparisons of the cloud mask algorithms for each scene of the study. Each Table displays pixel counts, cloud fractions, a contingency table, and the derived statistics described in Table 3.

| Total pixels | Cloudy pixels | | Cloud fraction | |
|---|---|---|---|---|
| | VCM | VIBCM | VCM | VIBCM |
| 49390823 | 28434565 | 21255906 | 0. 5757 | 0.4304 |

**Cloud in VCM**

| | | Yes | No | (totals) |
|---|---|---|---|---|
| **Cloud in** | Yes | ☐ 20474434 ☐ | 781472 | 21255906 |
| **VIBCM** | No | ☐ 7960131 | ■ 20174786 | 28134917 |
| | (totals) | 28434565 | 20956258 | 49390823 |

| Bias | Hit rate | Accuracy | FA rate | CSI | HSS | KSS |
|---|---|---|---|---|---|---|
| 0.7475 | 0.7201 | 0.8230 | 0.0373 | 0.7008 | 0.6533 | 0.6828 |

Table 4. Cloud fraction, contingency tables, and skill scores for Scene 1 (Hawaii).

Scene 1, which includes Hawaii in the lower left, is roughly split between cloudy and clear skies. In this scene, the VCM and the VIBCM largely agree on the locations of the cloudy and clear areas. This is supported qualitatively by the predominance of white and black pixels, respectively, in the lower panel of Figure 2. Quantitatively, Table 4 shows that this agreement on cloudy and clear pixels is born out by a high accuracy value. Table 4 also shows that the VIBCM compares favorably with the VCM by other measures; for example, CSI is much greater than zero, indicating a high number of correct events relative to false alarms and misses, and the false alarm rate is close to zero.

However, there are a significant number of misses by the VIBCM, as visually indicated by the blue pixels in the bottom panel of Figure 2. These misses cause a disparity in the computed cloud fraction, pull down skill scores like the hit rate, HSS, and KSS, and give a bias less than one because of the higher number of cloudy pixels identified by the VCM. Figure 2 shows that the misses tend to be concentrated around the edges of the cloudy regions. One explanation may be that the VIBCM does a poorer job of identifying clouds that aren't optically thick. Another is that the conditions we used to define a cloud in the VCM (i.e., a pixel marked as probably or confidently cloudy, with medium to high quality) may be too loose.

When we increased the Filter 4 threshold value to 440 K in this scene, the number of misses decreased by 46 percent, to 4303254, with concomitant increases in hit rate, accuracy, HSS and KSS skill scores. This hints at a nonlinear temperature dependency in this filter.

However, the VIBCM may not be entirely at fault for these misses: in the bottom panel of Figure 3, we noticed that the VCM is picking up the Mississippi and Missouri rivers. There is also an area of stratiform clouds over Lake Huron that, by inspection of the top panel of Figure 3, does not appear to cover as wide an area as indicated by the VCM. This strengthens that argument that we may not have been sufficiently careful in setting conditions for cloudy pixels in the VCM.

One difference between the current scene and Scene 1 is the doubling of the number of yellow false alarm pixels, where the VIBCM identifies a pixel as cloudy, while the VCM does not. Figure 3 shows that the false alarm pixels are grouped in the northeast corner of the scene. The location of the false alarm pixels again suggests an unattributed temperature dependency in the VIBCM.

| Total pixels | Cloudy pixels | | Cloud fraction | |
|---|---|---|---|---|
| | VCM | VIBCM | VCM | VIBCM |
| 49327866 | 30521583 | 25345629 | 0.6187 | 0.5138 |

| | | Cloud in VCM | | |
|---|---|---|---|---|
| | | Yes | No | (totals) |
| Cloud in | Yes | ☐ 23764738 ☐ 1580891 | | 25345629 |
| VIBCM | No | ☐ 6756845 ☐ 17225392 | | 23982237 |
| | (totals) | 30521583 | 18806283 | 49327866 |

| Bias | Hit rate | Accuracy | FA rate | CSI | HSS | KSS |
|---|---|---|---|---|---|---|
| 0.8304 | 0.7786 | 0.8310 | 0.0841 | 0.7403 | 0.6597 | 0.6946 |

Table 5. Cloud fraction, contingency tables, and skill scores for Scene 2 (Eastern United States).

Figure 3. VIIRS I3-I2-I1 false color composite (above) and cloud mask comparison (below) images for Scene 2 (Eastern United States).

In Scene 2, covering the Eastern United States and Canada, about half of the area is cloud-covered, with a mid-latitude cyclone exiting to the northeast, and a line of scattered thunderstorms along a trailing cold front to the south, as shown in the top panel of Figure 3.

As in Scene 1, there is good qualitative agreement between the cloud masks, as evidenced by the prevalence of white (cloudy) and black (clear) pixels in the bottom panel of Figure 3. The Figure also shows that the VIBCM is still missing clouds picked up by the VCM. The blue pixels indicative of these misses again tend to be found around the edges of deeper cloud banks. Although this occurs throughout the scene, it is especially noticeable in the southern portion.

Table 5 shows that, for this scene, like Scene 1, the total number of cloudy pixels, as well as the cloud fraction, are higher in the VCM. This is due to the number of misses (blue) by the VIBCM. However, the current scene actually has fewer misses than Scene 1, which gives a higher hit rate.

As in Scene 1, the accuracy in this scene is high. The count of correct cloudy pixels (white) and correct clear areas (black) are each an order of magnitude larger than the miss and false alarm (yellow) values. Likewise, the CSI score remains high because the number of correct cloudy pixels far outnumber misses and false alarms.

The bias score is in favor of the VCM, again because of the misses in the VIBCM. The bias is less than that in Scene 1, though, because of higher number of false alarm pixels in the current scene. The false alarm rate is higher than in Scene 1, but it is still close to zero.

The HSS and KSS scores for this scene are high, and are similar to those in Scene 1. Both are closer to one (a perfect prediction) than zero (random chance).

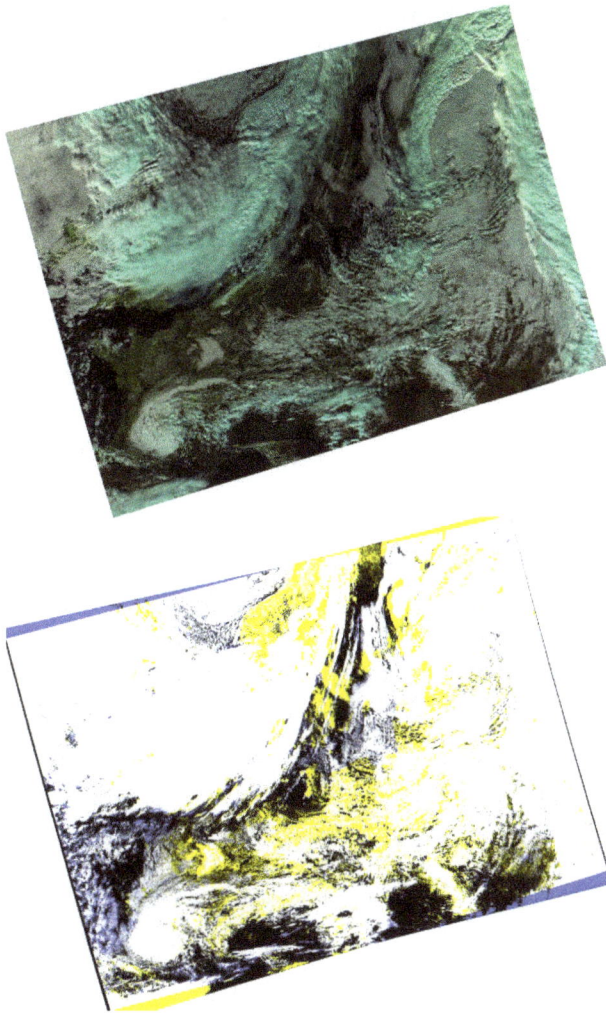

Figure 4. VIIRS I3-I2-I1 false color composite (above)
and cloud mask comparison (below) images for Scene 3
(Northern Europe).

Two storm systems dominate the weather over Northern Europe in Scene 3, with snow on the ground in Norway, in the northern half of Sweden, and in the Alps (NCDC, 2015). Clouds (white pixels) prevail in this scene.

Note that, in the bottom panel of Figure 4, there are far fewer blue pixels, denoting misses by the VIBCM, than in the previous two scenes. There are, however, more yellow pixels, indicating false alarms by the VIBCM. The VIBCM misclassifies the snow on the ground in Norway and Sweden as cloud. On the other hand, the VIBCM appears to correctly identify two low cloud banks to the southwest of Stockholm, and one to the north of Berlin, none of which are picked up by the VCM.

Table 6 shows that the number of false alarm pixels (yellow) is approximately three times as high as Scene 2. However, the number of misses (blue) is nearly three times lower. The false alarm rate is five times as high as in Scene 2. Note that this is due, in part, to the clouds missed by the VCM described above, so this number may be inflated.

Overall, the VIBCM compares well with the VCM in this scene, with the highest hit rate, accuracy and CSI scores of the three

scenes, and the lowest bias. These high scores are the result of the large number of cloudy pixels classified by both masks.

| Total pixels | Cloudy pixels | | Cloud fraction | |
|---|---|---|---|---|
| | VCM | VIBCM | VCM | VIBCM |
| 49436935 | 36149374 | 39745887 | 0.7312 | 0.8040 |

**Cloud in VCM**

| | | Yes | No | (totals) |
|---|---|---|---|---|
| Cloud in | Yes | ☐ 34155952 | ☐ 5589935 | 39745887 |
| VIBCM | No | ☐ 1993422 | ■ 7697626 | 9691048 |
| | (totals) | 36149374 | 13287561 | 49436935 |

| Bias | Hit rate | Accuracy | FA rate | CSI | HSS | KSS |
|---|---|---|---|---|---|---|
| 1.0995 | 0.9449 | 0.8466 | 0.4207 | 0.8183 | 0.5732 | 0.5242 |

Table 6. Cloud fraction, contingency tables, and skill scores for Scene 3 (Northern Europe).

When we lowered the Filter 4 threshold to 390 K, it resulted in a decrease of 23 percent in the number of false alarm pixels, to 4280523. This, in turn, decreased in the false alarm rate and the bias. Accuracy, along with HSS and KSS, increased. CSI remained approximately the same.

## 5. CONCLUSIONS

We developed a cloud mask algorithm (the VIIRS I-Band Cloud Mask, or VIBCM) based on the Landsat 7 ETM+ Automatic Cloud Cover Assessment. To assess its ability to identify clouds, we compared it, both qualitatively and quantitatively, with a cloud mask derived from the VIIRS Cloud Mask (VCM) Intermediate Product using a case study of three VIIRS scenes that varied in location and season. The results indicate, by various skill scores, a quantitatively good match between the two cloud masks; for example, the accuracy, defined as the sum of cloudy and clear pixels classified by both masks, divided by the total number of pixels in a scene, is above 80 percent in each of the scenes. However, there remains room for improvement.

The VIBCM provides the following advantages:

- It can quickly be computed from the five bands of a VIIRS Imagery EDR. No outside sources are needed.
- It computes a mask at the Imagery EDR resolution of 375 m nadir instead of the SDR resolution of 750 m.

The VIBCM also has disadvantages:

- It is not as accurate as the VCM: cloudy pixels are frequently missed or misidentified.
- It is not as detailed as the VCM: there are no confidence flags for cloudy pixels, and no distinction between high clouds, low clouds, fog, smoke, and shadow (JPSS, 2015).

There are three unresolved issues in the VIBCM that merit future work. The first is an investigation of what appears to be a temperature dependency in the threshold value of Filter 4. As the I5 brightness temperatures decreased from Scene 1 (Hawaii, warm) through Scene 3 (Europe, cold), the number of misses by the VIBCM decreased, and the number of false alarms increased. When we experimented with different threshold values in each scene, some of misses and false alarms were converted into hits. It would be better to perform an extended case study, where the use of many scenes might help quantify an empirical relationship between the threshold temperature in Filter 4 and its response. Alternately, the two-pass technique used by Irish (2000), which we chose not to implement in the VIBCM, could help address this issue.

The second unresolved issue lies in what we have chosen to define as a cloud in the VCM; that is, any pixel that is probably or confidently cloudy, with medium to high mask quality. As described above, this definition produces false alarms in Scene 2, and misses in Scene 3. We want to be careful in stating that we do not fault the VCM for this issue; rather, we may need to be more careful in our use of the quality flags produced by the VCM. In future work, we will explore how seasonal and latitudinal conditions on how the VCM is used affect the comparison with the VIBCM.

The third issue that merits further study is differentiating between clouds and snow. Only one scene in our case study had snow, and the VIBCM failed to identify it. Improving this behavior will require additional work with a range of VIIRS scenes that contain snow, cold surface temperatures, and cold clouds.

## ACKNOWLEDGMENTS

MP gratefully acknowledges the help of Curtis Seaman (CIRA/Colorado State University) in dealing with bowtie deletion issues in the VIIRS imagery.

## REFERENCES

Irish, R.R., 2000. Landsat 7 automatic cloud cover assessment. *Proceedings of SPIE*, 4049, pp. 348–355.

Irish, R.R., Barker, J.L., Goward, S.N., Arvidson, T., 2006. Characterization of the Landsat-7 ETM+ Automated Cloud-Cover Assessment (ACCA) algorithm. *Photogrammetric Engineering & Remote Sensing*, 72 (10), pp. 1179–1188.

Joint Polar Satellite System (JPSS), 2015. Joint Polar Satellite System (JPSS) Operational Algorithm Description (OAD) Document for VIIRS Cloud Mask (VCM) Intermediate Product (IP) Software. http://npp.gsfc.nasa.gov/sciencedocuments/2015-03/474-00062_OAD-VIIRS-Cloud-Mask-IP_H.pdf. (10.03.2015)

Joint Polar Satellite System (JPSS), 2011. Joint Polar Satellite System (JPSS) VIIRS Snow Cover Algorithm Theoretical Basis Document (ATBD). http://npp.gsfc.nasa.gov/sciencedocuments/ATBD_122011/474-00038_VIIRS_Snow_Cover_ATBD_Rev-_20110422.pdf. (10.03.2015)

National Climatic Data Center (NCDC), 2015. Europe/Asia Snow Cover. http://www.ncdc.noaa.gov/snow-and-ice/snow-cover/ea/20121105-20121107. (23.03.2015)

Piper, M., 2015. The VIIRS I-Band Cloud Mask. https://github.com/mdpiper/viirs-cloudmask. (22.03.2015)

Stanski, H.R., Wilson, L.J., Burrows, W.R., 1989. Survey of common verification methods in meteorology, *World Weather Watch Tech. Rept.*, 8, *WMO/TD*,.358, *WMO*, Geneva, 114 pp.

# 26

# A LABORATORY PROCEDURE FOR MEASURING AND GEOREFERENCING SOIL COLOUR

Á. Marqués-Mateu [a],[*] M. Balaguer-Puig[a], H. Moreno-Ramón[b], S. Ibáñez-Asensio[b]

[a] Universitat Politècnica de València (UPV), Department of Cartographic Engineering, Geodesy, and Photogrammetry, 46022 València, Spain – amarques@cgf.upv.es, balaguer@cgf.upv.es
[b] Universitat Politècnica de València (UPV), Department of Plant Production, 46022 València, Spain – hecmora@prv.upv.es, sibanez@prv.upv.es

**KEY WORDS:** Soil, Colorimetry, CIE, GIS, Geomatics

**ABSTRACT:**

Remote sensing and geospatial applications very often require ground truth data to assess outcomes from spatial analyses or environmental models. Those data sets, however, may be difficult to collect in proper format or may even be unavailable. In the particular case of soil colour the collection of reliable ground data can be cumbersome due to measuring methods, colour communication issues, and other practical factors which lead to a lack of standard procedure for soil colour measurement and georeferencing. In this paper we present a laboratory procedure that provides colour coordinates of georeferenced soil samples which become useful in later processing stages of soil mapping and classification from digital images. The procedure requires a laboratory setup consisting of a light booth and a trichromatic colorimeter, together with a computer program that performs colour measurement, storage, and colour space transformation tasks. Measurement tasks are automated by means of specific data logging routines which allow storing recorded colour data in a spatial format. A key feature of the system is the ability of transforming between physically-based colour spaces and the Munsell system which is still the standard in soil science. The working scheme pursues the automation of routine tasks whenever possible and the avoidance of input mistakes by means of a convenient layout of the user interface. The program can readily manage colour and coordinate data sets which eventually allow creating spatial data sets. All the tasks regarding data joining between colorimeter measurements and samples locations are executed by the software in the background, allowing users to concentrate on samples processing. As a result, we obtained a robust and fully functional computer-based procedure which has proven a very useful tool for sample classification or cataloging purposes as well as for integrating soil colour data with other remote sensed and spatial data sets.

## 1. INTRODUCTION

Vegetation maps are typical end products of remote sensing workflows that have multiple potential uses in agricultural and environmental engineering as well as in soil science. Early authors on the subject of vegetation mapping from remote sensed images were already aware of the influence of soil spectral characteristics on the final maps due to the spectral mixing of signal coming from the vegetation response together with signals from the background soil surface (Richardson and Wiegand, 1977, Tucker and Miller, 1977). The subject became a field of study on its own and its development led to the concept of 'soil line' which is common nowadays (Baret et al., 1993, Gitelson et al., 2002).

There exist several methods to model the influence of soils in remote sensing applications. All those methods are somewhat based on underlying assumptions that may or may not be met in real applications. In particular soil lines derived entirely from the images are limited to points or areas located on bare soil pixels which is clearly a practical limitation.

It is also well accepted that soil type, together with roughness, water content and other factors, are soil parameters that can distort the mathematical definition of soil lines. While roughness and water content parameters are highly variable in time and space, soil type is defined by a set of long-term characteristics that are routinely recorded in soil surveys. There are numerous applications of survey data sets for quantitative analyses (Bouma, 1989) which include using those data in combination with remote sensed images and other spatially distributed datasets.

Soil survey records include a number of physical, chemical, management, and environmental characteristics such as soil colour. Colour is recorded with three attributes that can be related with the reflectance of soils in the visible region of the spectrum, and thus it has a clear relationship with remote sensed data as reported in previous research (Baumgardner et al., 1985).

There are two general approaches to computing soil and vegetation fractions based either on visible and infrared data or on visible data only (Gitelson et al., 2002). In any case, soil is not just a gray background and previous studies state that soil processing in remote sensing applications requires specific collection of soil spectral characteristics (Escadafal and Huete, 1992). Therefore, a procedure for collecting soil colour data in spatial format appears to be useful for methods based on data from the visible spectrum.

There is still another class of experimental studies where soil reflectance data are used to find relationships between colour and soil properties. These studies include laboratory experiments (Torrent et al., 1983, Torrent and Barrón, 1993), agricultural landscape studies (Sánchez-Marañón et al., 1997, Gunal et al., 2008), and spatially based studies (Ibáñez-Asensio et al., 2013, Moreno-Ramón et al., 2014).

It seems obvious that remote sensing and GIS users could benefit from soil colour data sets if a precise procedure to transfer georeferenced data were available. This is the principal point in this paper which focuses on the particular subject of georeferencing soil colour data in the laboratory. Specifically, the goal herein is to present a rigorous laboratory procedure that provides a means for collecting georeferenced soil colour datasets which can be used in studies like those referenced above.

---

[*]Corresponding author

## 2.  SOIL COLOUR

Standard texts on soil colour often begin stating that colour is the most obvious physical characteristic of soil (Simonson, 1993, Thwaites, 2006) which has practical applications in classification tasks (FAO, 2006, Soil Survey Staff, 2010), remote sensing (Baumgardner et al., 1985, Metternicht and Zinck, 2009), and mapping (Boettinger et al., 2010).

The physical and numerical framework for processing colour information was established by the *Commission Internationale de l'Éclairage* (CIE) in 1931 when a set of resolutions were published (Schanda, 2007). The CIE resolutions set the principles of modern colorimetry by gathering all the technical and scientific knowledge on colour of the time, and are still in use with some modifications. What follows is a brief summary of the resolutions from the soil laboratory standpoint.

Colour is conveniently represented in a three dimensional space such as the CIE RGB system. This is a physically-based system, and therefore is considered as a device independent system in contrast to digital devices that output RGB data in their particular colour spaces. The CIE RGB system has obvious theoretical interest, but it is not used in practice. Instead, a number of derivative spaces such as the CIE XYZ, CIE Yxy or CIELAB are preferred. There are closed formulas to convert colour data between those three colour spaces (CIE, 2004, Malacara, 2011).

The CIE 1931 XYZ colour space represents a colour stimulus with three numbers XYZ called tristimulus values, where Y represents the luminance, that is, the total radiation reflected in the visible spectrum.

Tristimulus values can be converted to the so-called chromaticity coordinates using simple formulas:

$$x = X/(X + Y + Z); \quad y = Y/(X + Y + Z) \qquad (1)$$

where    $x, y$ = chromaticity coordinates
$X, Y, Z$ = tristimulus values

Chromaticity coordinates are normalised as a function of tristimulus values that allows positioning colour stimuli in the Yxy colour space by means of the chromaticity diagram (Figure 1). The values of $xy$ carry the chromatic content of a colour stimulus, whereas the third dimension contains the achromatic component Y. Although chromaticity coordinates allow trained users to know a colour, they are not psychophysical correlates of human vision. There are, however, geometric formulas that give estimates of such correlates which are named dominant wavelength and excitation purity in the Yxy space (CIE, 2004).

The chromaticity diagram suffers from a lack of uniformity over its domain, which is a known issue when computing colour diferences with euclidean distances in the Yxy space. To overcome this problem, the CIE published a new color space called CIELAB in 1976 that is supossed to be quasi uniform. In the CIELAB space a colour is depicted with three coordinates named $L^*$ (lightness), $a^*$ (red-green axis), and $b^*$ (yellow-blue axis) in a solid based on the theory of opponent colours. $L^*$ is a psychophysical correlate of human perceived lightness. However, $a^*$ and $b^*$ are not correlates of human perceived hue and saturation, but the CIELAB space provides formulas to calculate estimates of hue ($h_{ab}$) and chroma ($C^*_{ab}$) (CIE, 2004). The $L^*C^*h$ space is therefore a psychophysical counterpart of the CIELAB space.

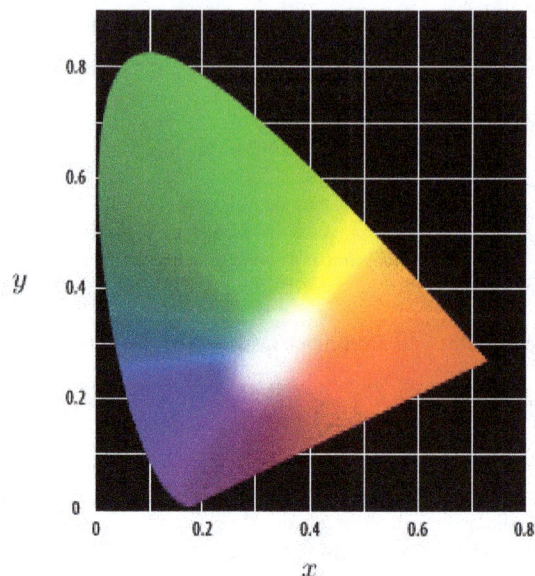

Figure 1: CIE chromaticity diagram

The CIELAB coordinates are computed as follows (CIE, 2004):

$$\begin{aligned}
L^* &= 116 \cdot (Y/Y_n)^{1/3} \\
a^* &= 500 \cdot \left[ (X/X_n)^{1/3} - (Y/Y_n)^{1/3} \right] \qquad (2) \\
b^* &= 200 \cdot \left[ (Y/Y_n)^{1/3} - (Z/Z_n)^{1/3} \right]
\end{aligned}$$

where    $L^*\, a^*\, b^*$ = CIELAB coordinates
$X, Y, Z$ = tristimulus values of the observed sample
$X_n, Y_n, Z_n$ = tristimulus values of reference white

The CIELAB space also provides formulas to compute colour differences. Given two colours $(L^*_1, a^*_1, b^*_1)$ and $(L^*_2, a^*_2, b^*_2)$ the difference is (CIE, 2004):

$$\Delta E^*_{ab} = \sqrt{(L^*_2 - L^*_1)^2 + (a^*_2 - a^*_1)^2 + (b^*_2 - b^*_1)^2} \qquad (3)$$

All the previous colour spaces were originally defined by means of visual experiments with human observers. The implementation of those physically-based colour spaces in modern instruments are done with a spectrophotometric approach that calculates tristimulus values with the sum of the product of three functions across the visible range of the electromagnetic spectrum. The expressions are (CIE, 2004):

$$\begin{aligned}
X &= \int_\lambda \rho_\lambda \cdot E_\lambda \cdot \overline{x}_\lambda \;\; d\lambda \\
Y &= \int_\lambda \rho_\lambda \cdot E_\lambda \cdot \overline{y}_\lambda \;\; d\lambda \qquad (4) \\
Z &= \int_\lambda \rho_\lambda \cdot E_\lambda \cdot \overline{z}_\lambda \;\; d\lambda
\end{aligned}$$

where    $\rho_\lambda$ = reflectance function of the specimen
$E_\lambda$ = light spectral function
$\overline{x}_\lambda\, \overline{y}_\lambda\, \overline{z}_\lambda$ = colour matching functions

The spectrophotometric formulas take into account the three elements of colour, that is, the object that reflects light energy, the light source characteristics, and the observer (or sensor that detects the light). The colour matching functions represent a human observer with normal colour vision. These functions are published by the CIE (CIE, 2004) and must be somehow embedded in modern instrumentation. It is worth noting that in engineering applications the integrals in Equation (4) are replaced with sums.

Although CIE colour spaces define a rigorous framework to process colour data, the study of colour in soil science followed a different path. The first references to color in soil surveys date back to the last decade of the 19th century. The approach to communicating colour in soil science was a visual procedure that allowed field surveyors to match the colours of soil samples with a collection of colour chips arranged in the so-called colour books (Figure 2). Although precursors of the CIE spaces already existed at that time, the technological development of instruments together with the limited computing power made the CIE spaces unsuitable for practical uses. In 1941, the United States Department of Agriculture (USDA) published the first soil colour charts (Simonson, 1993) in a format which has remained to the present day (Munsell Color, 2000).

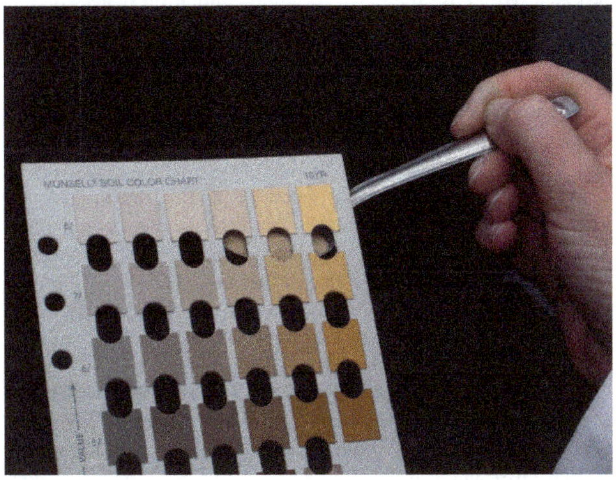

Figure 2: Visual assessment of soil colours with Munsell charts

The USDA published the colour charts in collaboration with the Munsell Color Company where a team of arstists and scientists leaded by A.F. Munsell developed a physical implementation of a colour solid. The colour solid was supposed to represent the whole domain of colours that were physically realizable with the colour technology available.

In the Munsell system, colours are arranged following an order in the three colour components: hue, value and chroma. It is therefore a colour order system that is specially suited to making fast visual comparisons with suitable training. In the Munsell system the term 'value' is used to denote an equivalent to luminance or lightness in the CIE spaces. There are clear similarities between the Munsell order system and the CIE spaces. For instance, the Munsell hue circle is very similar to the CIE chromaticity diagram. Note however that hues are arranged in opposite order in both representations (Figures 1 and 3).

Colour communication with Munsell charts is done with descriptions called Munsell notations. Notations are alphanumeric codes that describe a colour stimulus in terms of hue, value and chroma exactly in this order. Hue contains numbers and letters (R=red, Y=Yellow, B=Blue, and so on). Value and chroma are numbers. Each notation has an associated colour name. As an example a

notation 10YR6/4 has a colour name 'light yellowish brown. The three components are hue (10YR), value (6) and chroma (4).

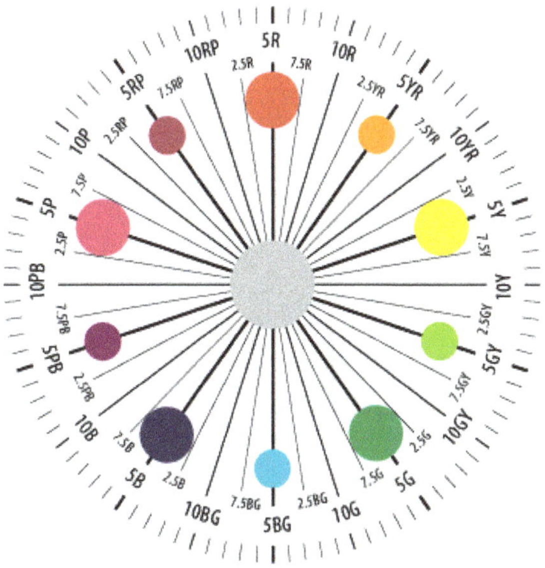

Figure 3: Munsell hue circle

The use of CIE spaces is well suited to laboratory work since they allow high degree of automation and efficient processing of soil colour data. However, the laboratory setup must be established very carefully to reach maximum accuracy in the measurements (Torrent and Barrón, 1993). The Munsell system, on the other hand, is still common use in soil science as observed in standard manuals (Soil Survey Division Staff, 1993). The coexistence of two colour systems poses a problem of transformation between two different spaces. This problem is not yet solved with analytic formulas (Malacara, 2011) although there are a number of different approximate methods in the literature. We face this problem in Section 3.4 with a simple machine learning technique.

## 3. PROCEDURE OUTLINE

The requirements considered when designing the procedure were twofold. First, we sought a procedure such as not to interfere with standard soil analyses. Secondly, we needed to create ready to use datasets in spatial format.

The first requirement was met by allocating a specific room in the Soils Laboratory of the Universitat Politècnica de València (UPV) as a dedicated colour laboratory. With this infrastructure, a small fraction of every soil sample can be processed to collect colour data in parallel with regular soil analyses.

The second requirement demanded a configurable computer environment in order to automate the measurement process as much as possible, and eventually to create the spatial databases. After assessing several options, we decided to write a program that fitted to our specific needs from the beginning.

Both criteria were taken into account when setting the processing routines of the procedure as outlined in Figure 4. The flowchart separates tasks done in the colour laboratory from those carried out in the soils laboratory.

The colour laboratory processing takes three steps. The first step is to read colour coordinates in a CIE space. The specific colour

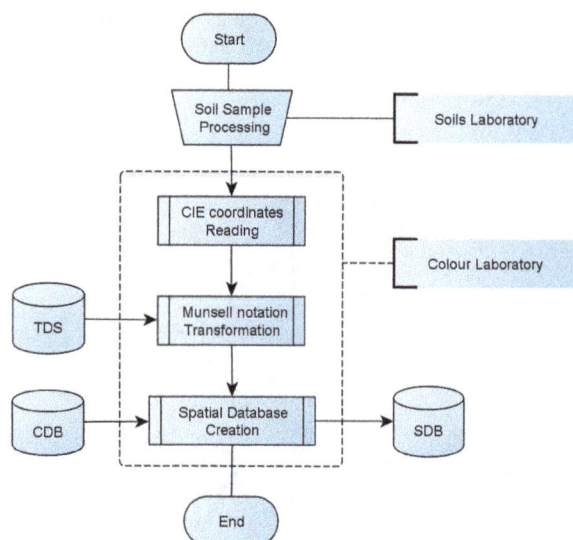

Figure 4: Flowchart of the procedure

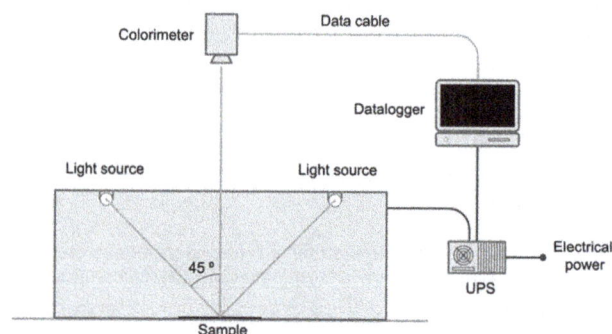

Figure 5: Laboratory setup

space used in this step depends on instrumentation characteristics. In our setup we used a trichromatic colorimeter that outputs coordinates in the CIE Yxy space. The coordinates are automatically converted to CIELAB coordinates provided that white reference readings are available in the data set.

In the second step, CIE coordinates are transformed to Munsell notations using records contained in the training data set file which is labeled as 'TDS' in Figure 4. For a description of the Munsell transformation see Section 3.4. This step is optional and depends on the existence or availability of the training data set.

Finally, the third step creates spatial data files on disk. This operation requires a coordinate database ('CDB' in Figure 4) containing spatial coordinates of the sample points. It is the responsibility of the user to create and maintain that database. The reference system of the coordinates must be consistent since the computer program that reads the file makes no assumption on this matter. This step requires the coordinate database to work properly. If, for any reason, the coordinate database is not available, the ouput will be a simple data table without spatial information. The final spatial database is labeled 'SDB' in Figure 4.

### 3.1 Laboratory setup

The laboratory setup is a critical point to achieve maximum accuracy and efficiency in soil colour processing (Torrent and Barrón, 1993). The components of our setup are: colorimeter, light booth, datalogger, and uninterruptible power supply (UPS).

We used a non-contact trichromatic colorimeter that fits well to measuring granular specimens. The instrument was the Chroma Meter CS-100A by Konica Minolta (Konica Minolta, 1992). This class of instruments require careful mounting to keep the distance to the specimen constant. If that distance were not constant luminance readings could not be compared among samples.

Another critical point is the geometry of illumination. We used a configuration known as 45°/0° geometry where light reaches the sample being observed at 45° measured from the normal to the sample surface. That geometry avoids specular reflection, glare, and other unwanted optical effects, and ensures that recorded colours belong to soil samples rather than to light. The colour

sensor is mounted at 0° from the normal to sample plane as seen in Figure 5.

We followed recommendations given by the instrument maker and by the CIE. Besides geometry, the recommendations also included the white plate and the illuminant used in the measurement process. We chose the CS-A20 white calibration plate which is part of the complementary equipment of the CS-100A colorimeter (Konica Minolta, 1992). Regarding illuminant, we used D65 (Daylight 6500K) simulators as proposed by the CIE for colour experiments (CIE, 2004).

The setup is completed with the datalogger which takes care of communicating with the colorimeter by sending messages to the instrument and collecting back colour data. In our setup the datalogger was a regular PC computer running a computer program (Section 3.2) that also performed other tasks such as computing average values from raw measurements, reading auxiliary files and writing the final spatial databases.

The communication is done over a RS-232 serial line which was available in the computer. We used a special RS-232 cable called LS-A12 which has two connectors. One connector is the well-known DB9 which has 9 pins and was the most common connector until the advent of USB ports. The other connector is a non-standard one called RP17-13RA-12SD created by Konica Minolta to fit the serial port located in the colorimeter frame.

It is important to note here that a time period of 30 minutes should be established before proceeding to measuring tasks. This period allows the light sources to reach their operating temperature which is necessary to avoid measurements deviations due to variations in ambient light. The UPS contributes to maintain such uniform light conditions avoiding voltage variations.

### 3.2 Computer program

The computer program described in this section was codenamed CS-100A and is one major element of the procedure to measure and georeference soil samples. We developed this program with laboratory user's needs in mind, and imposed ourselves several requirements such as multiplatform support, seamless integration in laboratory workflow, and connectivity with the colorimeter.

The program was developed in the Python programming language (van Rossum and Drake, 2011). This language uses a modular structure that allows new functionalities by importing modules or packages. There are two modules in our program that stand out out from the rest: Tkinter and pyserial. Tkinter is the default module for designing graphical user interfaces (GUI) in Python and is always available in standard installations. The module py-serial provides an interface to communicate with serial devices. This module is not part of the standard Python and must be installed prior to running the application.

It should be stressed here that proper communication with the colorimeter was possible because the communication protocol was available (Konica Minolta, 2000). The documentation was very precise and allowed us to write the communication functions without problems. The protocol is asynchronous, so that first the program sends a message to the colorimeter, then the colorimeter does something and sends some data back to the program. The program should finally interpret and decode the received data in a convenient way. The three important aspects of the protocol are the communication parameters, the list of valid commands and the format of the received data.

The CS-100A instrument has a fixed configuration with the following parameters:

- Baud rate: 4800

- Parity: even

- Data length: 7 bits

- Stop bits: 2 bits

The complete list of commands is shown in Table 1 although the most important for the program is 'MES' which measures one time. The commands sent over the serial line have two fields, the command name and the delimiter that marks the end of the message. In this protocol the delimiter is a two character string with CR (carriage return) and LF (line feed). Therefore, in order to take a measure the program must send the message:

```
MES + <CR> + <LF>
```

The data sent back by the instrument consist of text strings with fields separated by commas. A typical string returned after sending a 'MES' command looks like the following:

```
"MES: OK00, 121.5, .3325, .3565"
```

The string contains the command name and the message that carries the measurement. The measurement has four fields. The first field (OK00) is an error code that in this particular string means that there were no errors. The three remaining are $Yxy$ coordinates. The second field (121.5) is the luminance in units of $cd/m^2$. The units of luminance variable can be changed with a switch located in the battery chamber of the instrument. The third (.3325) and fourth (.3565) fields are chromaticity coordinates $x$ and $y$ respectively.

| Command | Description |
|---------|-------------|
| MES | Measures one time |
| MDS | Carries out various mode settings |
| CLE | Clears memory |
| RCR | Recalculates according to the mode settings |
| TDR | Returns target value for colour differences |
| UCR | Returns current standard calibration value |
| TDS | Stores new target value for colour differences |
| TDW | Receives new target value for colour differences |
| UCW | Stores new standard calibration value |

Table 1: Commands of CS-100A protocol

The first task performed by the program is to check the operating system it is running on. This is a simple but important step to guarantee multi-platform support. We successfully tested the program on Linux, Mac OS and Windows with exactly the same source code.

Next, the program reads a configuration file containing several variables. Those variables specify several disk locations to store data files, and other parameters such as the output spatial format or the serial port that provides colour data from the colorimeter. If the configuration file does not exist, then a default configuration is set by the program. After reading the configuration file into memory, the main graphical user interface is displayed on the screen (Figure 6).

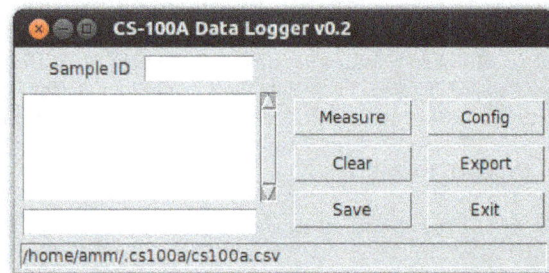

Figure 6: Main window of the program

Figure 7: Configuration window

Before proceeding with the measurement tasks, users can change the configuration of the working environment. Configurable items include: output data files, auxiliary files, spatial format, and serial port (Figure 7). There are two output files to store measurements and a log of the session. The measurements file contains a table of $Yxy$ coordinates in comma separated values (CSV) format. This file is suitable to be processed in other computing environments. The log file contain more information such as timestamps and raw text strings recovered from the colorimeter.

The auxiliary files are the coordinates and the Munsell files. The former contains spatial coordinates of the samples to be processed, whereas the latter contains CIE coordinates of Munsell chips. Both files must be provided by the user and are optional in some sense, that is, if they are not available, the program can still run and generate meaningful outputs, but they are required to obtain full functionality. For instance, if the coordinates file is not available the output will not be in spatial format.

The output spatial format can be selected from a list of three well-known formats: Shapefile, GML and GeoJSON. The Shapefile format is probably the most common format in the fields of geographic information systems (GIS) and geomatics. It was introduced by a commercial company and has become a *de facto* standard probably because the specification of the format is publicly available (ESRI, 1998). The geography markup languaje (GML) format, on the contrary, is a standard proposed by the Open Geospatial Consortium (OGC) and is "an XML grammar for expressing geographical features" whose specification is available in several official documents (OGC, 2010). Finally, the Geo-JSON is the most recent format of the three. GeoJSON is based on the JavaScript object notation (JSON) and is defined as "a format for encoding a variety of geographic data structures." This format can be a valid option to send geographic data over computer networks or in mobile devices and use the concept of dictionary or associative array to pack the data (Sriparasa, 2013).

The last configurable item is the serial port that will be the entry point of colorimetric data. The names of such ports are platform dependent and is one of the points that our program manages at start time to show correct port labels regardless of the platform.

### 3.3   Program operation

As described in the previous section, the GUI is very simple and consists of two windows that allow measuring and configuring tasks. The main window has two sections as seen in Figure 6. The left half of the window has three text boxes and the right half contain a number of buttons with specific functions. A brief outline of the program operation and some recommendations are given below.

The first action of the user is to type the Sample ID in the entry located in the upper left section. The ID allows joining colour measurements with spatial coordinates and other data, therefore it must be unique. There are two special IDs ('w' and 'white') reserved to indicate measurements on the white calibration plate. It is common to start a measuring session with a reading on the reference white plate.

Below the sample ID entry there are two text boxes where colour measurements are shown as they are read from the colorimeter. The upper (and bigger) box contains all the measurements done on the current sample, whereas the bottom box contains mean values of the coordinates measured up to that point. The mean values of luminance $Y$ and chromaticity coordinates $xy$ are updated after every new measurement.

In the right section there is a group of six buttons that allow actions such as measuring colour coordinates, clearing the data to start a new measurement series, saving measured data, configuring the working environment, exporting the data to spatial format, and stopping the program.

The 'Measure' button sends a 'MES' command to the colorimeter to request a new measurement. After receiving the command, the instrument takes a new measurement and sends it back to the other side of the serial link. In the meantime, the program stays listening for any incoming data. When the new coordinates arrive, they are displayed on the measurement text box and the mean values are updated.

The natural way of measuring the colour of a sample is by reading a series of values. The program allows this and shows all the individual measurements and the mean values for each colour coordinate coordinate ($Y$, $x$ and $y$). If for any reason it is necessary to reject any previous measures, the user should click on button 'Clear'. This button removes any contents both in the text boxes and in memory, and awaits for a new measurement series.

If the data are correct the user should click on button 'Save' to store the data on disk files. This button also removes any previous content from the text boxes and awaits for a new series. The mean values are stored in the file specified in the configuration window under the item 'Measurement' of the 'Output files' area. The individual raw measurements are recorded in the log file with timestamps just in case the user needs to check the whole session.

The 'Config' button shows the configuration window (Figure 7) where the user can customise the environment. As described in the previous section, there are four configurable items: output files, auxiliary files, spatial format, and serial port.

Once the whole laboratory session has finished, the user should click on 'Export' to create a spatial database containing colour information about the processed samples. When clicking this button, several actions take place in the background in a transparent fashion for the user. First, the program searches for the spatial coordinates database. If this file does not exist, it will not be possible to create the spatial database. In this event, the program will print a warning message. Next, the program searches for the Munsell file with the CIE coordinates of the Munsell colour chips to transform CIE coordinates into Munsell notations. If both files exist, the the user ends up with a spatial database containing point geometries and an attribute table with CIE coordinates as well as with Munsell notations. It should be noted that the program also calculates CIELAB coordinates and writes them to the attribute table if white reference readings are found in the data file.

It is convenient to highlight a couple of key points before closing this section. As stated above, there are several CIE colour spaces used in soil science, the Yxy and CIELAB spaces being the most commonly used in practice. It is worth noting that program CS-100A does not check the colour space of the Munsell data. It is the responsibility of the user to ensure that both, samples coordinates and Munsell chips coordinates, are expressed in the same colour space. Otherwise, the results will contain errors. Likewise, it is necessary to measure soil samples under the same environmental light conditions of the Munsell charts measurements to obtain consistent results.

### 3.4   CIE to Munsell transformation

As mentioned above, one point of interest in soil colorimetry is to report colours as Munsell notations to ensure compatibility with common practice. While there are closed formulas to convert between CIE XYZ, CIE Yxy, and CIELAB spaces (CIE, 2004), in the case of transforming from CIE coordinates into Munsell notations such formulas do not exist. Instead, a number of approximate methods can be found in the colorimetry literature. We addressed this problem with a non-conventional approach in soil colorimetry based on the k nearest neighbours (k-NN) technique (Steinbach and Tan, 2009).

The goal of the k-NN method is to assign class labels to unknown objects. Those objects are just points in a multidimensional coordinate system. In our particular case, the obvious choice is to define the coordinate system with the three colour dimensions, $Yxy$ in the CIE Yxy space or $L^*a^*b^*$ in the CIELAB space.

There are two datasets named test and training sets. The test dataset contains records that represent unknown objects, that is, objects whose classes are to be defined. The training dataset, on the contrary, contains objects with known classes that have been somehow assigned. Each record in both datasets holds the coordinates of a single object.

The classification of a test (or unknown) object is done by computing distances from the test object to all the training objects.

Then, the k nearest neighbours are selected and their labels retained. The label to be assigned to the test is chosen using a voting strategy, that is, the label with more votes or occurrences among the k neighbours will be selected.

The key points to be defined in the k-NN method are therefore (Steinbach and Tan, 2009):

- The set of labeled or known objects

- A distance or similarity metric

- The value of $k$

- The method used to determine the class of the test objects

It is necessary to adapt those key elements to the problem of converting from CIE coordinates to Munsell notations. The set of labeled objects must be obviously the Munsell chips, that have to be observed with the colorimeter using the laboratory setup described above. The point here is that our labels are Munsell notations, so that classifying test objects is analogous to set Munsell notations.

The metric used is the euclidean distance in the CIE space. In this respect, it is better to use CIELAB coordinates rather than $Yxy$ coordinates for colour uniformity reasons. The program described here will always use CIELAB coordinates if there are white calibration readings available.

The value of $k$ is one in our case since each class, that is each Munsell notation, contains only one member that corresponds to a particular colour chip. The method to determine the class of test objects makes no sense with a value of $k = 1$.

In summary, it is possible to convert CIE coordinates to Munsell notations using the k-NN method with a value $k = 1$. This is equivalent to select the notation of the nearest Munsell chip to the test sample. In spite of the simplicity of this method, there are several drawbacks that can be limiting in certain circumstances and have been studied in the literature (Steinbach and Tan, 2009).

Although the k-NN classification method can be easily implemented, we used the `KNeighborsClassifier` class from the Scikit-learn module (Pedregosa et al., 2011). This classifier provides an interface to execute the k-NN method in a few lines of code.

The process requires importing the package:

```
>>> from sklearn import neighbors
```

The k-NN classification is a three-step process. First, a classifier object must be created specifying the number of neighbours that must be one in our problem:

```
>>> knn = neighbors.KNeighborsClassifier(1)
```

In the previous line it is possible to pass optional parameters to indicate the weighting scheme with `weights='uniform'` or the metric with `metric='euclidean'`.

The second step requires two lists that contain the training dataset records (`training`) and the class labels (`labels`). The two lists are joined with the `fit` method:

```
>>> knn.fit(training, labels)
```

Finally, the classification of new data points is done with `predict`:

```
>>> test_class = knn.predict(test)
```

where `test` is a list that contains test records and `test_class` is an array of labels that allow classifiying the unknown records.

The parameters of the previous methods can be lists or arrays that should match in their dimensions. For instance, `training` has dimensions m×n, where m is the number of training records and n is the number of dimensions of the space. The number of items in list `labels` must be m and the dimensions of `test` must be u×n, where u is the number of test records and n the number of coordinates or dimensions of those test records.

### 3.5  Measurement routine

After setting up the laboratory and assembling all the components (light booth, colorimeter, computer, and so on) we defined a measurement routine based on user's experience. The goals of the routine are to speed up the laboratory work sessions and to minimise errors.

The routine consists of the following steps:

1. Fill up a Petri plate with soil

2. Shake the plate to obtain a 'flat' and 'homogeneous' surface

3. Put the plate into the light booth

4. Measure colour coordinates

5. Take the plate from the booth and mix soil material

6. Repeat steps 2, 3 and 4

7. Repeat step 5

8. Repeat steps 2, 3 and 4

9. Save measurements to disk file

This routine provides three measurements per sample which allows being aware of possible deviations across measurements. It should be noted that small deviations are possible due to the granular nature of soil samples. It is for this reason that samples should be shaken before repeating a measure. When the user accepts the measured values, the average is calculated and reported in the final data files.

The measurement routine can be complemented with reference white readings that should be inserted at regular intervals during the laboratory work session. These measurements are mandatory if CIELAB coordinates are needed.

## 4.  CONCLUSIONS

The procedure described herein is an effective tool for users interested in studying soil colour in combination with other environmental and agricultural variables.

The idea of developing a tailored application allowed fitting the procedure to specific laboratory requirements and integrating colour processing in the laboratory workflow.

The potential of the k-NN method to convert CIE coordinates into Munsell notations was demonstrated. Moreover, the k-NN approach emulated reasonably well the visual matching process experienced by human observers that consists of assessing the minimum chromatic distance between a sample and a collection of colour chips.

In summary, the procedure presented in this paper covers all needs for soil colorimetry and creates spatial databases that can be used in environmental, agricultural and remote sensing applications.

## REFERENCES

Baret, F., Jaquemoud, S. and Hanocq, J., 1993. About the Soil Line Concept in Remote Sensing. Advanced Space Research 13(5), pp. 281–284.

Baumgardner, M., Silva, L., Biehl, L. and Stoner, E., 1985. Reflectance properties of soils. In: N. Brady (ed.), Advances in Agronomy, Vol. 38, pp. 2–44.

Boettinger, J., Howell, D., Moore, A., Hartemink, A. and Kienast-Brown, S. (eds), 2010. Digital Soil Mapping. Bridging Research, Environmental Application, and Operation. Springer.

Bouma, J., 1989. Using Soil Survey Data for Quantitative Land Evaluation. In: B. Stwart (ed.), Advances in Soil Science, Vol. 9, Springer-Verlag, pp. 177–213.

CIE, 2004. Colorimetry. Commission Internationale de l'Éclairage.

Escadafal, R. and Huete, A., 1992. Soil optical properties and environmental applications of remote sensing. International Archives of Photogrammetry and Remote Sensing 29(B7), pp. 709–715.

ESRI, 1998. ESRI Shapefile Technical Description. Environmental Systems Research Institute.

FAO, 2006. World reference base for soil resources 2006. World Soil Resources Reports No. 103. Food and Agriculture Organization of the United Nations.

Gitelson, A., Stark, R., Grits, U., Rundquist, D., Kaufman, Y. and Derry, D., 2002. Vegetation and soil lines in visible spectral space: a concept and technique for remote estimation of vegetation index. International Journal of Remote Sensing 23(13), pp. 2537–2562.

Gunal, H., Ersahin, B., Yetgin, B. and Kutlu, T., 2008. Use of chromameter-measured color parameters in estimating color-related soil variables. Communications in Soil Science and Plant Analysis 39, pp. 726–740.

Ibáñez-Asensio, S., Marqués-Mateu, A., Moreno-Ramón, H. and Balasch, S., 2013. Statistical relationships between soil colour and soil attributes in semiarid areas. Biosystems Engineering 116(2), pp. 120–129.

Konica Minolta, 1992. Chroma Meter CS-100A. Instruction Manual. Konica Minolta.

Konica Minolta, 2000. Chroma Meter CS-100A. Communication Manual. Konica Minolta.

Malacara, D., 2011. Color Vision and Colorimetry: Theory and Applications. 2nd edn, SPIE Press.

Metternicht, G. and Zinck, J. (eds), 2009. Remote Sensing of Soil Salinization. Impact on Land Management. CRC Press.

Moreno-Ramón, H., Marqués-Mateu, A. and Ibáñez-Asensio, S., 2014. Significance of soil lightness versus physicochemical soil properties in semiarid areas. Arid Land Research and Management 28, pp. 371–382.

Munsell Color, 2000. Munsell Soil Color Charts. GretagMacbeth.

OGC, 2010. OpenGIS Implementation for Geographic Information - Simple feature access - Part 1. Open Geospatial Consortium.

Pedregosa, F., Varoquaux, G., Gramfort, A., Michel, V., Thirion, B., Grisel, O., Blondel, M., Prettenhofer, P., Weiss, R., Dubourg, V., Vanderplas, J., Passos, A., Cournapeau, D., Brucher, M., Perrot, M. and Duchesnay, E., 2011. Scikit-learn: Machine Learning in Python. Journal of Machine Learning Research (12), pp. 2825–2830.

Richardson, A. and Wiegand, C., 1977. Distinguishing Vegetation from Soil Background Information. Photogrammetric Engineering and Remote Sensing 43(12), pp. 1541–1552.

Sánchez-Marañón, M., Delgado, G., Melgosa, M., Hita, E. and Delgado, R., 1997. CIELAB color parameters and their relationships to soil characteristics in Mediterranean red soils. Soil Science 162(11), pp. 833–842.

Schanda, J. (ed.), 2007. Colorimetry. Understanding the CIE system. John Wiley & Sons.

Simonson, R., 1993. Soil color standards and terms for field use - history of their development. In: J.M. Bigham and E.J. Ciolkosc (ed.), Soil color, Soil Science Society of America, pp. 21–34.

Soil Survey Division Staff, 1993. Soil Survey Manual. USDA Handbook 18. Soil Conservation Service. U.S. Department of Agriculture.

Soil Survey Staff, 2010. Keys to Soil Taxonomy. 11th edn, USDA-Natural Resources Conservation Service.

Sriparasa, S., 2013. JavaScript and JSON Essentials. Packt Publishing.

Steinbach, M. and Tan, P., 2009. kNN: k Nearest Neighbors. In: X. Wu and V. Kumar (ed.), The Top 10 Algorithms in Data Mining, CRC Press, pp. 151–161.

Thwaites, R., 2006. Color. In: R. Lal (ed.), Encyclopedia of Soil Science, Taylor & Francis, pp. 303–306.

Torrent, J. and Barrón, V., 1993. Laboratory measurements of soil color: theory and practice. In: J.M. Bigham and E.J. Ciolkosc (ed.), Soil color, Soil Science Society of America, pp. 21–34.

Torrent, J., Schwertmann, U., Fetcher, H. and Alférez, F., 1983. Quantitative relationships between soil color and hematite content. Soil Science 136(6), pp. 354–358.

Tucker, C. and Miller, L., 1977. Soil spectra contributions to grass canopy spectral reflectance. Photogrammetric Engineering and Remote Sensing 43(6), pp. 721–726.

van Rossum, G. and Drake, F., 2011. Python Languaje Reference Manual. Network Theory.

# 27

# 3D RECONSTRUCTION OF CULTURAL TOURISM ATTRACTIONS FROM INDOOR TO OUTDOOR BASED ON PORTABLE FOUR-CAMERA STEREO VISION SYSTEM

Zhenfeng Shao [a], Congmin Li [a,*], Sidong Zhong [b], Bo Liu [c], Honggang Jiang [c] , Xuehu Wen [d]

[a] State key Laboratory for Information Engineering in Surveying, Mapping and Remote Sensing, Wuhan University, Wuhan, 430079, China
[b] Wuhan University, Wuhan, 430079, China
[c] Geomatics Center of Guangxi, 5 Jianzheng Road, Nanning China,530023
[d] The Third Surveying and Mapping Engineering Institute of Sichuan, Chengdu Sichuan ,610500
shaozhenfeng@whu.edu.cn, cminlee@whu.edu.cn, sdzhong@whu.edu.cn,676802836@qq.com, 19743408@qq.com, 398169068@qq.com

**\*cminlee@whu.edu.cn**

**KEY WORDS:** Cultural Tourism Resources Protection, Building Information Modelling, Indoor-outdoor Seamless Modelling, Four-camera Stereo Photographic Measurement System, 3D Reconstruction, Image-based reconstruction

**ABSTRACT:**

Building the fine 3D model from outdoor to indoor is becoming a necessity for protecting the cultural tourism resources. However, the existing 3D modelling technologies mainly focus on outdoor areas. Actually, a 3D model should contain detailed descriptions of both its appearance and its internal structure, including architectural components. In this paper, a portable four-camera stereo photographic measurement system is developed, which can provide a professional solution for fast 3D data acquisition, processing, integration, reconstruction and visualization. Given a specific scene or object, it can directly collect physical geometric information such as positions, sizes and shapes of an object or a scene, as well as physical property information such as the materials and textures. On the basis of the information, 3D model can be automatically constructed. The system has been applied to the indoor-outdoor seamless modelling of distinctive architecture existing in two typical cultural tourism zones, that is, Tibetan and Qiang ethnic minority villages in Sichuan Jiuzhaigou Scenic Area and Tujia ethnic minority villages in Hubei Shennongjia Nature Reserve, providing a new method and platform for protection of minority cultural characteristics, 3D reconstruction and cultural tourism.

## 1. INTRODUCTION

With the rapid development of Chinese tourism, cultural tourism resources become a new direction. It is widely believed that the development of cultural tourism resources brings prosperity to local social, economic and cultural, while it may cause changes in local ethnic culture. Traditional cultural in some areas is dramatically changing in clothing, architecture, customs and the way of life, which are rapidly converging with the foreign culture. In fact, local culture is facing the danger of losing individuality and characteristics. For example, in the recent years, new residential construction in Jiuzhaigou is mainly made of brick or reinforced concrete structure, while traditional cedar chips roof is replaced by tile roof, and rammed earth walls were changed to brick. This type of building has a richer exterior decoration and colour and integrates more Han Chinese architectural elements. Actually, these changes enhance the comfort and convenience of living in Tibetan villages. However, the unique architectural culture of minorities is disappearing. In order to alleviate the contradiction between cultural resource protection and utilization in cultural tourism, three-dimensional reconstruction technologies have become an important means to protect and present cultural tourism resources.

At present, the common used method of three-dimensional reconstruction of a scene or object is mainly depended on existing interactive modelling software. Although the modelling accuracy can meet the requirements of practical applications, it needs heavy manual operation, and thus has lower efficiency in modelling and higher cost. Among the automatic 3D reconstruction techniques, laser scanners and image-based approach have attracted widespread attention. The former is widely used in the digitization of cultural heritage, because it is able to quickly scan the surface of the target area, and generate high-density 3D point cloud data. But these devices are relatively expensive and are generally used for surface modelling of buildings. While the latter has lower hardware requirement and can construct models at lower cost. Moreover, 3D models obtained by the latter can basically meet certain requirements. A real 3D model should contain not only the description of the appearance of the building, but also its internal structure. As respect of data acquisition, modelling for indoor lacks of effective 3D data acquisition equipment at present.

Based on the above requirements and analysis, a portable four-camera stereo vision system is developed and implemented, which has been successfully applied to 3D reconstruction of architectural features from indoor to outdoor in two typical cultural tourism resources in China. The system can directly

---

* Corresponding author

measure the geometry and attribute information by the acquired images, and automatically model based on this information. It solves the problems of indoor-outdoor data acquisition and integrated modelling, and provides a new method and platform for the protection of minority cultural characteristics, 3D reconstruction and cultural tourism.

## 2.  RELATED WORK

Compared with 3D range scanning, Image-based 3D reconstruction techniques have been considered as a low-cost and effective tool for producing high-quality 3D models of real world in terms of hardware requirements, knowledge backgrounds and man-hours (Koutsoudis, et al, 2014). Researchers have done considerable work to create high-quality 3D models, especially in computer vision community. Among these techniques, the Structure-From-Motion (SFM) (Robertson& Cipolla, 2009) and Dense Multi-View 3D Reconstruction (DMVR) (Furukawa& Ponce, 2010) are the most popular algorithms. These methods can be used to recover a scene or object by processing numerous unordered images taken from perspective viewpoints with large overlapping areas. The pipeline of image-based 3D reconstruction is generally composed of different steps that can be automatically performed. First, camera parameters can be obtained by matching corresponding features or calibrated in a control field. According to the camera imaging model, 3D geometry information as well as texture can be restored. A number of software based on the two algorithms has been made available, such as Bundler (Snavely, 2008), Visual SFM(Wu, 2011), etc.. Based on these basic theories, image-based 3D reconstruction techniques have been widely used. However, these applications mainly focus on surface modelling of a scene or building, which can hardly meet the requirements of fine interior modelling.

In the research of stereo vision, according to the number of cameras, stereo vision measurement methods can be mainly divided into three categories: monocular stereo vision measurement method, binocular stereo vision measurement method and multi-view stereo vision measurement method. Due to self-occlusion and shadows of the measured objects, as well as the limited field of view of a camera and depth of field, neither monocular nor binocular stereo measurement system can get the whole surface data of the measured object through a single measurement. Usually, it needs to increase the times of measurement from a different angle range. When the number of cameras reaches certain level, it will add the difficulty of subsequent processing, which can be a big challenge for data mosaic and integration. Therefore, the developed portable four-camera measurement system can be of great significance.

## 3.  PROTABLE FOUR-CAMERA BASED STEREO VISION SYSYTEM

Four-camera Stereo photogrammetry system is a type of image-based non-contact measurement. Firstly, it obtains four images of the same object or scene simultaneously from four different perspectives by using four relatively fixed cameras. Then, measurements can be carried out on these images, thus, physical geometric information such as 3D description of the object coordinates, sizes and shapes of an object or a scene, as well as physical property information such as the materials and textures can be automatically achieved. On the basis of the information, 3D model can be automatically constructed. Section 3.1 and 3.2 respectively introduces the hardware design and methods of 3D model acquisition of the developed system.

### 3.1  Hardware Platform

As illustrated in figure 1, the system is mainly composed of four digital cameras, rigid connection rod and synchronous controller. Those four cameras are divided into two groups on average and each group is fixed at ends of the rigid connecting rod within the camera cover, thus, the relative position and orientation of the camera can be maintained during shooting. The most important part of the measurement system is the synchronous controller. It is capable of receiving signals from a handheld remote shooting and then transmits high precision synchronization signals to cameras, ensuring synchronization of four cameras shooting.

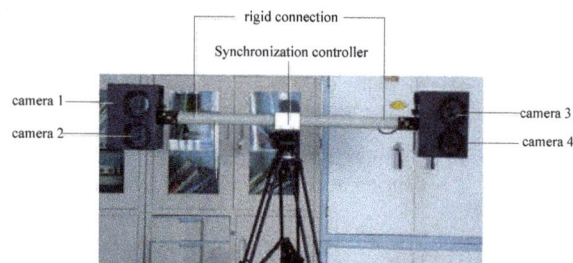

Figure 1. Hardware structure of portable four-camera stereo vision system

### 3.2  Methodology of fully automatic 3D models acquisition

Stereoscopic image collected by the four cameras can be transmitted to computers in real-time. When measuring, three-dimensional coordinates of the target can be obtained by the computer automatically calculate and applied directly to the object three-dimensional positioning, geometric measurement and object reconstruction. Procedures of acquiring 3D models through the proposed system are described as Figure 2.

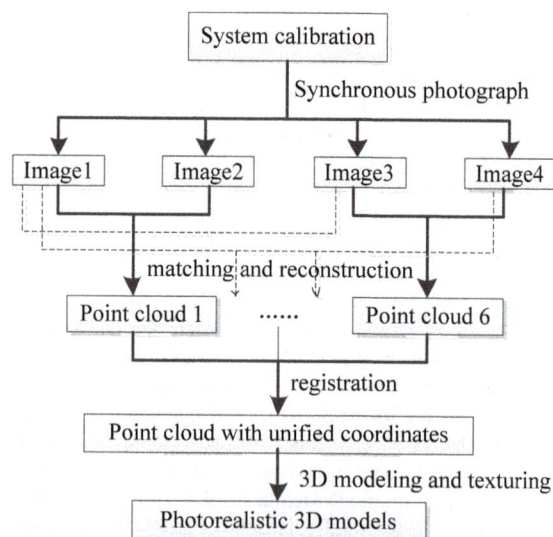

Figure 2. General pipeline for delivering photorealistic 3D models from collections of portable four-camera stereo vision system.

**3.2.1** **Camera calibration**: Calibrating this system to obtain camera parameters is one of prerequisite for the measurement. Because of the portability of the system, only digital cameras or digital video cameras are needed, not requiring professional cameras. However, considering the accuracy of cameras and precision of 3D reconstruction, the camera calibration of the system is necessary. Images of the control field acquired by four cameras existing in this system can be utilized to calculate camera parameters, including internal and external orientation elements of cameras, and camera distortion parameters etc.. Thus, there is no need to providing rulers or setting known points. Once the system is calibrated in the professional calibration field, its parameters can be kept.

**3.2.2** **Data acquisition**: When acquiring indoor-outdoor data, the special synchronous controller must be used to control shutter switch signal of four cameras, ensuring synchronization photography of four cameras to prevent camera ego motion and parameters failure caused by dynamic scenes and the movement of objects. In order to obtain high accuracy of texture data in the measured area as well as providing guidance for the automatic matching of the whole scene, UAVs will be applied to assist modelling.

**3.2.3** **Feature extraction and image matching**: After acquiring data, a series of pre-processing operations for images are needed, such as feature extraction and stereo matching. In the paper, SIFT algorithm (Lowe, 2004) is used for image feature extraction because of its scale and rotation invariant characteristics. As for matching, matching strategy based on the principle of nuclear lines intersecting is proposed. Combining with camera imaging model, each feature point can be constructed into six imaging binocular stereo vision pairs when using four-camera stereo vision system. As is shown in figure 3, a reality scene point can be imaged on four pictures, assuming to be $p_1$, $p_2$, $p_3$, $p_4$, and $l_{12}$, $l_{13}$, $l_{14}$ $l_{23}$, $l_{24}$, $l_{34}$, $l_{14}$ respectively stands for the nuclear line of the six stereo pairs. When extracting the feature point $p_1$ in the upper left picture, it is easy to find the corresponding point $p_3$ in the lower left image with high accuracy. According to the epipolar constraint, nuclear line $l_{12}$ and $l_{23}$ can intersect at point $p_{2'}$, which is corresponding point of $p_1$ in the upper right image. Similarly, it is easy to find corresponding points $p_2$, $p_4$ and $p_{4'}$, which is the intersection point of the two nuclear lines $l_{24}$, $l_{34}$.

Therefore, whether the match of point $p_2$ and $p_4$ is correct can be verified by judging whether the two points $p_4$ and $p_{4'}$ coincide. If the two points do not coincide, it indicates a mismatch and needs to find another point until all of the matches are correct.

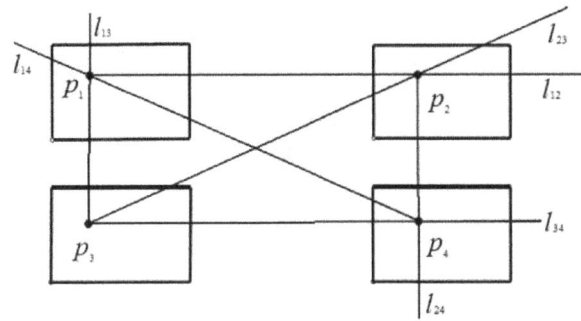

Figure 3. Corresponding points matching based on the principle of nuclear lines intersect.

The biggest advantages of the four-camera system is that it can not only greatly reduce the searching area for feature point matching, but also eliminate the gross error and providing redundant observation for checking and ensuring the accuracy of matching.

**3.2.4** **3D modelling and texturing**: Once stereo images are transmitted over the network into the computer, the 3D spatial information can be automatic obtained by professional software. From 3D space information, 3D point cloud data and texture data can be extracted. Then 3D point cloud data can be used for automatic construction of triangulation. At last, finer precise 3D models will be built through dense matching. Detail implementation can refer to the open-source software bundler (Robertson& Cipolla, 2009).

**3.2.5** **Scene modelling and texturing**: After extracting 3D models of indoor and outdoor areas, unmanned aerial vehicles images are used to assist scene modelling, such as helping for extracting top texture of architecture and providing guidance for the whole scene matching.

## 4. APPLICATIONS IN Reconstruction CULTURAL TOURISM RESOURCES

### 4.1 Experimental Area

The system is applied to the indoor-outdoor seamless modelling of distinctive architecture existing in two typical cultural tourism zones in China, that is, Tibetan and Qiang ethnic minority villages in Sichuan Jiuzhaigou Scenic Area and Tujia ethnic minority villages in Hubei Shennongjia Nature Reserve.

Jiuzhaigou Scenic Area is located on the edge of the Tibetan Himalayan Plateau in Northern Sichuan Province, which is named for nine Tibetan villages and it was declared a UNESCO World Heritage Site in 1992. Historically external blocking traffic, specific historical and closed conditions make Jiuzhaigou retain the original nature of traditional Tibetan culture. Building facilities in Jiuzhaigou Scenic Area are mainly nine Tibetan villages and temples. Among the nine villages, Zezawa Village, Shuzheng Village and Heye Village are well-preserved and have frequent tours.

Shennongjia Nature Reserve is located in the northwest of hubei province. In 1990, it was incorporated in the UNESCO World Network of Biosphere Reserves. People living in Shennongjia forest region are mainly Tujia ethnic minority. Diaojiaolou with red lanterns and Tunkou have become a distinctive architectural culture.

For the unique cultural buildings and landscape in the two cultural tourism zones, 3D reconstruction work has been conducted by the developed system, so as to preserve the unique cultural tourism resources as much as possible. The workflow is described as figure 4.

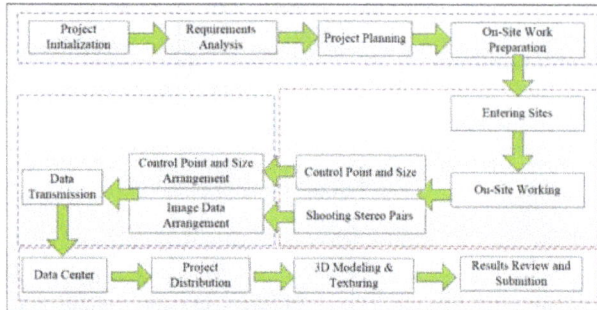

Figure 4. Workflow of 3D modelling on-site.

### 4.2 Results and Discussions

This study mainly focuses on the 3D reconstruction of the outdoor surface and indoor structure for two typical cultural tourism resources. Experiment results show that it is feasible to precede indoor-outdoor reconstruction with the use of the portable four-camera stereo photographic measurement system.

The measurement system is a type of image-based non-contact measurement. As for image collection, the specialised four-camera device can be manipulated by one person to obtain both facade and interior structure data of the buildings. And unmanned aerial vehicles images are used to help for extracting top texture of architecture. Once these images are collected, they will be transmitted to the computer to be measured through the Internet in real-time. Then 3D models can be automatically constructed by the professional software. Because of the distribution characteristics of the four-camera, image matching can be greatly efficient and accurate. Finally, 3D models containing detailed descriptions of both its appearance and its internal structure can be constructed. For the purpose of visualization of cultural tourism resources, 3D models can be presented and manipulate in a specialized software, meanwhile, they can be saved as AVI form. The final photorealistic 3D models of the distinctive architecture are shown in figure 5-8.

Figure 5. 3D reconstruction of Heye Village in Jiuzhaigou Scenic Area

Figure 6. Internal-outdoor reconstruction of residential building with Tibetan characteristics in Jiuzhaigou Scenic Area

Figure 7. 3D reconstruction of distinctive building hanging with red lanterns in Shennongjia Nature Reserve

Figure 8. Internal structure model of the distinctive building in Shennongjia Nature Reserve

### 5. CONCLUSIONS

In this paper, a portable four-camera stereo photographic measurement system is developed and implemented, which can provide a professional solution for fast 3D data acquisition, processing, reconstruction and visualization. Compared to traditional methods, it is a non-contact measurement method and generates 3D models containing detailed descriptions of both its appearance and its internal structure, including architectural components.

The system is characterized by: (1) it is portable and can be carried and operated by one person conveniently; (2) multi-source data can be integrated to fully automatically generate 3D models; (3) high-efficient modelling, once images of the scene are captured by the four cameras, they will be transmitted to computers to generate models through the Internet in real-time;

(4) indoor-outdoor seamless modelling, it reconstructs 3D models containing detailed descriptions of both its appearance and its internal structure.

The system is applied to the indoor-outdoor seamless modelling of distinctive architecture existing in two typical cultural tourism zones, that is, Tibetan and Qiang ethnic minority villages in Sichuan Jiuzhaigou Scenic Area and Tujia ethnic minority villages in Hubei Shennongjia Nature Reserve, providing a new method and platform for protection of minority cultural characteristics, 3D reconstruction and cultural tourism.

## ACKNOWLEDGEMENTS

This research was financially supported by the National Science & Technology Specific Projects under Grant 2012YQ16018505, 2013BAH42F03, by National Natural Science Foundation of China under Grant 61172174, the Basic Research Program of Hubei Province(2013CFA024), Special Project on the Integration of Industry, Education and Research of Guangdong Province (2012B090500016),Shenzhen science and Technology Development Foundation (JCYJ20120618162928009)，public research fund on surveying and mapping (201412010),and project of Sichuan Provincial Bureau on Surveying and mapping(J2013ZH02,and J2014ZC03).

## REFERENCES

Furukawa, Y. and J. Ponce, 2010. Accurate, dense, and robust multiview stereopsis. Pattern Analysis and Machine Intelligence, IEEE Transactions on. 32(8): p. 1362-1376.

Lowe, D. G., 2004. Distinctive image features from scale-invariant keypoints. International journal of computer vision, 60(2), 91-110.

Koutsoudis, Anestis, et al, 2014. Multi-image 3D reconstruction data evaluation. Journal of Cultural Heritage, 15(1), 73-79.

Robertson, D.P. and R. Cipolla, , 2009. Structure from Motion. *Practical Image Processing and Computer Vision*. John Wiley, Hoboken, NJ, USA,p. 49.

Snavely, N., 2008. Bundler: Structure from motion for unordered image collections.

Wu, C., 2011. 9. VisualSFM: A visual structure from motion system. http://homes. cs. washington. edu/~ ccwu/vsfm.

# 28

# POTENTIAL IMPROVEMENT FOR FOREST COVER AND FOREST DEGRADATION MAPPING WITH THE FORTHCOMING SENTINEL-2 PROGRAM

L. Hojas-Gascón [a, d], A. Belward [a], H. Eva [a*], G. Ceccherini [a], O. Hagolle [b], J. Garcia [c], P. Cerutti[d]

[a] European Commission Joint Research Centre, Via E. Fermi 2749, 21027 Ispra (VA), Italy - hugh.eva@jrc.ec.europa.eu
[b] CESBIO / CNES, Toulouse, France - olivier.hagolle@cnes.fr
[c] Unidad de Investigación de Teledetección, Dr. Moliner 50, 46100 Valencia, Spain - j.garcia.haro@uv.es
[d] CIFOR Center for International Forestry Research, Bogor Barat 16115, Indonesia - p.cerutti@cgiar.org

**KEY WORDS:** Forest cover, Forest degradation, SPOT, Sentinel-2, REDD+, Tropical Dry forest.

**ABSTRACT:**

The forthcoming European Space Agency's Sentinel-2 mission promises to provide high (10 m) resolution optical data at higher temporal frequencies (5 day revisit with two operational satellites) than previously available. CNES, the French national space agency, launched a program in 2013, 'SPOT4 take 5', to simulate such a dataflow using the SPOT HRV sensor, which has similar spectral characteristics to the Sentinel sensor, but lower (20m) spatial resolution. Such data flow enables the analysis of the satellite images using temporal analysis, an approach previously restricted to lower spatial resolution sensors. We acquired 23 such images over Tanzania for the period from February to June 2013. The data were analysed with aim of discriminating between different forest cover percentages for landscape units of 0.5 ha over a site characterised by deciduous intact and degraded forests. The SPOT data were processed by one extracting temporal vegetation indices. We assessed the impact of the high acquisition rate with respect to the current rate of one image every 16 days. Validation data, giving the percentage of forest canopy cover in each land unit were provided by very high resolution satellite data. Results show that using the full temporal series it is possible to discriminate between forest units with differences of more than 40% tree cover or more. Classification errors fell exclusively into the adjacent forest canopy cover class of 20% or less. The analyses show that forestation mapping and degradation monitoring will be substantially improved with the Sentinel-2 program.

## 1. INTRODUCTION

### 1.1 Monitoring deforestation and forest degradation

For more than a decade the monitoring of deforestation has successfully been carried out at regional levels using moderate spatial resolution satellite data, predominantly from the Landsat sensor (Achard et al. 2009, INPE 2014, FAO, JRC, SDSU and UCL 2009), which has 30 m spatial resolution and a revisit frequency of 16 days. More recently the University of Maryland, in conjunction with Google, have produced global forest change maps (the Global Forest Maps) based on a synthesis of the Landsat archive for the years 2000-2012 (Hansen et al. 2013).

The activities proposed in 2011 under the Reduced Emissions from Deforestation and forest Degradation (REDD+) framework, brought new requirements for monitoring deforestation and forest degradation at national levels and finer scales (UNFCCC 2011). Deforestation was defined as a direct human-induced decrease in tree crown cover below 10-30% of forest areas with a minimum size of 0.05-1 ha (UNFCCC 2001), and degradation as a loss of carbon stock in forest areas with a decrease in the tree crown cover not below the 10-30% threshold (IPCC 2003).

Participatory countries should implement forest monitoring systems that use an appropriate combination of remote sensing and ground-based forest carbon inventory approaches, with a focus on estimating anthropogenic forest area changes and forest carbon stocks (UNFCCC 2009, p.12). Preliminary comparisons of the Global Forest Maps with very high resolution satellite data has already highlight that the spatial resolution of the Landsat sensor is too coarse for monitoring deforestation with high accuracy in the context of REDD+, and therefore higher (c. 5-10m) spatial resolution satellite data should be employed (CIFOR 2015).

The addition of forest degradation in the program implies that the estimations of forest carbon stock changes need to be based, not only on monitoring transitions of land cover classes (e.g. forest to non-forest), but also on transitions within the forest class when there is a loss of carbon sequestration (e.g. forest with more than 30% crown cover into forest with less than 10% crown cover). In this study carried out over a test site in Tanzania, we have considered forest as an area of land with at least 0.5 ha and a minimum tree crown cover of 10%, with trees which have, or have the potential, to reach a minimum height of 5 meters at maturity *in situ*, according to the definition adopted by the Tanzanian national REDD+ strategy (UN-REDD 2013).

### 1.2 The Sentinel-2 program

The European Union's first Earth Observation programme, Copernicus, is building a series of technologically advanced satellites (the Sentinels), which includes the Sentinel-2 satellites. Sentinel-2 aim to contribute providing inputs for services relying on multi-spectral high-resolution optical observations over global land surfaces, like SPOT and Landsat satellites, but also attempt to cover current limitations with the addition of the technical needs for new requirements. These include higher revisit frequencies, more spectral bands with narrower bandwidths and finer spatial resolutions, in order to improve services as vegetation monitoring (ESA, 2010).

---

* Corresponding author.

The design of the Sentinel-2 platform benefited from the experience and lessons learned from other satellites building on their technology. The selection of the spectral bands has been guided by the Landsat, SPOT-5, MERIS and MODIS heritage (ESA, 2010). The Multi Spectral Instrument (MSI)'s 13 spectral bands' centre range from 0.433 to 2.19μm. There are four visible and near-infrared bands at 10 m spatial resolution, three red edge, one near-infrared and two SWIR at 20 m, and three channels to help in atmospheric correction and cloud screening at 60 m (Drusch et al., 2012). When complete, the Sentinel-2 program will have two satellites offset in orbit operating simultaneously on opposite sides (Sentinel-2A and Sentinel-2B), each carrying the same instruments. Sentinel-2A is scheduled for launch in June 2015 and Sentinel-2B in late 2016. Together these two satellites will provide coverage every five days at the equator with a 290 km field of view (ESA, 2010).

Forest monitoring is one of the priority services of the Global Monitoring for Environment and Security (GMES) programme for which Sentinel-2 has been tailored. In fact, the revisit requirements were driven by vegetation monitoring, for which those of Landsat and SPOT were not enough. Sentinel-2 observations are explicitly intended to develop key inputs required for Kyoto protocol reporting. Potentially, they could contribute to the Baseline Mapping Service for the REDD+ programme (ESA, 2010). The Copernicus plans also aim for multiple global acquisitions and a free and open access data policy (European Commission, 2013), similarly to Landsat.

### 1.3 Time series for forest cover mapping

Short revisit periods are potentially important to monitor forest at national/regional scales. Firstly, the increased coverage provides more opportunity for acquiring cloud-free images, particularly important in tropical regions (Beuchle et al., 2011). Secondly, because they should allow us to exploit seasonal differences in canopy reflectance characteristics as a means of discriminating between forest cover types and different forest conditions (e.g. closed and open forests, or deciduous and degraded forests), which is especially important for the dry forest.

With the advent of the Sentinel-2, data availability over target areas will increase, allowing temporal analysis previously restricted to moderate (>100m) spatial resolution satellite data (such as MODIS), to be employed in the monitoring of forests at finer spatial resolutions. Sentinel-2 will bring an improvement in the spatial resolution (with the three visible and a near infrared bands at 10m), which will allow a more accurate assessment of deforestation and forest degradation areas taking the minimum scales defined by the UNFCC, and in the spectral sampling (i.e. higher amount of bands with narrower width), with the inclusion of three bands in the red edge, which has shown to be useful for quantitative assessment of vegetation status (Frampton et al., 2013).

### 1.4 SPOT4 Take 5

Whilst the Sentinel have not been launched yet, in order to prepare for the use of its data, on the 29th of January 2013 the French space agency CNES lowered the orbit of SPOT4 to put it on the same repeat cycle of Sentinel-2 until 19th June of the same year. During this period, SPOT passed over by the same 45 selected places every 5 days, one of them in the dry forest in Tanzania as requested by JRC. SPOT4 records in 5 spectral bands: three visible, one near-infrared and one SWIR at 20 m

spatial resolution (Hagolle et al., n.d.). This experiment, SPOT4 Take 5, does not simulate the full spectral and radiometric capabilities of Sentinel-2, but does simulate the revisit frequency and the spatial resolution of Sentinel-2.

For the selected place in Tanzania we obtained 23 SPOT images from 6th Feb to 19th June 2013 at level 2A (ortho-rectified surface reflectance data provided with a cloud mask). They cover an area of 360.000 ha and a period ranging from the end of the wet season to deep into the dry season.

This paper examines whether improved temporal sampling at high spatial resolution (20m) with satellite data actually improves our knowledge of deforestation and forest degradation in dry forest ecosystems, such as those found in Tanzania. For this we will: 1) estimate the increment of data availability, by comparing the cloud free image area of SPOT4 Take 5 with that of Landsat for the same period; 2) evaluate the improvement of the temporal resolution, by comparing the time series of SPOT4 Take 5 with that from MODIS; and 3) estimate the improvement of forest classification accuracy, by calculating the separability of forest classes, with Sentinel-2 A and B and only with (the most proximate) Sentinel-2 A.

## 2. METHODOLOGY

### 2.1 Study area and reference data

The study area is located in the Somalia-Masai ecoregion, in the dry highlands of Central Tanzania. The climate is semiarid; the rainfall is less than 500mm per year with high interannual variation, the mean monthly temperature between 20 and 25 ºC, and it has a well-defined arid season from beginning of May to end of November. Most of the region is covered with deciduous bushland and thicket (Acacia-Commiphora is the climax vegetation), which grade into evergreen and semi-evergreen bushland and thicket on the lower slopes of the mountains. At higher altitude in the mountains dry forests dominate.

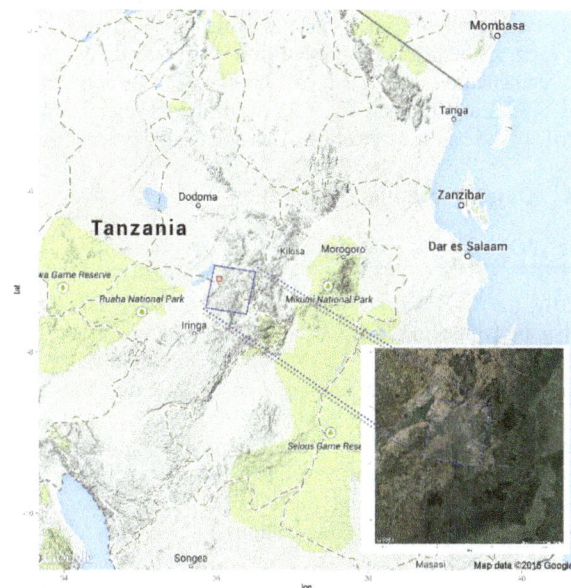

Figure 1. Study area

The area covered by the SPOT images (blue polygon, Figure 1) is centred at Lat Long (-7.226 36.182), between the Dodoma and the Iringa regions. To the north there are agriculture fields and to the South a mosaic of degraded forest, which was fully covered by forest in 1980. From the Landsat archive we can see that severe deforestation and degradation took place between 1983 and 1994, and in less degree between 1994 and 2011.

A 0.5 m spatial resolution pan-sharpened multispectral image acquired by the WorldView-2 satellite on the 4th September 2010 was used as reference data, along with field data collected in 2012 (Hojas Gascón and Eva, 2014). It covers an area of 5.000 ha centred at Lat Long (-7.092, 36.035) (red polygon, Figure 1).

## 2.2 Data availability

From the cloud occurrence maps provided with the data we summed up the cloud free data frequency for each pixel in the SPOT scene area simulating 10 day (Sentinel-2A) and 5 day (Sentinel-2A and 2B) frequency acquisition.

We also acquired the available Landsat-8 images for the same period comprised by the SPOT4 Take5 data and produced the cloud free data frequency map for comparison.

## 2.3 SPOT image segmentation

To divide the images in land units we created two 'seasonal' mosaics, using SPOT images from the wet and dry season respectively. These were then segmented in combination to create polygons of a minimum mapping unit (MMU) of 0.5 ha and 1 ha size. This method was employed so as to retain and discriminate land features that may be distinct in either season.

NDVI, SAVI and MSAVI indices were calculated for the polygons. Analysis showed no significant difference between NDVI and SAVI trends, and the he MSAVI was not found to be effective at discriminating between woody and non-woody vegetation. For easy of processing we reduced the data to the NDVI series.

The NDVI was calculated for each single data image and for the layer stack. From the layer stack we extracted the average NDVI for each of the segments created from the seasonal mosaic. In the segmentation at a MMU of 1 ha, the average polygon size was 5.5 ha and the maximum 25 ha. In the segmentation at a MMU of 0.5 ha the average was 1 ha and the maximum 5 ha.

## 2.4 Object vegetation classification

The SPOT data was classified in a two steps processing.

Firstly, the WorldView-2 image was segmented so as to obtain polygons with a mean of 0.1 ha. Areas of bare soil and grass were identified using a 5% reflectance threshold in the red channel. Woody vegetation was then divided into tree cover (woody vegetation higher than 5 m) and shrub cover (woody vegetation lower than 5 m). Field data provided information on the ratio of woody vegetation height to crown width, which was found to be around 1. This was effected by classifying crown width less than 5 m as shrub formations.

Secondly, the segments from the SPOT data containing the NDVI profiles were then cross tabulated with the very high resolution (VHR) reference data. Therefore the segments from

the SPOT data contain proportion of tree cover, shrub cover and non-woody land cover (grass or bare soil). As the data come from different dates, fine spatial resolution RapidEye data of 2013 were screened, so as to remove any areas that had undergone major land cover changes between the acquisition of the Worldview-2 data and the SPOT data. We then classified each segment by its proportions of the three elements with 6 category levels (0-10, 10-20, 20-40, 40-60, 60-80 and 80-100 cover percentages). For easy of nomenclature we combine the woody and tree proportions in a simple concatenation, for example, class W100F00 is all woody vegetation, of which trees (forest) is the only component, class W40F20 has 40% of woody vegetation, of which 20% is trees and 20% shrubs, hence 60% non-woody vegetation.

## 2.5 Extraction of NDVI profiles

The NDVI profiles were examined by vegetation classes. Cloud-affected dates were removed from the series by averaging proximate date values.

Figure 2 shows that generally the NDVI of the land units falls from a peak at the start of the observation period (end of the wet season) until the end of the period (deep into the dry season). We also noted that the mean NDVI from the different classes pass from being very divergent at the end of the wet season to be very similar in the dry season.

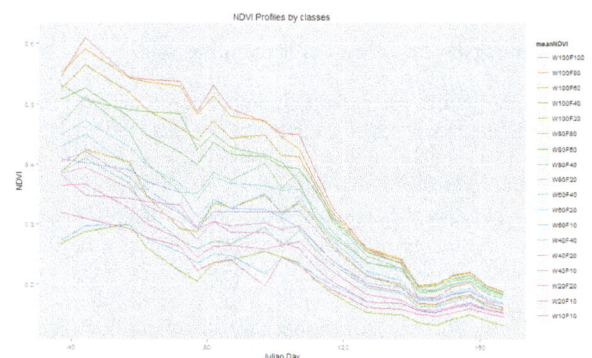

Figure 2. NDVI profiles from by vegetation classes.

## 2.6 Class separability

### 2.6.1 Jeffries-Matusita Distance

The Jeffries-Matusita distance (hereafter JM distance) is a statistical measure of distance between two distributions (Swain, 1972). The JM distance is defined as:

$$JM = \sqrt{2(1 - e^{-\alpha})} \qquad (1)$$

$$\alpha = \frac{1}{8}(\mu_a - \mu_b)^T \left(\frac{C_a - C_b}{2}\right)^T (\mu_a - \mu_b) + \frac{1}{2}\ln\left(\frac{\frac{1}{2}|C_a + C_b|}{\sqrt{|C_a| \times |C_b|}}\right) \qquad (2)$$

where    a and b are the two distributions
         C is the covariance matrix
         μ is the mean vector
         T is the transposition function

The JM distance is asymptotic to √2 and as such, a value of √2 suggests that the two distributions are very separable. The JM distance is widely used in remote sensing applications (Ghiyamat et al., 2013; Padma and Sanjeevi, 2014) to determine how distinct, and thus separable, different land cover classes or spectral signals are from each other.

Our goal is to assess the potential of the SPOT time series (simulating Sentinel-2) to discriminate between different forest classes. We calculated the JM taking full advantage of the Sentinel-2 repetitive observations (i.e. 5 days), and also simulating the revisit frequency with just one Sentinel (i.e. 10 days) and the current status (i.e. 16 days), by excluding half and two third of NDVI measurements, respectively.

### 2.6.2 Random Forest

Random Forest is a learning algorithm widely used in the statistical community to cluster data in different classes (among other analysis), constructing a multitude of decision trees at training time (Breiman, 2001). It has been shown to be effective at land cover classification (Rodriguez-Galiano et al., 2012). Our data were examined in Random Forest with a training of 150 of the 832 sample sites, also simulating 5 days, 10 days and 16 days revisit frequency.

## 3. RESULTS AND CONCLUSIONS

### 3.1 Cloud free image area

For the full SPOT scene 90% of the area is acquired cloud free at least once during the five month period of the Spot4 Take 5 experiment simulating one Sentinel-2 (an image every 10 days). This rises to 99% with Sentinel-2 A and B (acquiring images every 5 days). Figure 3 shows the increment of cloud free data frequency with both Sentinel-2 (SPOT) with respect to Landsat-8 during the same period. Combining both satellites (image not shown) the scene coverage is of 100%.

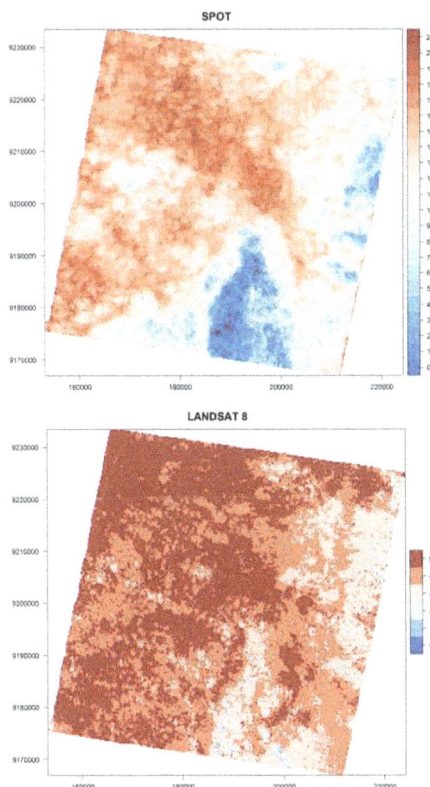

Figure 3. Cloud free data frequency maps with SPOT (top) and Landsat-8 (bottom).

### 3.2 NDVI time series

The averaged SPOT NDVI time series was compared to that obtained with the 250 m spatial resolution MODIS sensor (Figure 4). The latter has been produced from mean NDVI records from 2000 to 2012 in 16-days periods across the study area. Taking into account that the SPOT time series is a one single acquisition composite, we can say that it corresponds well with the smoothed MODIS product, except in the transitional period between the wet and the dry season. The dip in the profile at this period could be caused by an anomaly in temperature or rainfall regime or to the low quality of the cloud flagging.

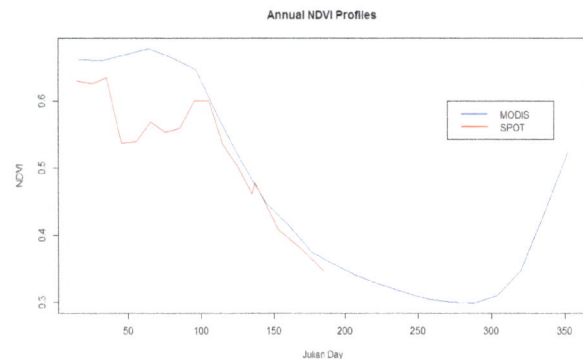

Figure 4. NDVI profile from SPOT and MODIS.

### 3.3 Forest classification

#### 3.3.1 Jeffries-Matusita Distance

**Sentinel A+B:** We present in the appendix the matrix of the JM distance between the different forest classes (Figure 4). For the sake of simplicity, JM distance ranges from 0 to 1414. The bigger the distance, the better the separability. We can see that JM distance within classes which exhibit small differences in forest cover (i.e. 20%) is small. Conversely, when the forest cover falls from 40% on, the distance increases, thus rending easier the discrimination between the corresponding classes. For example, this is true in the case of W100F80 where JM distance from W100F60 is only 78, whereas the distance from W100F40 is 339.

**Sentinel A**: We have computed the matrix of the JM distance between the different forest classes also in the case of Sentinel A (not shown here). For this configuration we used half of the observations, in order to simulate the 10 day revisit time. To wrap up, if we compute the mean of the ratio between the JM distance with all the observations (i.e. Sentinel A+B) and half of the observations, we obtain a positive increment of 5.7% in the distance when using both satellites. This means that, overall, there is an improvement of separation capabilities by increasing the frequency of observations. However, this small value of the increment might be due to the fact that many forest classes are undistinguishable (see Figure 2), and reducing or increasing the frequency does not always help to differentiate them. Also note that the relatively high values of the standard deviation of NDVI (not shown in the graphic) do not always allow the complete separability within two similar classes, resulting often in a partial overlap.

**Current acquisition**: The same procedure has been applied also for the case of the current frequency acquisition of satellite data of similar spatial resolution (i.e. 16 days), and so only one third of the observations were retained. In this case, the mean of the ratio of the corresponding JM distances indicates an improvement up to 14.5% when using Sentinel A and B, with respect to the current acquisition frequency. Again, there is always the issue of the separability between close classes that might lower this statistic.

### 3.3.2 Random Forest

**Sentinel A+B:**

|     |      | SPOT |      |      |      |      |      |
| --- | ---- | ---- | ---- | ---- | ---- | ---- | ---- |
|     |      | F100 | F80  | F60  | F40  | F20  | F10  |
| WV2 | F100 | 242  | 151  | 40   | 10   | 6    | 5    |
|     | F80  | 77   | 567  | 176  | 25   | 3    | 16   |
|     | F60  | 16   | 209  | 442  | 147  | 19   | 44   |
|     | F40  | 4    | 17   | 158  | 342  | 119  | 72   |
|     | F20  | 2    | 0    | 18   | 138  | 232  | 190  |
|     | F10  | 2    | 1    | 7    | 7    | 121  | 670  |

Table 1. Confusion matrix of forest classification with SPOT simulating Sentinel-2 A and B in comparison to Worldview-2.

From the confusion matrix (Table1) we can assess the class separability and hence assess by what percentage forest cover would have to fall to be correctly identified. We note that a change of two classes (i.e. 40% fall in forest cover) is required to ensure correct identification (e.g. from F100 to F60), with a probability of 95% approximately. The detection of 20% forest cover loss has a risk of misclassification by 20% approximately.

**Sentinel A:**

|     |      | SPOT |      |      |      |      |      |
| --- | ---- | ---- | ---- | ---- | ---- | ---- | ---- |
|     |      | F100 | F80  | F60  | F40  | F20  | F10  |
| WV2 | F100 | 240  | 151  | 41   | 10   | 7    | 5    |
|     | F80  | 78   | 556  | 183  | 27   | 5    | 15   |
|     | F60  | 15   | 217  | 442  | 144  | 21   | 38   |
|     | F40  | 5    | 19   | 158  | 337  | 116  | 77   |
|     | F20  | 2    | 1    | 18   | 141  | 233  | 185  |
|     | F10  | 1    | 1    | 8    | 8    | 129  | 664  |

Table 2. Confusion matrix of classification with SPOT simulating Sentinel-2A in comparison to Worldview-2.

The confusion matrix from the SPOT data with a 10 days acquisition frequency does not give very different results to the previous one. Random Forest analysis upholds our previous findings using the Jeffries-Matusita Distance: close classes (cover difference smaller than 20%), are difficult to discriminate; conversely, when the forest cover difference between two classes is greater or equal to 40%, the separability is easy.

**Current acquisition:**

|     |      | SPOT |      |      |      |      |      |
| --- | ---- | ---- | ---- | ---- | ---- | ---- | ---- |
|     |      | F100 | F80  | F60  | F40  | F20  | F10  |
| WV2 | F100 | 225  | 151  | 39   | 23   | 12   | 4    |
|     | F80  | 81   | 513  | 214  | 40   | 3    | 13   |
|     | F60  | 10   | 229  | 405  | 150  | 34   | 49   |
|     | F40  | 3    | 24   | 170  | 293  | 114  | 108  |
|     | F20  | 2    | 1    | 26   | 130  | 207  | 214  |
|     | F10  | 2    | 2    | 16   | 16   | 156  | 631  |

Table 3. Confusion matrix of classification with the current acquisition in comparison to Worldview-2.

In the case of the current acquisition frequency, Random Forest results give smaller improvement respect to the higher frequencies than the JM distance. The detection of 20% forest cover loss has a risk of misclassification of 25% approximately, while the detection of 40% forest cover loss of only a bit more than 5%.

Both JM distance and Random Forest analysis were done at a 0.5 and 1 ha MMU. However little differences were found between them and so only the tables with the 0.5 ha are presented.

### 3.4    Conclusions

The increased provision of medium (~10m) spatial resolution data acquisition from its current (c.16 days) to 10 days with one Sentinel platform, and 5 days with two operating platforms, promises to bring higher potential for detecting and quantifying forest degradation. Using the 20m resolution SPOT4 Take 5 data, processed to a simple vegetation index (NDVI) we have shown that forest degradation can be detected when a reduction of 40% canopy cover or more occurs in 0.5 ha land units. This is valid for both 5 and 10 day acquisitions. Lower reductions in canopy cover are also detectable, however, with a higher (~ 20-5%) chance of misclassification.

Deforestation and forest degradation monitoring in the context of REDD+ require change detections of 10% and less than 10% respectively in the forest cover of land units of 0.05-1 ha. Here we could only discriminate classes with forest cover with more than 40% difference. However, the results should underestimate the potential of Sentinel, which has a finer (10m) spatial resolution and finer band widths. At the same time data were available only for a limited period (5 months) of the year.

The development of better indices and the employment of wave analysis (e.g. Fourier) to characterize the vegetation changes over the full growing season, and eventually to historical data, should provide more robust results. Further improvement could be made by the integration of Landsat - and even MODIS data.

## ACKNOWLEDGEMENTS

This work was partly funded under a grant from CIFOR, the Center for International Forestry Research.

The authors would like to thank Cesbio for providing of the Spot4 Take 5 dataset needed to perform this research.

## REFERENCES

Achard F., Beuchle R., Bodart C. *et al.* 2009. Monitoring forest cover at global scale: the JRC approach. *Proceeedings of the 33rd International Symposium on Remote Sensing of Environment (ISRSE), 4-8 May 2009, Stresa,Italy,* p. 1-4.

Breiman, L., 2001. Random forests. Mach. Learn. 45, 5–32.

Beuchle, R., Eva, H.D., Stibig, H.-J., Bodart, C., Brink, A., Mayaux, P., Johansson, D., Achard, F., Belward, A., 2011. A satellite data set for tropical forest area change assessment. Int. J. Remote Sens. 32.

Drusch, M., Del Bello, U., Carlier, S., Colin, O., Fernandez, V., Gascon, F., Hoersch, B., Isola, C., Laberinti, P., Martimort, P., Meygret, A., Spoto, F., Sy, O., Marchese, F., Bargellini, P., 2012. Sentinel-2: ESA's Optical High-Resolution Mission for GMES Operational Services. Remote Sens. Environ. 120, 25–36.

ESA, 2010. GMES Sentinel-2 Mission Requirements Document.

European Commission, 2013. Commission Delegated Regulations (EU) No 1159/2013 of 12 July 2013. Off. J. Eur. Union.

FAO, JRC, SDSU and UCL 2009. The 2010 Global Forest Resources Assessment Remote Sensing Survey: an outline of the objectives, data, methods and approach. Forest Resources Assessment Working Paper 155. FAO,Rome, Italy.

Frampton, W.J., Dash, J., Watmough, G., Milton, E.J., 2013. Evaluating the capabilities of Sentinel-2 for quantitative estimation of biophysical variables in vegetation. ISPRS J. Photogramm. Remote Sens. 82, 83–92.

Ghiyamat, A., Shafri, H.Z.M., Amouzad Mahdiraji, G., Shariff, A.R.M., Mansor, S., 2013. Hyperspectral discrimination of tree species with different classifications using single- and multiple-endmember. Int. J. Appl. Earth Obs. Geoinformation 23, 177–191.

Hagolle, O., Huc, M., Dedieu, G., Sylvander, S., n.d. SPOT4 (Take 5) Times series over 45 sites to prepare Sentinel-2 applications and methods.

Hansen, M.C., Potapov P.V., Moore R. et al. 2013. High-resolution global maps of 21st-Century forest Cover change. Science, 342 (6160), 850-853.

Hojas Gascón, L., Eva, H., European Commission, Joint Research Centre, Institute for Environment and Sustainability, 2014. Field guide for forest mapping with high resolution satellite data. Publications Office, Luxembourg.

Hojas Gascón L., Eva H., Cerutti Paolo 2015. Policy report: Lessons learnt in monitoring deforestation and forest degradation in the context of REDD+. For publication.

INPE 2014. Projeto Prodes - Monitoramento Da Floresta Amazônica Brasileira Por Satélite, INPE, São José dos Campos, Brazil. http://www.obt.inpe.br/prodes/index.php

IPCC 2003. Definitions and methodological options to inventory emissions from direct human-induced degradation forest and vegetation of other vegetation types. Ninth session of the Conference of the Parties (COP 9). http://unfccc.int/meetings/milan_dec_2003/session/6271/php/view/reports.php

OECD 2007. Financing mechanisms to reduce emissions from deforestation: issues in design and implementation. Retrieved on 20th February 2014 from: http://www.oecd.org/env/cc/39725582.pdf

Padma, S., Sanjeevi, S., 2014. Jeffries Matusita based mixed-measure for improved spectral matching in hyperspectral image analysis. Int. J. Appl. Earth Obs. Geoinformation 32, 138–151.

Rodriguez-Galiano, V.F., Ghimire, B., Rogan, J., Chica-Olmo, M., Rigol-Sanchez, J.P., 2012. An assessment of the effectiveness of a random forest classifier for land-cover classification. ISPRS J. Photogramm. Remote Sens. 67, 93–104.

Swain, P.H., 1972. Pattern recognition: a basis for remote sensing data analysis.

UNFCCC 2011. Report of the Conference of the Parties on its sixteenth session, held in Cancun from 29 November to 10 December 2010. http://unfccc.int/resource/docs/2010/cop16/eng/07a01.pdf

UNFCCC 2009. Report of the Conference of the Parties on its fifteenth session, held in Copenhagen from 7 to 19 December 2009. http://unfccc.int/resource/docs/2009/cop15/eng/11a01.pdf

UNFCCC 2001. Report of the Conference of the Parties on its seventh session, held at Marrakesh from 29 october to 10 november 2001. http://unfccc.int/resource/docs/cop7/13a01.pdf

UN-REDD 2013. Roadmap for Development of a Reference Emission Level / Reference Level for the United Republic of Tanzania. Discussion draft – v1.

UN-REDD 2009. Tanzania final UN-REDD National Joint Programme. Retrieved on 7th January 2014 from: http://www.unredd.org/UNREDDProgramme/CountryActions/Tanzania/tabid/1028/language/en-US/Default.aspx

# APPENDIX

Table 4. JM distance matrix between the different forest classes
from the SPOT data (simulating Sentinel-2 A and B)
classification.

| | W10F10 | W100F10 | W80F10 | W60F10 | W20F10 | W40F10 | W20F20 | W100F20 | W80F20 | W40F20 | W60F20 | W40F40 | W60F40 | W100F40 | W80F40 | W60F60 | W100F60 | W80F60 | W80F80 | W100F80 | W100F100 |
|---|---|---|---|---|---|---|---|---|---|---|---|---|---|---|---|---|---|---|---|---|---|
| W100F100 | 759 | 724 | 720 | 653 | 670 | 534 | 629 | 723 | 544 | 487 | 416 | 521 | 343 | 322 | 231 | 635 | 76 | 204 | 394 | 36 | 0 |
| W100F80 | 781 | 747 | 742 | 675 | 691 | 554 | 645 | 745 | 566 | 505 | 433 | 539 | 357 | 339 | 243 | 654 | 78 | 212 | 415 | 0 | 36 |
| W80F80 | 464 | 419 | 409 | 317 | 339 | 166 | 298 | 412 | 178 | 118 | 66 | 158 | 125 | 97 | 207 | 295 | 352 | 245 | 0 | 415 | 394 |
| W80F60 | 669 | 632 | 621 | 538 | 558 | 390 | 487 | 624 | 411 | 331 | 249 | 369 | 157 | 151 | 39 | 504 | 136 | 0 | 245 | 212 | 204 |
| W100F60 | 741 | 706 | 699 | 626 | 644 | 496 | 590 | 702 | 510 | 444 | 367 | 479 | 286 | 270 | 168 | 602 | 0 | 136 | 352 | 78 | 76 |
| W60F60 | 255 | 234 | 194 | 102 | 115 | 134 | 85 | 195 | 135 | 194 | 278 | 156 | 371 | 372 | 471 | 0 | 602 | 504 | 295 | 654 | 635 |
| W80F40 | 638 | 600 | 589 | 505 | 524 | 354 | 455 | 592 | 375 | 295 | 212 | 333 | 122 | 113 | 0 | 471 | 168 | 39 | 207 | 243 | 231 |
| W100F40 | 547 | 505 | 494 | 404 | 425 | 248 | 361 | 497 | 268 | 188 | 105 | 229 | 54 | 0 | 113 | 372 | 270 | 151 | 97 | 339 | 322 |
| W60F40 | 563 | 525 | 508 | 415 | 436 | 251 | 349 | 511 | 279 | 186 | 101 | 223 | 0 | 54 | 122 | 371 | 286 | 157 | 125 | 357 | 343 |
| W40F40 | 379 | 343 | 317 | 214 | 235 | 43 | 146 | 320 | 92 | 42 | 127 | 0 | 223 | 229 | 333 | 156 | 479 | 369 | 158 | 539 | 521 |
| W60F20 | 475 | 435 | 417 | 320 | 342 | 152 | 265 | 421 | 180 | 87 | 0 | 127 | 101 | 105 | 212 | 278 | 367 | 249 | 66 | 433 | 416 |
| W40F20 | 405 | 366 | 344 | 242 | 265 | 68 | 187 | 347 | 106 | 0 | 87 | 42 | 186 | 188 | 295 | 194 | 444 | 331 | 118 | 505 | 487 |
| W80F20 | 309 | 264 | 247 | 146 | 170 | 49 | 174 | 251 | 0 | 106 | 180 | 92 | 279 | 268 | 375 | 135 | 510 | 411 | 178 | 566 | 544 |
| W100F20 | 63 | 56 | 6 | 110 | 88 | 283 | 275 | 0 | 251 | 347 | 421 | 320 | 511 | 497 | 592 | 195 | 702 | 624 | 412 | 745 | 723 |
| W20F20 | 333 | 317 | 275 | 187 | 199 | 148 | 0 | 275 | 174 | 187 | 265 | 146 | 349 | 361 | 455 | 85 | 590 | 487 | 298 | 645 | 629 |
| W40F10 | 342 | 303 | 280 | 176 | 199 | 0 | 148 | 283 | 49 | 68 | 152 | 43 | 251 | 248 | 354 | 134 | 496 | 390 | 166 | 554 | 534 |
| W20F10 | 150 | 120 | 85 | 24 | 0 | 199 | 199 | 88 | 170 | 265 | 342 | 235 | 436 | 425 | 524 | 115 | 644 | 558 | 339 | 691 | 670 |
| W60F10 | 172 | 137 | 107 | 0 | 24 | 176 | 187 | 110 | 146 | 242 | 320 | 214 | 415 | 404 | 505 | 102 | 626 | 538 | 317 | 675 | 653 |
| W80F10 | 65 | 52 | 0 | 107 | 85 | 280 | 275 | 6 | 247 | 344 | 417 | 317 | 508 | 494 | 589 | 194 | 699 | 621 | 409 | 742 | 720 |
| W100F10 | 65 | 0 | 52 | 137 | 120 | 303 | 317 | 56 | 264 | 366 | 435 | 343 | 525 | 505 | 600 | 234 | 706 | 632 | 419 | 747 | 724 |
| W10F10 | 0 | 65 | 65 | 172 | 150 | 342 | 333 | 63 | 309 | 405 | 475 | 379 | 563 | 547 | 638 | 255 | 741 | 669 | 464 | 781 | 759 |

# 29

# MONITORING OF RAPID LAND COVER CHANGES IN EASTERN JAPAN USING TERRA/MODIS DATA

I. Harada [a,] *, K. Hara [a], J. Park [a], I. Asanuma [a], M. Tomita [a], D. Hasegawa [a], K. Short [a], M. Fujihara [b]

[a] Department of Informatics, Tokyo University of Information Sciences,
4-1 Onaridai Wakaba-ku, Chiba, 265-8501 Japan – *iharada@rsch.tuis.ac.jp
[b] Graduate School of Landscape Design and Management, University of Hyogo/Awaji Landscape Planning and Horticulture
Academy, 954-2 Nojimatokiwa, Awaji, Hyogo Prefecture, 656-1726 Japan – fujihara@awaji.ac.jp

**Commission VIII, WG VIII/8**

KEY WORDS: Vegetation Map, National Survey of the Natural Environment, Landscape Changes, Terra/MODIS, Landsat, LTS, the Great Eastern Japan Earthquake

**ABSTRACT:**

Vegetation and land cover in Japan are rapidly changing. Abandoned farmland in 2010, for example, was 396,000 ha, or triple that of 1985. Efficient monitoring of changes in land cover is vital to both conservation of biodiversity and sustainable regional development. The Ministry of Environment is currently producing 1/25,000 scale vegetation maps for all of Japan, but the work is not yet completed. Traditional research is time consuming, and has difficulty coping with the rapid nature of change in the modern world. In this situation, classification of various scale remotely sensed data can be of premier use for efficient and timely monitoring of changes in vegetation.. In this research Terra/MODIS data is utilized to classify land cover in all of eastern Japan. Emphasis is placed on the Tohoku area, where large scale and rapid changes in vegetation have occurred in the aftermath of the Great Eastern Japan Earthquake of 11 March 2011. Large sections of coastal forest and agricultural lands, for example, were directly damaged by the earthquake or inundated by subsequent tsunami. Agricultural land was also abandoned due to radioactive contamination from the Fukushima nuclear power plant accident. The classification results are interpreted within the framework of a Landscape Transformation Sere model developed by Hara et al (2010), which presents a multi-staged pattern for tracking vegetation changes under successively heavy levels of human interference. The results of the research will be useful for balancing conservation of biodiversity and ecosystems with the needs for regional redevelopment.

## 1. INTRODUCTION

Vegetation and land cover in Japan are rapidly changing. Over the past half century, for example, coppices and other managed secondary woodlands, which had formerly provided various vital ecosystem services, were abandoned when fossil fuels replaced firewood and charcoal as the main source of energy. These abandoned secondary woodlands revert to the natural forest cycle, and are now slowly moving back to their original climax stage. Urbanization, as well as depopulation in rural areas, also produce wide spread shifts in land cover. Natural disasters can result in dramatic changes. The Great East Japan Earthquake of March, 2011 and the subsequent tsunami, for example, caused massive abandonment of farmland and damage to various vegetation communities, especially in the heavily-damaged Tohoku Region.

In this environment, accurate, comprehensive and up-to-date data on changes in land cover and vegetation are required for proper planning and management of land and resources. In Japan, basic national level data for ecosystem and biodiversity conservation were first produced by the Ministry of Environment (MOE) National Survey of the Natural Environment project, which lasted from 1973 to 1998. These 1/50,000 scale maps provided a crucial understanding of Japan's vegetation patterns, and for several decades were the standard reference work for ecological studies and environmental impact assessments.

This first series of national level vegetation maps, however, gradually became out-dated, and in 1999 MOE began a new project to update them with 1/25,000 scale maps. Unfortunately, this work has proved costly and time-consuming, and as of March 2013 only 68% of the national land area had been completed. As can be seen in Figure 1, the new maps as they stand are still insufficient for following rapid vegetation changes, especially in mountainous areas. Furthermore, accurate, detailed continuously updated land cover data is now required for assessing this damage caused by the 2011 earthquake and tsunami, and also for designing reconstruction strategies that balance social and economic recovery with conservation of biodiversity and ecosystems.

Remote sensing research shows a great potential to respond to this need for timely land cover data. Remote sensing enjoys relatively low financial and labor costs; and can be implemented safely and effectively even on steep mountainsides and in disaster areas where access is limited. In addition, the research can be conducted over a wide region on an immediate or periodic basis.

Worldwide remote sensing research on vegetation distribution and change has developed rapidly since Global Vegetation Index (GVI) data produced by the NOAA/AVHRR weather satellites became available in the 1980s. Dehrich et al.

---

* Corresponding author. This is useful to know for communication with the appropriate person in cases with more than one author.

(1994) and Loveland et al. (2000), for example, employed this remote sensing data to map global vegetation. In 1999, NOAA launched the Terra/MODIS satellite, which was designed as the successor to NOAA AVHRR. MODIS, covers a 2330 km wide swath, and passes over the same spot twice daily. Improved functions include spatial and spectral resolution, orbital data, and on-board luminance correction（Justice et al. 1998）.

MODIS researches focusing on vegetation patterns in East Asia, however, have been few. To quantify general trends in land cover change over the past century, Harada et al (2011, 2014) combined MODIS data with digitalized versions of historic maps and maps of potential vegetation as predicted by Warmth Index. The ultimate goal of this continuing research is to take advantage of the wide scale and regularity of MODIS data to develop a system for timely monitoring of land cover for all of Japan. This system can supplement the MOE vegetation research, and can be updated periodically as required. This paper reports on the first step in this project, which is to generate a land cover map for eastern Japan that can be put to immediate use in the areas impacted by the earthquake and tsunami of 2011.

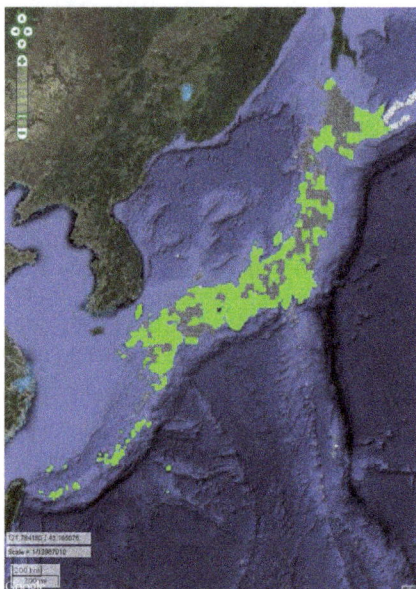

Figure 1. Area of Japan currently covered by National Survey of the Natural Environment 1/25,000 scale vegetation maps (green) (http://www.vegetation.biodic.go.jp)

## 2. DATA AND MONTHLY COMPOSITE IMAGES

At Tokyo University of Information Sciences (TUIS), MODIS data from the Terra and Aqua satellites has been directly acquired since November of 2000, and VIIRS data from the Suomi-NPP satellite since May of 2001. 500m resolution atmospherically-corrected MODIS data sets have been archived since 2001. Since these data sets cover the time periods both before and after the 2011 earthquake and tsunami, they can be utilized to analyse land cover changes caused by the disaster.

Multi-temporal composite Terra/MODIS data sets with the effects of cloud cover removed were constructed, and used to generate vegetation maps of the Tohoku Region before and after the disaster. Compositing generally results in loss of vegetative phonological information. In this research, however, a system proposed by Park et al (2009), which utilizes the White Index (proportion of cloud in the image) and MODIS MOD35 data, were employed. This system allowed for

generation of monthly composite data without the loss of vegetative phonological information.

In order to remove the effects of cloud cover from 500 m resolution data sets, the data must be acquired at high frequency. NASA does provide atmospherically-corrected composite images with cloud cover removed, but to assemble these on a scale such as all Japan would be exceedingly difficult. TUIS, however, acquires MODIS data at three spots; Miyake Island in the southern Ryukyus, the main campus near Tokyo in central Honshu, and Abashiri along the Sea Of Okhotsk in eastern Hokkaido. This allows high frequency data acquisition over an area stretching from the Philippine Sea to the Okhotsk Sea. As a result, monthly composite data sets with the effects of cloud cover removed can be used to construct time sequences that follow seasonal changes in vegetation (Hara et al. 2010 and Harada et al. 2014).

This research employs 500 m resolution MOD09 and MOD03 data acquired at TUIS from April to November 2013. 171 images from April, 139 from May, 144 from June, 124 from July, 165 from August, 142 from September, 134 from October and 155 from November, were used to construct multi-temporal monthly composite data sets with the effects of cloud cover removed. In order to eliminate distorted images, only data from a sensor zenith angle of 40 degrees or less were used. Monthly composite maps were generated for April to November 2013 (Figure 2). Data for July was less than for other months, so the results for this month were not included in the subsequent analyses. The composite data for the seven remaining months was then integrated into image data. Nine bands (Surface Reflectance Bands1-7+NDVI+RVI) for 7 months resulted in a total of 63 bands.

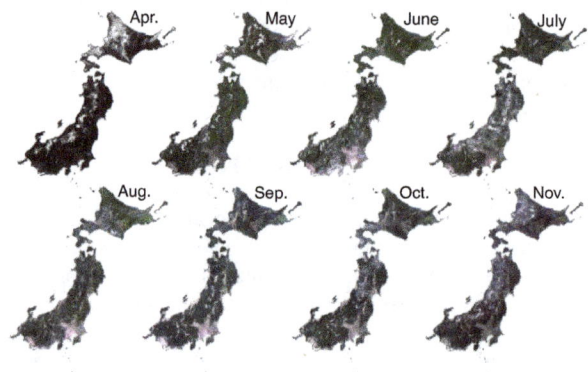

Figure 2. Monthly Composite Images from Terra/MODIS data for 2013

## 3. LAND COVER CHANGES IN EASTERN JAPAN

The flowchart for constructing vegetation maps from Terra/MODIS data is shown in Figure 3. To begin with, vegetated areas and non-vegetated areas were separated using NDVI threshold (NDVI=0.35) from the composite data for August and September, with cloud cover effect removed. Further, monthly composite data with cloud cover effect removed were used to generate image data (Surface Reflectance Bands1-7+NDVI+RVI); and lake and marsh data from the National land Numerical Download Service were employed to mask the surrounding seas, isolating an image of the land mass of eastern Japan, including the island of Hokkaido and Honshu from the northern tip south to the Chubu and Kanto regions.

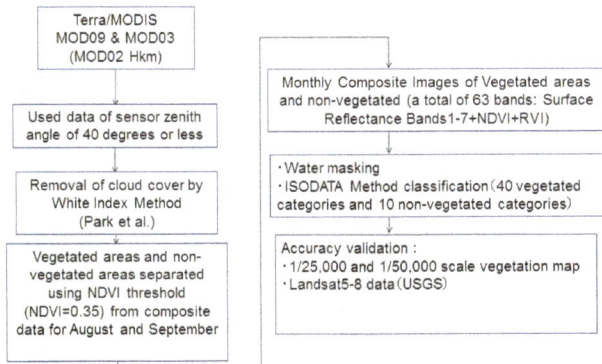

Figure 3. Flowchart for constructing vegetation maps from Terra/MODIS data

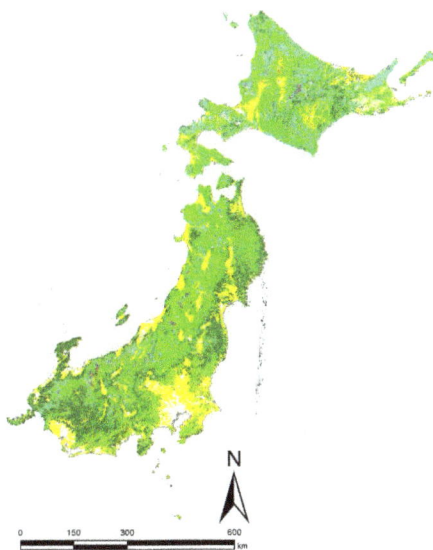

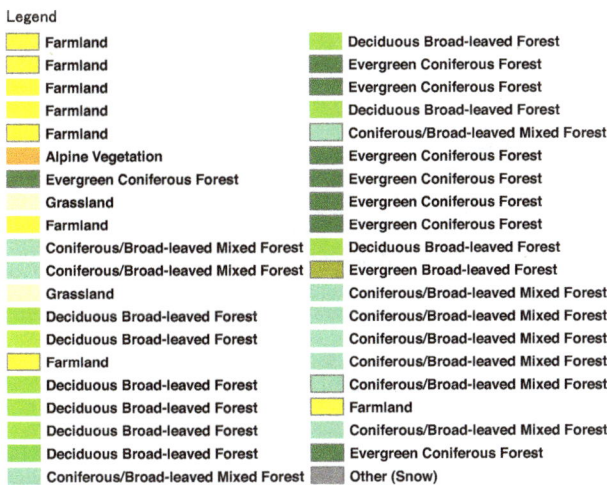

Figure 5. 10 non-vegetated categories by ISODATA method classification of Terra/MODIS data

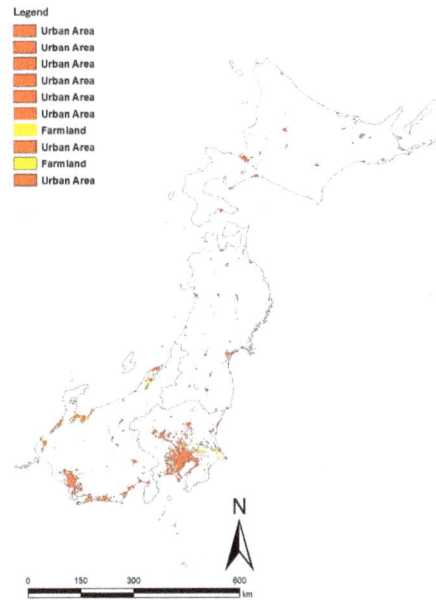

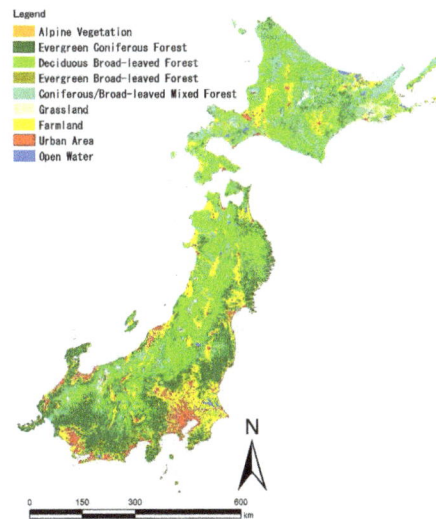

Figure 4. 40 vegetated categories by ISODATA method classification of Terra/MODIS data

Figure 6. Land Cover Map in eastern Japan based on Terra/MODIS data for 2013

A total of 50 initial classification categories were established using the ISODATA method (non-hierarchical cluster analyses). These included 40 vegetated categories and 10 non-vegetated categories. The distribution of the 40 vegetated categories are shown in Figure 4, and the 10 non-vegetated categories in Figure 5.

These initial categories were then consolidated into nine final categories employed in the subsequent analysis. Lakes and marshes were restored and classified as Open Water, and the vegetated and non-vegetated Farmland categories were consolidated. As a result, the 50 initial vegetated categories were consolidated into the following categories: Alpine Vegetation; Evergreen Coniferous Forest; Deciduous Broad-leaved Forest; Evergreen Broad-leaved Forest; Coniferous/Broad-leaved Mixed Forest; Grassland; Farmland; Urban Area and Open Water.

The accuracy of the map was evaluated using ENVI5.1 (Exelis VIS) image analyses software. References used included GIS data from the 1:25,000 scale MOE maps (incomplete), the 1:50,000 scale MOE maps (complete) and Landsat5-8 satellite data (USGS). To implement this MOE comparison the 9 classification categories adopted here were collated with the 58 categories employed on the MOE maps. The results of the classification are shown in Figure 6.

## 4. LAND COVER CHANGES IN THE DISASTER REGION

One purpose of this research was to evaluate the changes in land cover and vegetation in the area of Tohoku severely damaged by the earthquake and tsunami of 2011. To accomplish this, the results generated here, after the disaster, were compared with those of a similar MODIS-based land cover classification for the Tohoku region implemented in 2001, before the disaster (Hara et al. 2010 and Harada et al., 2014).

The results of his comparison are indicated in Figure 7. As can be seen, large areas of farmland were damaged and abandoned. Loss of farmland is especially apparent along the coast at the bottom right hand corner of the maps. This area is where the Fukushima Number One Nuclear Power Plant is located. The plant was damaged by the tsunami and leaked radioactive substances into the surrounding atmosphere. These substances were carried by the wind, and the area within a 30 km radius of the plant was designated as an Evacuation Zone, with about 30,000 residents being evacuated. Today entry to the zone remains closed or restricted.

In addition, analysis of the Evacuation Zone is being implemented using MODIS (500 m resolution), Landsat (30 m resolution) and GIS data. The purpose of this part of the research is to clarify the general land cover changes discussed above in greater detail, and also to develop a system for continual monitoring of this area.

As a baseline, the land cover in the target area was analysed using USGS Landsat data (http://www.usgs.gov/) downloaded for 3 July 2000, 3 March 2002 and 9 May 2003. The ISODATA (non-hierarchical clustering) using ENVI5.1 was employed to classify the data into 50 initial categories, which were then consolidated using ArcGIS into 9 land cover categories. These categories are basically the same as those employed in the MODIS classifications, but Grassland has been eliminated, and Farmland has been divided into Upland Field (non-irrigated) and Rice Paddy (irrigated). The results of this classification are shown in Figure 8. The only vegetation reference data available for this area was the MOE 1:50,000 map.

In addition, Landsat-5 TM data for 2 June 2009 was downloaded from USGS (http://www.usgs.gov/). Figure 9 presents this image in natural color (RGB=band3, band5, band1). This image was then compared with the classification results shown in Figure 8. In both images large areas of rice paddy filled with water, dark red in Figure 9 and yellow in Figure 8, can be seen along the coastal plain.

To further understand these changes in land cover caused by the disaster, a Landsat-8 OLI image for 31 May 2014 was downloaded from USGS. This image is shown in natural color (RGB=band4, band6, band2) in Figure 10. As can be clearly

seen, rice paddies along the coastal plain remain only at the top of the image, around 40 km from the power station. The paddies closer to the station are no longer filled with water, indicating that they have been abandoned.

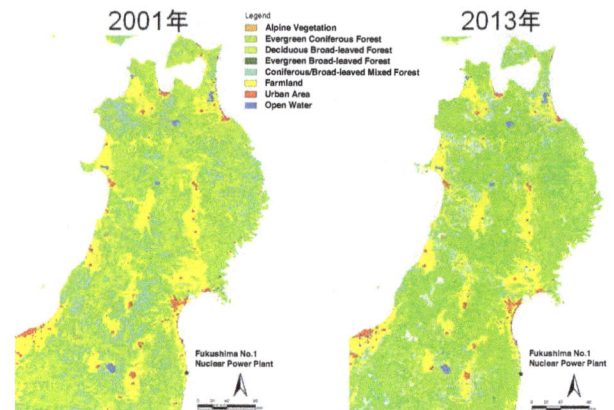

Figure 7. Land cover change in eastern Japan based on Terra/MODIS data for 2001 and 2013

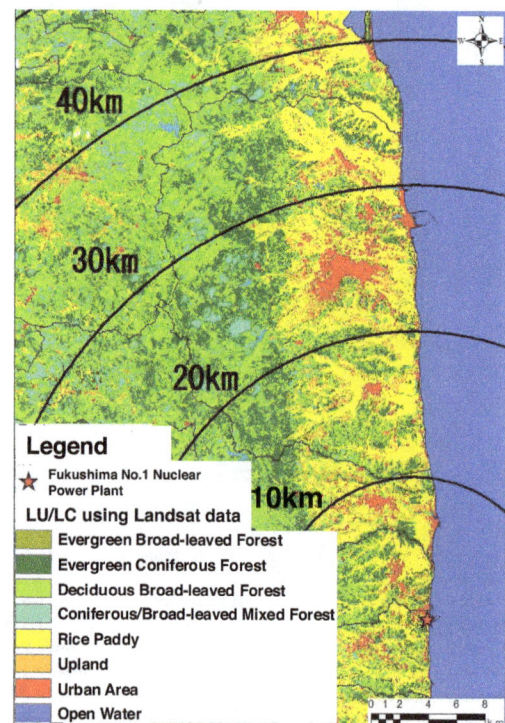

Figure 8. Land Cover Map in Fukushima Prefecture based on Landsat TM data during 2000 and 2003

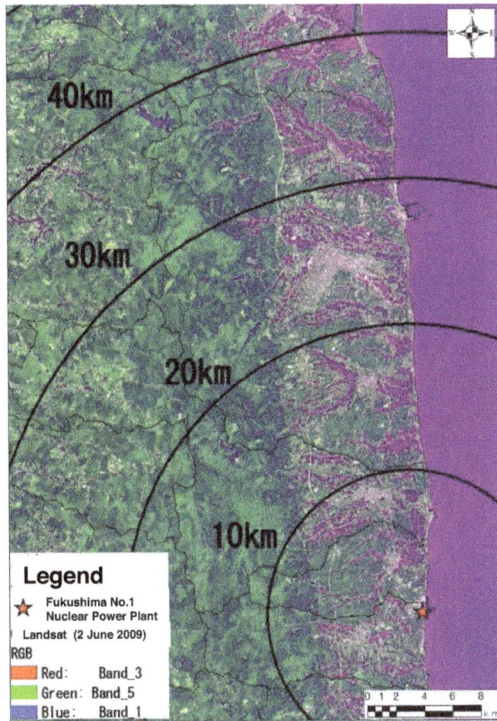

Figure 9.
Natural Color Image using Landsat-5 TM data
(2 June 2009, RGB : 3,5,1)

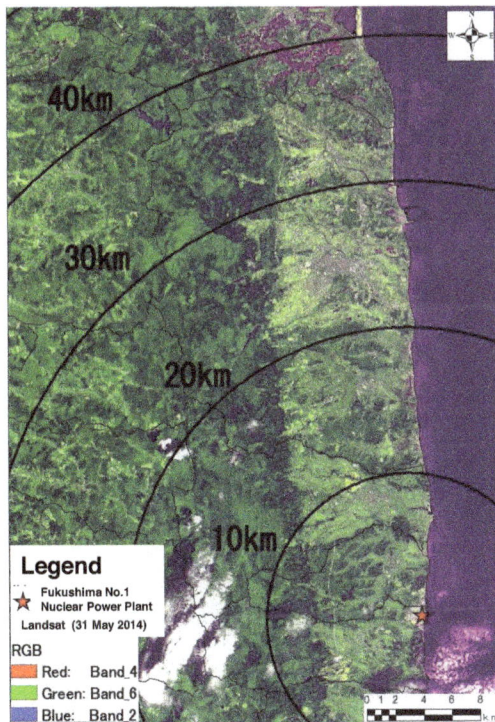

Figure 10. Natural Color Image using Landsat-8 OLI data
(31 May 2014, RGB : 4,6,2)

## 5. CONCLUSIONS

The amount of abandoned farmland in Japan remained steady until around 1985 (130,000 ha). After 1990, however, the pace of farmland abandonment increased rapidly, and by 2010 had reached 396,000 hectares (Ministry of Agriculture, Forestry and Fisheries, 2011). The earthquake and tsunami of 2011 inundated rice paddies and dealt a powerful blow to agricultural production in the Tohoku Region. Four years later the industry is yet to recover. Restoration efforts are hampered by a loss of population and aging of the agricultural workforce, and also by negative consumer images generated by the nuclear fallout.

Abandoned farmland first reverts to weed-field, but the subsequent changes in vegetation vary according to factors such as climate, altitude and topography. Hara et al (2010) have compared actual land cover based on multi-temporal Terra/MODIS data to natural land cover as predicted by climate zone; and used the results to develop a Landscape Transformation Sere (LTS) model for predicting how vegetation and land cover change under increasing levels of human intervention. The basic pattern is presented in Figure 11; and involves change from primary forest to managed secondary woodland, then on to grassland, agricultural land and finally non-vegetated residential or industrial area. Abandoned farmland may be expected to change in the opposite direction.

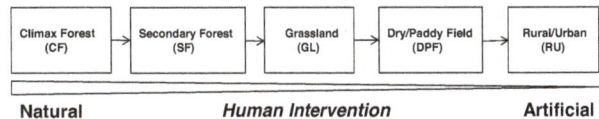

Figure 11. Typical Landscape Transformation Sere (LTS)
model (after Hara et al.,2009)

This research clearly demonstrates that wide-scale remote sensing data such as MODIS and Landsat are effective tools for supplementing the Ministry of Environment's vegetation mapping, and also for evaluating and monitoring rapid changes in land cover such as those caused by large-scale infrequent disasters.

## ACKNOWLEDGEMENTS

This research was supported in part by the MEXT Strategic Research Infrastructure Formation Project "Research project for sustainable development of economic and social structure dependent on the environment of the eastern coast of Asia (TUIS)"; by the Environment Research and Technology Development Fund (1-1405) of Ministry of Environment, Japan; and by a MEXT Japan grant-in-aid for scientific research (No. 26350403).

## REFERENCES

Ehrlich, D., Estes, J., E. & Singh, A., 1994. Applications of NOAA-AVHRR 1 km data for environmental monitoring, *International Journal of Remote Sensing*, 15(1), pp. 145-161.

Ministry of Agriculture, Forestry and Fisheries, 2010. Report on Results of 2010 Census of Agriculture and Forestry in Japan.

Justice, C., et al., 1998. The Moderate Resolution Imaging Spectroradiometer (MODIS): land remote sensing for global change research. *IEEE Trans*. On Geosci. And Remote Sens., 36(4), 1228-1249.

Hara, K., Harada, I., Tomita, M., Short M., Park J., Shimojima, H., Fujihara, M., Hirabuki, Y., Hara, M. and Kondoh, A., 2010. Landscape transformation sere: in which directions will our landscape move and how can we monitor these changes,

Landscape ecology - methods, applications and interdisciplinary approach, 165-172.

Harada, I., Matsumura, T., Hara, K. and Kondoh, A., 2011. Spatial analysis of changes in Japanese forests during the process of modernization, Landscape Ecology and Management, 16(1), 17-32. (in Japanese with English abstract)

Harada, I., Hara, K., Tomita, M., Short, K. and Park, J. 2014. Monitoring landscape changes in Japan using classification of MODIS data combined with a Landscape Transformation Sere (LTS) model, Journal of Landscape Ecology, 7(3), 23-38.

Loveland, T. R., Reed, B. C., Brown, J. F., Ohlen, D. O., Zhu, z,, Yang, L., and Merchant, J.W., 2000. Development of a global land cover characteristics database and IGBP DISCover from 1 km AVHRR data, International Journal of Remote Sensing, 21, 1303-1330.

Park, J., Yasuda, Y., Sekine, H., Tateishi, R. and Susaki, J.,2009. Estimation of the cloud cover ratio in a pixel using MODIS data , Journal of the Japan Society of Photogrammetry  and Remote Sensing, 47(6), 30-37. (in Japanese with English abstract)

National Survey of the Natural Environment, 2013. http://www.vegetation.biodic.go.jp

# 30

# FOREST AND FOREST CHANGE MAPPING WITH C- AND L-BAND SAR IN LIWALE, TANZANIA

J. Haarpaintner [a], C. Davids [a], H. Hindberg [a], E. Zahabu [b], R.E. Malimbwi [b]

[a] Norut, P.O. Box 6434, Tromsø Science Park, N-9294 Tromsø, Norway – joergh@norut.no
[b] Sokoine University of Agriculture, Morogoro, United Republic of Tanzania

THEME: BIOD - Forests, Biodiversity and Terrestrial Ecosystems.

KEY WORDS: REDD+, Forest, Forest Change, SAR, Tanzania

**ABSTRACT:**

As part of a Tanzanian-Norwegian cooperation project on Monitoring Reporting and Verification (MRV) for REDD+, 2007-2011 C- and L-band synthetic aperture radar (SAR) backscatter data from Envisat ASAR and ALOS Palsar, respectively, have been processed, analysed and used for forest and forest change mapping over a study side in Liwale District in Lindi Region, Tanzania. Land cover observations from forest inventory plots of the National Forestry Resources Monitoring and Assessment (NAFORMA) project have been used for training Gaussian Mixture Models and k-means classifier that have been combined in order to map the study region into forest, woodland and non-forest areas. Maximum forest and woodland extension masks have been extracted by classifying maximum backscatter mosaics in HH and HV polarizations from the 2007-2011 ALOS Palsar coverage and could be used to map efficiently inter-annual forest change by filtering out changes in non-forest areas. Envisat ASAR APS (alternate polarization mode) have also been analysed with the aim to improve the forest/woodland/non-forest classification based on ALOS Palsar. Clearly, the combination of C-band SAR and L-band SAR provides useful information in order to smooth the classification and especially increase the woodland class, but an overall improvement for the wall-to-wall land type classification has yet to be confirmed. The quality assessment and validation of the results is done with very high resolution optical data from WorldView, Ikonos and RapidEye, and NAFORMA field observations.

## 1. INTRODUCTION

As part of the Group on Earth Observations Forest Carbon Tracking Task (GEO FCT) and following Global Forest Observations Initiative (GFOI), a Norwegian-Tanzania cooperation project was established. The goal is to support Measuring, Reporting and Verification (MRV) for REDD+ initiative, which aims to be a financial incentive for tropical countries to reduce deforestation and forest degradation (REDD+).

An important research issue inside GFOI is to investigate the role of synthetic aperture radar (SAR) in establishing national forest monitoring systems and develop robust methods to map forests and forest change (GFOI, 2013; Haarpaintner et al., 2012). Two study regions are investigated in this Norwegian-Tanzanian project by airborne LiDAR as well as by satellite remote sensing: the Amani forest reserve and a large study region of the Liwale district in the Lindi region in Tanzania. Here we focus on the use of C- and L-band SAR of the Liwale study site. L-band SAR is generally better suited than C-band SAR for forest monitoring as its longer wavelength better penetrates the forest canopy. L-band SAR data is provided by the Japanese ALOS satellite with its Palsar sensor. C-band SAR data is provided by the European Space Agency's Envisat A(dvanced)SAR. The project should complement the National Forestry Resources Monitoring and Assessment (NAFORMA), which is a nation-wide forest inventory program that has collected a total of more than 36000 forest plots (Tomppo, 2014).

This paper presents forest mapping results from both ALOS Palsar and Envisat ASAR individually as well as combined, and yearly forest change detection from ALOS Palsar only.

## 2. THE LIWALE STUDY SITE

The Liwale study site of about 15000 km$^2$ is located in the Lindi region in the south-east of Tanzania. There are large forested areas mainly in the north and west of the site, but the majority land type is scattered open to close Miombo woodland. The north-eastern quadrant of the study site has little vegetation, dominantly open (Miombo) woodland and non-forest areas.
Fig. 1 shows the location of the Liwale study site in Tanzania.

Figure 1. The location of the Liwale study site in Tanzania (© GoogleEarth).

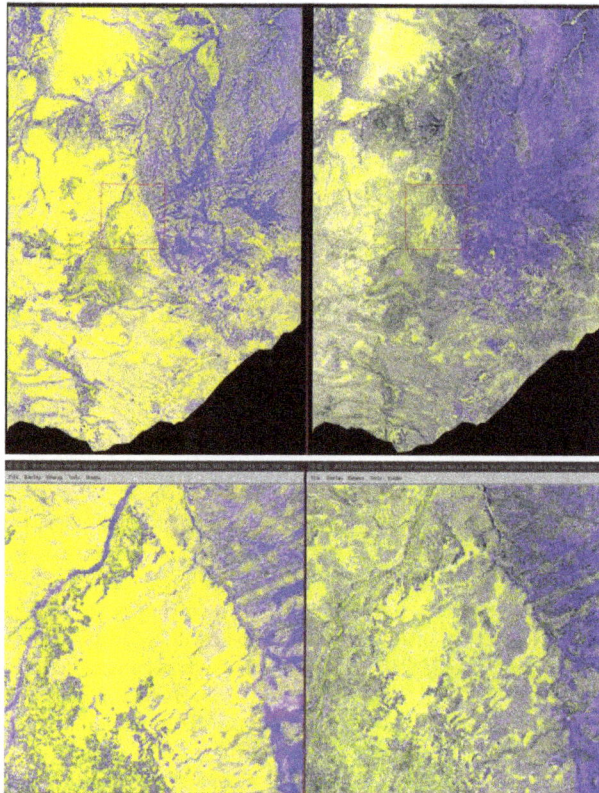

Figure 2. (Upper panels) 2007-2010 averaged ALOS Palsar image (left) and 2009-2011 averaged Envisat ASAR APS image (right) over the Liwale study site. (Lower panels) red rectangle zoom.

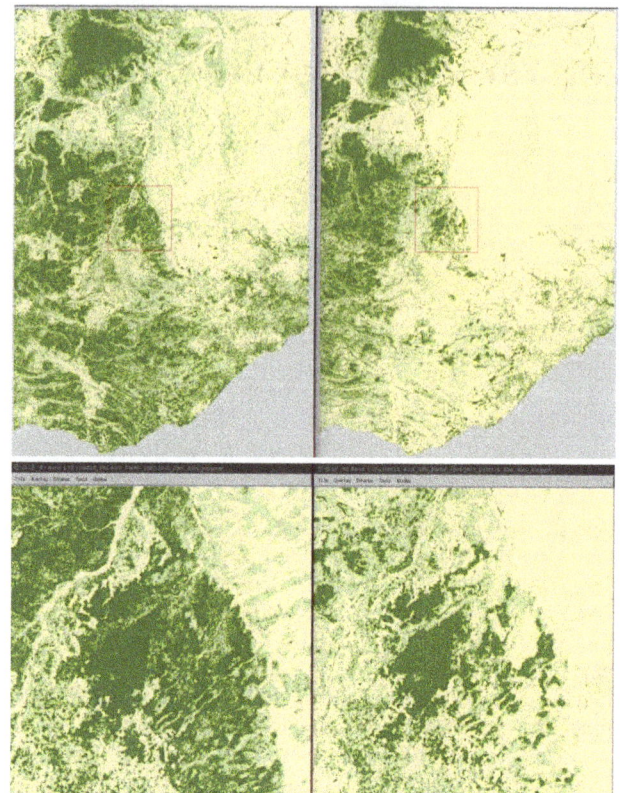

Figure 3. Forest/Woodland/Non-forest classification from averaged ALOS Palsar FBD (left) and Envisat ASAR APS (right). Lower panels show the red rectangle zoom. Forest in dark green, woodland in light green, and non-forest in beige.

### 3.  SAR PRE-PROCESSING

SAR pre-processing, i.e. georeferencing, radiometric calibration and slope correction, has been done with Norut's in-house developed GSAR software (Larsen et al., 2005) using the SRTM.v4 DEM. An advanced sigma nought ($\sigma°$) correction was introduced to the pre-processing in order to account for an angle dependency of the illuminated area, according to Shimada & Takahiro (2010) and Ulander (1996). The 2007-2010 ALOS Palsar Fine-Beam Dual (FBD) data in HH and HV polarization has been then processed into yearly dry season mosaics. Furthermore an average as well as a maximum backscatter (per pixel) mosaic for the dry season have been established for the four year period in order to build a maximum forest extent mask to filter out changes in non-forest areas (see section 5) for the forest change product. The same has been done for the 2009-2011 Envisat ASAR alternate polarization (APS) data in VV and VH polarizations. As Envisat ASAR APS data however is far more noisy and variable because of higher sensitivity for humidity and phenology effects, the Envisat ASAR APS average backscatter in VV and VH have been used only for forest mapping. The average backscatter images for ALOS Palsar FBD data and for Envisat ASAR APS data for the periods 2007-2010 and 2009-2011, respectively, are shown in Fig 2. The RGB channels are (R) co-polarization (HH for Palsar and VV for ASAR), (G) cross-polarization (HV for Palsar and VH for ASAR), and only for presentation purposes (B) the normalized difference index (NDI) describing the ration between the co-polarization XX (i.e. HH or VV) and cross-polarization XY (i.e. HV or VH) backscatter as:

$$NDI=(XX-XY)/(XX+XY).$$

### 4.  FOREST/WOODLAND/NON-FOREST MAPPING (FWNF)

Ground reference data has been collected as forest plots through the NAFORMA project providing vegetation types at each plot location. As it is unrealistic that each NAFORMA land/vegetation type is detectable in SAR data, the different types have been aggregated into three main classes: forest, dense woodland and non-forest; resulting in a higher number of training sites for each class for the classification algorithm. In order to reduce speckle of the SAR data, the data has been temporally averaged over the totality of available data, which are the dry seasons of the years 2007-2010 and 2009-2011 for ALOS Palsar FBD and Envisat ASAR APS data, respectively.

#### 4.1  Single sensor approach

Each of these averaged data sets have then been classified using the co- and cross-polarization band (HH/HV for ALOS Palsar and VV/VH for Envisat ASAR) and with these training sites into forest, woodland and non-forest areas using two methods that are part of Norut's GSAR software package: the Gaussian Mixture Model (GMM) and the k-means (KM) classifier. The two classification results, GMM and KM, have then been combined into one classification taking, by priority, the maximum extent of forest, the maximum extent of woodland and the rest as non-forest areas. Fig. 3 shows these classification results using the ALOS Palsar mean backscatter HH and HV, and the Envisat ASAR APS mean backscatter in VV and VH. Obviously, using this method, ALOS Palsar FBD detects more forest and woodland than Envisat ASAR APS.

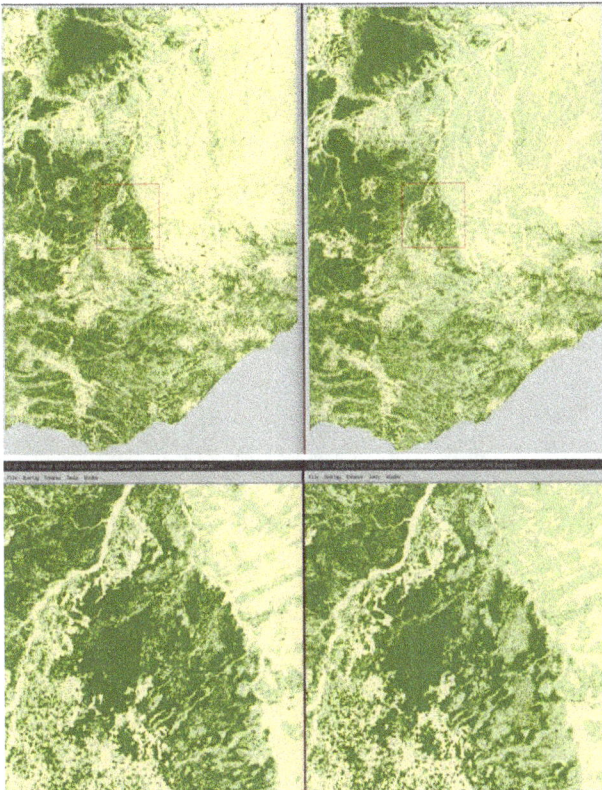

Figure 4. Single sensor classification with ALOS Palsar (left) versus multi-sensor classification using ALOS Palsar and Envisat ASAR (right). Lower panels are enlargements of the red rectangles. Forest in dark green, woodland in light green, and non-forest in beige.

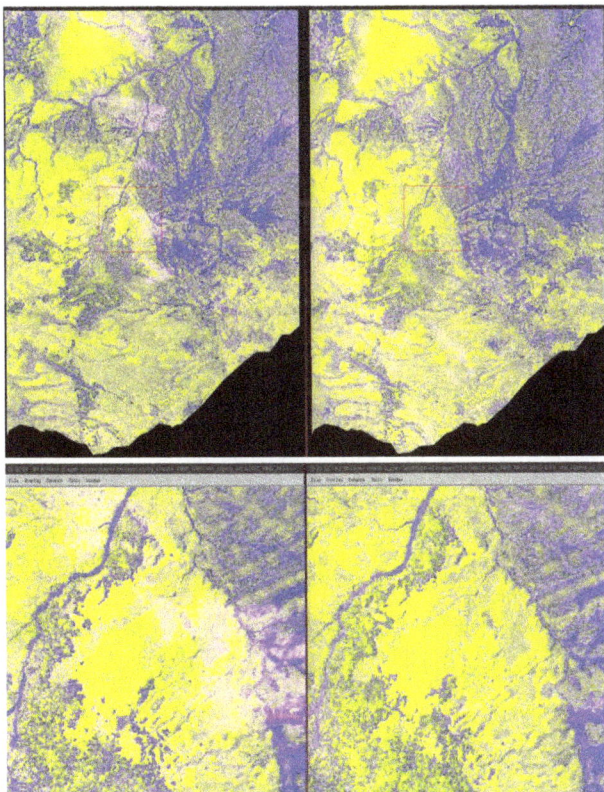

Figure 5. 2007 (left) and 2010 (right) ALOS Palsar mosaic. Rectangle enlargements in lower panel.

## 4.2 Multi-sensor approach (C- and L-band SAR combined)

Using the same classification approach than in 4.1., i.e. combing the GMM and KM classification, in the multi-sensor approach, we feed the classifier with all averaged backscatter channels, i.e. Palsar (HH), Palsar (HV), ASAR (VV) and ASAR (VH). The multi-sensor results from the C-and L-band SAR combination is shown in Fig. 4 in comparison with the individual ALOS Palsar classification result. The forest area extent is very similar in both classification, but the multi-sensor approach classifies a larger area into woodland especially in the north-eastern area, even though Envisat ASAR individually classifies this area in large majority into non-forest.

## 5. INTER-ANNUAL FOREST CHANGE FROM ALOS PALSAR HV

Inter-annual forest change has been detected by subtracting the ALOS Palsar HV backscatter image of year (y+1) from year y. All pixels with a decrease stronger than 3dB are considered as forest or woodland loss and all pixels with a variation of +3dB are considered as forest gain. Prior to this step the maximum ALOS Palsar HH and HV backscatter mosaics have been classified into forest, woodland and non-forest and the result is used to mask out changes in non-forest areas, which are mainly due to difference in ground humidity variation, precipitation or agriculture changes. Inter-annual changes have been detected for the consecutive years 2007-2008, 2008-2009, 2009-2010 as well as for the three year period 2007-2010. The strong agreement between the sum of the forest loss of the consecutive years and the directly detected three year loss from 2007 to 2010 indicates that this simple method is quit robust and reliable as long as the forest/woodland mask is accurate.

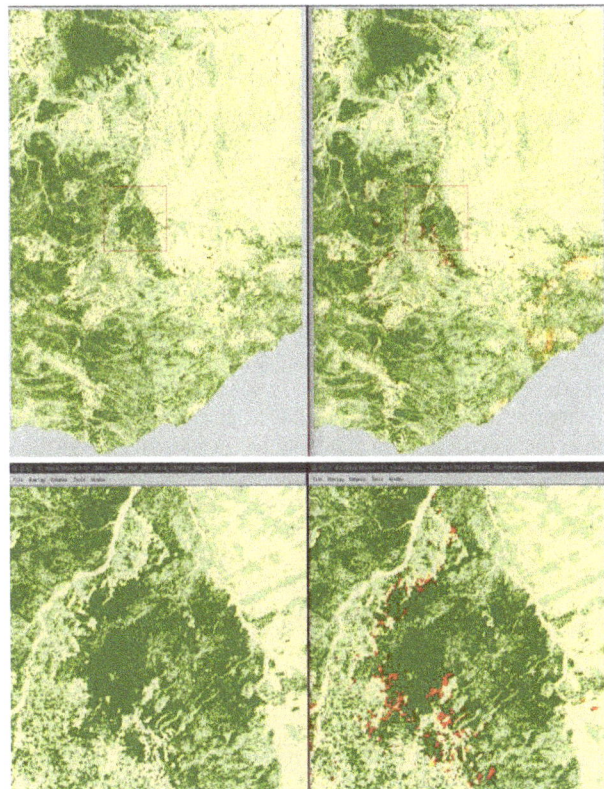

Figure 6. (Left) forest/woodland mask from classifying the maximum HH and HV backscatter mosaics. (Right) 2007-2010 forest and woodland loss in red and orange, respectively.

Single forest/woodland loss or forest gain pixels have been filtered out as they are often due to speckle effects.

## 6. VALIDATION

The accuracy of the forest-woodland-non forest classification was determined visually using a very high resolution (VHR) satellite image WorldView, with 4 m resolution, from 12.04.2009. In addition, the forest change mapping was visually checked against VHR optical images from different years (Ikonos, Digital Globe, Rapid Eye and WorldView).

For the accuracy assessment, a set of 250 random points was created across the WorldView image; of these, a total of 134 points were located in cloud and shadow free areas and were classified visually. The visual classification is based on an area of approximately 100×100 m around each point. Prior to visual classification, the observer was trained by comparing NAFORMA field observations with the WorldView image. The visual classification is divided into three classes: forest, woodland and other (non forest), where woodland includes all woodland with >10% tree crown cover.

The result of the visual accuracy assessment (Table 1) shows that the producer's accuracy of the three classes vary between 0.65 and 0.96. The overall accuracy of the classification is 73.1%. The main confusion is in the classification of woodland, of which 24% is classified as forest and 12% as non forest. Two examples (Fig. 7), comparing the classified image with the WorldView image, show that the classification correctly identifies the main landscape patterns.

Table 1. Confusion table for a forest/woodland/non-forest classification. PA: producer's accuracy. UA: user's accuracy.

| | ALOS PALSAR classes | | | total | PA |
|---|---|---|---|---|---|
| | forest | woodland | non forest | | |
| forest | 21 | 5 | 0 | 26 | 0.81 |
| woodland | 20 | 55 | 10 | 85 | 0.65 |
| non forest | 0 | 1 | 22 | 23 | 0.96 |
| total | 41 | 61 | 32 | 134 | |
| UA | 0.51 | 0.90 | 0.69 | | |

*Average accuracy (PA) 80.4%*
*Average reliability (UA) 70.0%*
*Overall accuracy 73.1%*

When combining the forest and woodland classes into a forest (= forest + woodland) - non forest classification, the overall accuracy increases to 91.8% (Table 2). About 9% of the reference forest pixels were classified as non-forest, indicating that the extent of forest could be underestimated in this classification.

Table 2. Confusion table for a forest/non-forest classification. PA: producer's accuracy. UA: user's accuracy.

| | ALOS PALSAR classes | | total | PA |
|---|---|---|---|---|
| | forest | non forest | | |
| forest | 101 | 10 | 111 | 0.91 |
| non forest | 1 | 22 | 23 | 0.96 |
| total | 102 | 32 | 134 | |
| UA | 0.99 | 0.69 | | |

*Average accuracy (PA) 93.3%*
*Average reliability (UA) 83.9%*
*Overall accuracy 91.8%*

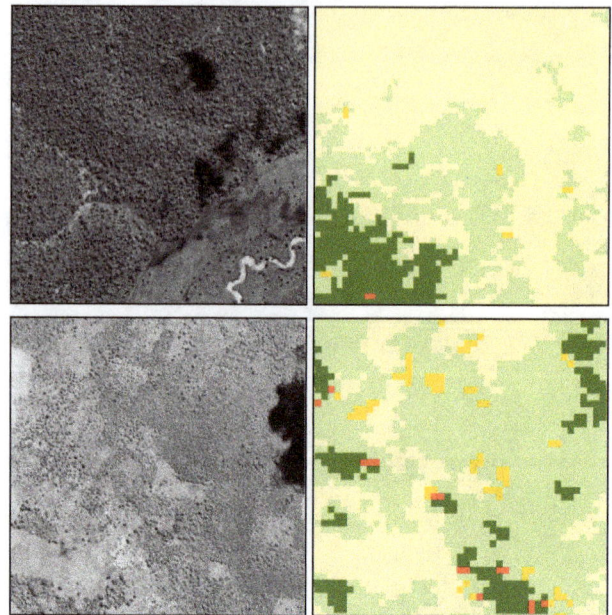

WorldView image                    Classified image
Figure 7. Two examples of the classified ALOS PALSAR image compared with the WorldView image. Dark green: forest. Light green: woodland. Beige: non-forest. Red: forest loss. Orange: woodland loss.

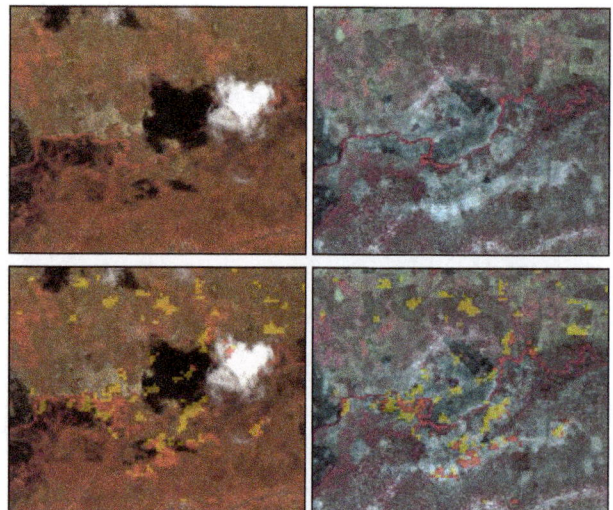

Ikonos 2008                    RapidEye 2012S
Figure 8. Example of an area where forest and woodland loss was identified. On the left an Ikonos image from 2008, on the right a RapidEye image from 2012. The forest and woodland loss classes (in red and orange, respectively) are overlain on the bottom two images.

The accuracy of the classified forest change between 2007 and 2010 is difficult to properly validate because of the limited availability of overlapping VHR images of the same years. The mapped deforested areas were visually checked against pairs of overlapping VHR images: Ikonos images from 2008 and 2010, the Ikonos image from 2008 against a RapidEye image from 2012, and a WorldView image from 2009 against a 2012 RapidEye image. The comparison showed that the main trends are correctly classified, although single pixels classified as forest or woodland loss tend to be mostly noise. Fig. 8 shows an example of an area where forest and woodland are changed into agricultural land. The images are from 2008 and 2012 and it is therefore not possible to identify the exact timing of deforestation. However, the forest/woodland loss classes appear to match well with areas where deforestation is clearly ongoing. One should keep in mind, however, that deforested areas in the RapidEye image from 2012 may have been deforested after 2010 and are therefore not picked up in the forest/woodland loss classes of 2007-2010.

## 7. CONCLUSION

C- and L-band SAR imagery have been analysed over a study site in Liwale district in Lindi region in Tanzania. At least L-band SAR imagery has been shown to be a suitable instrument to classify into forest, woodland and non-forest with an overall accuracy of about 73%. Aggregating the forest and woodland class into one forest class results in accuracies of about 92%. The classification method presented and used here is a combination of a Gaussian Mixture Model and a k-means classifier. C-band SAR alone seems to underestimate the woodland areas compared to the L-band classification, but combing C-and L-band inside the classifier increases the woodland area even compared to the single sensor L-band classification. C-band and multi-sensor classification results have not been validated at this stage. A forest/woodland/non-forest classification based on the maximum backscatter from the time period 2007 to 2010 has been used as a forest/woodland mask to detect forest changes from inter-annual HV backscatter variation higher than 3dB. Comparison of added single year forest loss with a 3 year loss and a visual assessment with VHR optical satellite data shows that the forest loss detection method seems to be robust. C-band SAR data has not yet been analysed for forest change detection, but is planned for future studies.

## ACKNOWLEDGEMENTS

Funding for this study has been provided by the Norwegian Embassy in Dar-Es-Salam in Tanzania through a contract with Sokoine University of Agriculture. Additional funding at Norut was provided by the Norwegian Research Council through grant nr. 204430/E40. Ground reference land cover type data has been provided by the NAFORMA project, funded by FAO and Finland. Envisat ASAR and ALOS Palsar satellite data has been provided by ESA and JAXA, respectively, though GEO FCT. Additional data sets were provided by ESA through a Category-1 project nr. 27689. VHR optical data has been provided by DigitalGlobe, Ikonos and RapidEye through the KSAT/GEO-FCT/GFOI.

## REFERENCES

GFOI, 2013. Review of Priority Research & Development Topics: R&D related to the use of Remote Sensing in National Forest Monitoring. *Pub. GEO, Switzerland, 2013 ISBN 978-92-990047-5-3.*

Haarpaintner, J., Ø. Due Trier, and J. Otieno, 2012. GEO FCT Product Development Team report – Tanzania. *Oral presentation at the GEO-FCT Science & Data Summit #3, Arusha, Tanzania, 6-10 Feb. 2012.*

Larsen, Y., G. Engen, T.R. Lauknes, E. Malnes, and K.A. Høgda, 2005. A generic differential interferometric SAR processing system, with applications to land subsidence and snow-water equivalent retrieval. *Proc. FRINGE 2005, Frascati, Italy, Nov 28 - Dec 2, 2005.*

Shimada, M., and O. Takahiro, 2010. Generating Large-Scale High-Quality SAR Mosaic Datasets: Application to PALSAR Data for Global Monitoring. *IEEE journal of selected topics in applied earth observations and remote sensing, 3(4), pp. 637-656.*

Tomppo, E., R. Malimbwi, M. Katila, K. Mäkisara, H.M. Henttonen, N. Chamuya, E. Zahabu, and J. Otieno, 2014. A sampling design for a large area forest inventory: case Tanzania. *Canadian Journal of Forest Research, 44(8), pp. 931-948.*

Ulander, L., 1996. Radiometric slope correction of synthetic aperture radar images. *IEEE Transactions on Geoscience and Remote Sensing,34(5), pp.1115-1122.*

# 31

# COMPARISON OF FIELD AND AIRBORNE LASER SCANNING BASED CROWN COVER ESTIMATES ACROSS LAND COVER TYPES IN KENYA

J. Heiskanen [a, *], L. Korhonen [b], J. Hietanen [a], V. Heikinheimo [a], E. Schäfer [a], P. K. E. Pellikka [a]

[a] Department of Geosciences and Geography, University of Helsinki, P.O. Box 68, FI-00014, Helsinki, Finland - (janne.heiskanen, jesse.hietanen, vuokko.heikinheimo, elisa.schafer, petri.pellikka)@helsinki.fi
[b] Department of Forest Sciences, University of Helsinki, P.O. Box 27, FI-00014, Helsinki, Finland – lauri.z.korhonen@helsinki.fi

**KEY WORDS:** Canopy cover, Crown relascope, Hemispherical photography, LiDAR

**ABSTRACT:**

Tree crown cover (CC) provides means for the continuous land cover characterization of complex tropical landscapes with multiple land uses and variable degrees of degradation. It is also a key parameter in the international forest definitions that are basis for monitoring global forest cover changes. Recently, airborne laser scanning (ALS) has emerged as a practical method for accurate CC mapping, but ALS derived CC estimates have rarely been assessed with field data in the tropics. Here, our objective was to compare the various field and ALS based CC estimates across multiple land cover types in the Taita Hills, Kenya. The field data was measured from a total of 178 sample plots (0.1 ha) in 2013 and 2014. The most accurate field measurement method, line intersect sampling using Cajanus tube, was used in 37 plots. Other methods included CC estimate based on the tree inventory data (144 plots), crown relascope (43 plots) and hemispherical photography (30 plots). Three ALS data sets, including two scanners and flying heights, were acquired concurrently with the field data collection. According to the results, the first echo cover index (FCI) from ALS data had good agreement with the most accurate field based CC estimates (RMSD 7.1% and 2.7% depending on the area and scan). The agreement with other field based methods was considerably worse. Furthermore, we observed that ALS cover indices were robust between the different scans in the overlapping area. In conclusion, our results suggest that ALS provides a reliable method for continuous CC mapping across tropical land cover types although dense shrub layer and tree-like herbaceous plants can cause overestimation of CC.

## 1. INTRODUCTION

Tropical landscapes show often great variation in tree crown cover (CC) depending on the land cover type and land use. Typically, the moist tropical forests have closed and multi-layered canopies whereas drier vegetation types such as savannah woodlands have low CC. In the natural vegetation types, CC can be altered by degradation due to various disturbances. Furthermore, the managed land cover types, such as croplands and agroforestry systems, can have relatively high CC in the tropics (Zomer et al., 2009).

CC is the single most important variable in the various definitions of forest, including the definitions of forest and other wooded land by Food and Agriculture Organization (FAO) of the United Nations (FAO, 2010). Cover of trees is also an elementary classifier in FAO Land Cover Classification System (LCCS) for natural and semi-natural vegetated areas (Di Gregorio, 2005). The monitoring of tropical forest area has gained increasing attention because it is needed for the implementation of climate change mitigation policies, such as United Nations collaborative initiative on Reducing Emissions from Deforestation and forest Degradation (REDD) in developing countries.

CC is defined as the proportion of ground covered by the vertical projection of tree crowns in percentage (Jennings et al., 1999; Korhonen et al., 2006; Gschwantner et al., 2009). Crown is defined by its outer perimeter and hence the small within crown gaps are considered to belong to the crown.

CC can be estimated in the field using vertical sighting tubes (Korhonen et al., 2006). The accurate measurement of CC is very time consuming, and thus several methods have been developed to decrease the measurement time. However, the alternative methods are usually biased and not necessarily applicable for complete range of CC variation. One method is to estimate CC based on the tree inventory data

Often CC estimate is needed for the sample plots with basic forest inventory data. If tree positions and crown diameters are available, it is possible to estimate CC based on these. However, the estimate of the crown area can be biased when assuming a circular shape. Hemispherical photography (HP) is commonly used method for estimating canopy gap fraction (GF) and leaf area index (LAI) (Jonckheere et al., 2004). If view zenith angle is restricted close to zenith, HP can provide a proxy of CC $(1-GF)$. However, images observe within crown gaps and hence the CC is underestimated. The additional methods for CC estimation include crown relascope (Stenberg et al. 2008), which can be used for very rapid CC assessment in sparsely stocked areas. Hence, it could be a useful method for CC estimation in savannah woodlands and croplands in the tropics.

Field measurements are viable only at sample plot scale and remote sensing is needed for mapping. Airborne laser scanning (ALS) has become a standard source of high-resolution remote sensing data for mapping forest attributes. The laser pulses are capable of detecting gaps in forest canopies, and hence offer three-dimensional information on canopy structure and sub-

---

* Corresponding author

canopy topography. In addition to being very useful for assessing forest attributes, such as tree height and aboveground biomass, ALS data has been shown to provide accurate CC estimates even without field calibration data (Korhonen et al., 2011). Several ALS cover indices have been proposed to estimate CC, GF and LAI from discrete return ALS data (Korhonen and Morsdorf, 2014). Basically, the indices differ in terms of considered return types (single, first-of-many, intermediate and last-of-many returns) and if they are strictly geometrical or take into account return intensity. In all the indices, a height threshold needs to be set to separate canopy and ground returns.

So far, the agreement of the field and ALS based CC estimates has been rarely assessed in the tropics. In this paper, our objective was to compare field based CC estimates (vertical sighting tube, tree inventory, crown relascope and HP) and ALS based CC estimates across multiple land cover types in the Taita Hills, Kenya.

## 2. MATERIAL AND METHODS

### 2.1  Study area

The Taita Hills are located in the northernmost part of the Eastern Arc Mountains of Kenya and Tanzania, and cover approximately 1000 km$^2$ (Figure 1). The hills rise from the lowlands at 600–900 m a.s.l. elevation to approximately 2200 m a.s.l. The hills are intensively cultivated and much of the forested land has been cleared for agriculture (Pellikka et al., 2009). Some remaining fragments of the indigenous cloud forest are restricted to the highest altitudes. In addition to the indigenous forest patches, plantations of exotic tree species, including eucalyptus (mostly *Eucalyptus saligna*), pines (*Pinus* spp.), cypress (*Cupressus lusitanica*) and black wattle (*Acacia mearnsii*) were established in the hills between the 1950s and 1970s. Mixed stands of indigenous and exotic species are also common (Pellikka et al., 2013). In the lower altitudes, the landscape is characterized by cultivated areas, open woodlands, shrublands and thickets with relatively low CC. Drought resistant tree species, such as *Commiphora* spp. *Acacia* spp. *and Albizia amara* are typical across the lowlands. Also fruit trees, such as mango (*Mangifera indica*) and cashew (*Anacardium occidentale*) are grown in the area.

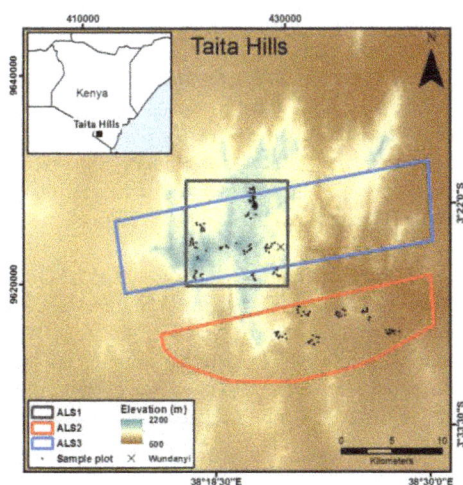

Figure 1. Location of the study area and sample plots, and extent of the three ALS data sets.

### 2.2  Field estimates of CC

We used CC estimates from a total of 177 sample plots (Figure 1). All the data were collected in January–February 2013 and January–February 2014. The circular sample plots had a size of 0.1 ha (radius = 17.84 m). The plot centres were positioned using a Trimble GeoXH GNSS receiver with an external antenna (Trimble Zephyr Model 2) mounted on a 2.6 m range pole. The positions were measured as long as we stayed in the plot. The differential correction was made using a GNSS base station located in the town of Wundanyi (Figure 1). The position of the base station was determined using Trimble RTX post-processing service (http://www.trimblertx.com).

The field data was combined from the several field campaigns in different parts of the study area and hence sampling strategy and methods of estimating CC varied between the plots. In 2013 measurements and in 2014 lowland measurements we selected sample plots randomly within 1 km$^2$ clusters (ten plots in each cluster). In 2014, we sampled additional plots from forest areas in order to cover variation in aboveground biomass and tree species composition. This sampling was designed subjectively based on the canopy height model generated from 2013 ALS data and AisaEAGLE imaging spectroscopy data (Schäfer, 2014; Heiskanen et al., in press).

We used four methods for estimating CC in the field depending on the land cover type and tree density. The reference method, Cajanus tube with line intersect sampling (Korhonen et al., 2006), was used in 37 sample plots. CC measurement by Cajanus tube is time consuming and hence the number of plots was relatively small. However, the measurements covered all the land cover types and CC range in the study area. Cajanus tube is a vertical sighting tube, which provides an unbiased CC estimate if sample size is sufficient (Korhonen et al., 2006). In each of the 37 sample plots, we established nine transects (length 15.8–35.7 m) with four meter distance starting from the plot center point. In each transect, we recorded the starting and ending points of the crowns and identified the exact position of the crown edge by the Cajanus tube. Trees smaller than 3 m in height were not included in CC. Finally, we computed the distance that was covered by crowns and divided it by the total length of transects to estimate CC. These estimates are called hereafter CC$_{Cajanus}$.

In 144 plots, CC was estimated based on tree inventory. The method was used in the plots outside forests (croplands, agroforestry, woodlands and shrublands), where it was feasible to measure position for each tree within the sample plot. The position (direction and distance from the plot center) and the diameter at breast height (DBH) were measured for all the trees having DBH ≥ 10 cm by using a measurement tape and precision compass. Furthermore, crown diameter (CD) was measured by measurement tape in two perpendicular directions for all the trees in 2013 and for at least three trees in 2014 (minimum, median and maximum DBH). The mean CD was predicted for the trees missing CD measurement using linear regression. Finally, CC was computed as a percentage of the plot area covered by the tree crowns (overlapping crowns were counted only once). These estimates are called hereafter CC$_{trees}$.

Crown relascope was used in 43 sample plots in the lowlands, where tree densities are relatively low. Our crown relascope consisted of a 30 cm long and 3.2 cm wide plastic sheet and a string, the length of which was adjusted according to the crown basal area factor (CBAF) (Stenberg et al., 2008). Two CBAFs were tested: 100 and 200. When CBAF is 100 (200), each tallied tree corresponds to 1% (2%) increase in CC. These estimates are called hereafter $CC_{cr1}$ and $CC_{cr2}$, respectively.

Furthermore, CC was estimated by HP for 30 plots. The majority of these plots were located in indigenous forests, but some plots were established in plantations (pine, cypress, eucalyptus and black wattle). We used Nikon D5000 camera and Sigma 4.5 mm 1:2.8 DC HSM fisheye lens. The lens was mounted on a tripod and levelled to the height of 1.3 m. The number of camera positions per plot varied between five and eight. In the case of five positions, the images were taken from the center and 9 m to each cardinal direction. In the case of eight positions, the images were taken from 3 m and 9 m distance to each cardinal direction. In order to determine the exposure setting, we followed the histogram method of Beckschäfer et al. (2013). The image processing included the classification of images to the canopy and sky pixels using the blue band and automatic thresholding algorithm (Nobis and Hunziker, 2005). If the thresholding resulted in clear classification errors according to the visual assessment, we used either algorithm of Ridler and Calvard (1978) or determined threshold manually. After the classification, we computed CC for each image as a percentage of canopy pixels in the 0–15° zenith angle range and averaged all the measurements in the plot for the plot-wise CC. These estimates are called hereafter $CC_{HP}$.

## 2.3 ALS data

We used three discrete return ALS data sets that covered different parts of the study area (Figure 1). The first scan (ALS1) was made 4−5 February 2013 and covered 10 km × 10 km area in the highest parts of the hills. The second scan (ALS2) was made 17 January 2014 and covered 150 km² in the lowlands. The third scan (ALS3) was made 4 February 2014 and covered 330 km², and partly overlapped with the first scan. In the first two scans, the sensor was Optech ALTM 3100 and in the third scan Leica ALS60. In the first scan, the flying height was relatively low and targeted pulse density relatively high in comparison to the 2014 scans. Further details are given in Table 1. The number of sample plots measured by each method and covered by the different ALS scans are summarized in Table 2.

All the ALS data sets were pre-processed by the data vendors (Topscan Gmbh, Ramani Geosystems) and delivered as georeferenced point clouds in UTM/WGS84 coordinate system with ellipsoidal heights. Ground returns in ALS1 and ALS2 data sets were filtered by the vendor using Terrascan software (Terrasolid Oy). Then, we used ground classified returns of ALS1 and ALS2 to produce digital elevation model (DEM) at 1 m cell size.

## 2.4 ALS based CC estimates

We extracted ALS data for the sample plots using several radii depending on the field measurement method. First, we extracted returns using 17.84 m radius corresponding to the area of 0.1 ha sample plot. Furthermore, in the crown relascope plots, we extracted returns for radius depending on the mean CD in the sample plot. The maximum radius (r) within which a tree crown is still tallied was computed as:

| Parameter | ALS1 | ALS2 | ALS3 |
|---|---|---|---|
| Date | 4−5 Feb 2013 | 17 Jan 2014 | 4 Feb 2014 |
| Sensor | Optech ALTM 3100 | Optech ALTM 3100 | Leica ALS60 |
| Mean range (m) | 760 | 1240 | 1460 |
| Pulse rate (kHz) | 100 | 70 | 58 |
| Scan rate (Hz) | 36 | 37 | 66 |
| Scan angle (°) | ±16 | ±18 | ±16 |
| Pulse density (pulses m⁻²) | 9.6 | 2.9 | 3.0 |
| Return density (returns m⁻²) | 11.4 | 3.3 | 3.4 |
| Beam divergence at 1/e² (mrad) | 0.3 | 0.3 | 0.22 |
| Footprint diameter (cm) | 23 | 37 | 32 |

Table 1. Survey and sensor specifications for 2013 and 2014 scans. All sensors recorded a maximum of four returns per pulse.

| Method | ALS1 | ALS2 | ALS3 | Total |
|---|---|---|---|---|
| Cajanus tube | 23 | 14 | 17 | 37 |
| Tree inventory | 83 | 61 | 66 | 144 |
| Crown relascope | | 43 | | 43 |
| HP | 30 | | 21 | 30 |

Table 2. The number of sample plots measured by the different method and covered by different ALS scans.

$$r = \frac{50CD}{\sqrt{CBAF}} \qquad (1)$$

where     CD = mean CD in the sample plot
          CBAF = crown basal area factor (100 or 200)

Finally, we extracted returns also for larger fixed radii of 25 m in the plots measured by HP because of the non-zero view zenith angle.

When extracting ALS data for the sample plots, we also normalized return heights to the heights from the ground level by using DEM. Furthermore, as some plots were covered by several flight lines, we removed overlap between the adjacent flight lines based on minimum scan angle using lasoverage tool in LAStools software (rapidlasso GmbH). This was done in order to minimize bias in the ALS based CC estimates due to the scan angle (Korhonen et al., 2011).

Then, we computed two ALS cover indices for the sample plots. First return cover index (FCI) (e.g., Solberg et al., 2006) has been found to be a good proxy of CC and relatively robust index between sensors and scans (Korhonen et al., 2011; Korhonen and Morsdorf, 2014). However, indices that incorporate intermediate and last returns, such as all return cover index (ACI) (e.g., Morsdorf et al., 2006) have been shown to provide better estimates of canopy GF, because they include information concerning both between crown and within crown canopy gaps (Korhonen et al., 2011). FCI and ACI were computed as:

$$FCI = \frac{\sum Single_{canopy} + \sum First_{canopy}}{\sum Single_{all} + \sum First_{all}} \times 100 \qquad (2)$$

$$ACI = \frac{\sum All_{canopy}}{\sum All} \times 100 \qquad (3)$$

where     $Single_{canopy}$ = single return from canopy
          $First_{canopy}$ = first return from canopy
          $Single_{all}$ = any single return
          $First_{all}$ = any first return
          $All_{canopy}$ = any return from canopy
          All = any return (i.e. single, first, intermediate of last)

In the computation of ALS cover indices, canopy refers to the returns above a given height threshold that separates canopy returns from the understory and ground returns. Here, we computed indices using height thresholds of 3, 4 and 5 m.

## 2.5 Analysis

The best agreement with the field and ALS based CC (i.e. radii and height thresholds) was searched for each field measurement method. The agreement between the CC estimates was assessed using root mean square difference (RMSD) and average difference (AD):

$$RMSD = \sqrt{\frac{\sum_{i=1}^{n}(y_i - \hat{y}_i)^2}{n}} \qquad (4)$$

$$AD = \frac{\sum_{i=1}^{n}(y_i - \hat{y}_i)}{n} \qquad (5)$$

where     $y_i$ = field based CC estimate
          $\hat{y}_i$ = ALS based CC estimate

## 3. RESULTS

First, we compared $CC_{Cajanus}$ and cover indices from ALS2 and ALS3 scans. ALS1 was not considered in this comparison as it was acquired in 2013 and there was one year difference with regard to the field measurements made in 2014. Furthermore, ALS covers approximately the same plots as ALS3 scan (Figure 1). $CC_{Cajanus}$ showed good agreement with FCI and AD were small (Figure 2a, Table 3). In the indigenous forest plots, $CC_{Cajanus}$ was 100% or close to it, which describes the closed and multi-layered canopies in those forests. The optimal height threshold to separate canopy returns was different between the areas and scans. In the lowland areas (ALS2), 3 m height threshold provided the best agreement but in the hills 5 m was the best (ALS3). FCI provided better agreement with $CC_{Cajanus}$ than ACI, which underestimated $CC_{Cajanus}$ in the plots of high canopy density where multiple returns were produced (Figure 2b, Table 3).

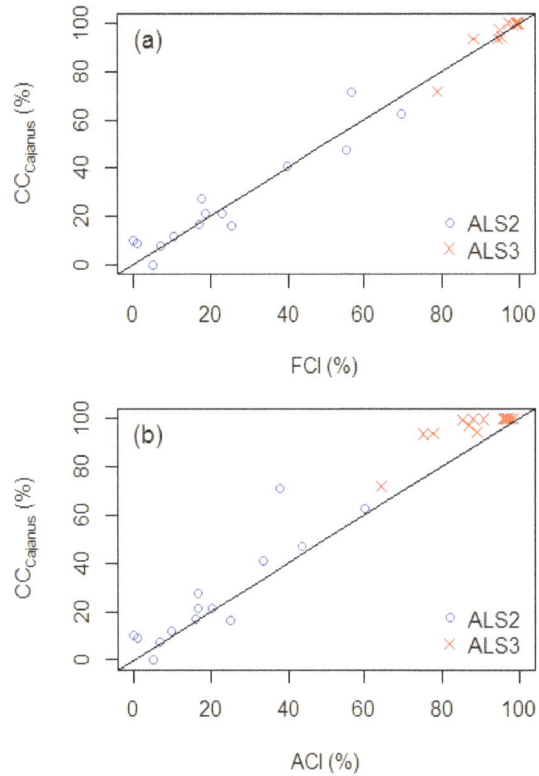

Figure 2. (a) FCI and (b) ACI against $CC_{Cajanus}$.

| Scan | Index | HT | RMSD | AD |
|------|-------|-----|------|------|
| ALS2 | FCI | 3 m | 7.1% | 1.3% |
| ALS2 | ACI | 3 m | 10.8% | 5.2% |
| ALS3 | FCI | 5 m | 2.7% | 0.1% |
| ALS3 | ACI | 3 m | 6.4% | 4.6% |

Table 3. Summary of the comparison between $CC_{Cajanus}$, and ALS cover indices. HT = height threshold for separating canopy and ground returns.

Next, we compared $CC_{trees}$ and FCI from ALS1 and ALS2 scans, which corresponded to the years when tree inventories were made (Figure 3). Only FCI was tested because it showed the best agreement with $CC_{Cajanus}$, which we considered the most accurate method for CC estimation. In general, $CC_{trees}$ had good agreement with FCI until 20%. However, the values larger than 20% were underestimated in $CC_{trees}$ and AD were negative. The height threshold of 5 m was the best for ALS1 (hills) and 4 m for ALS2 (lowlands).

Next, we compared crown relascope measurements and FCI from ALS2 because those measurements were available only from the lowland area. $CC_{cr1}$ and $CC_{cr2}$ rarely exceeded 20% although FCI values were greater than that (Figure 4). The agreement was somewhat better for $CC_{cr2}$. The mean CD had a range of 3.5–14.0 m (mean 7.0 m). Furthermore, the radii for $CC_{cr1}$ had a range of 17.5–70.2 m (mean 34.8 m) and $CC_{cr2}$ 12.4–49.6 m (mean 24.6 m). The height threshold of 3 m was used as it gave the best agreement between $CC_{Cajanus}$ and FCI in this area (Table 3). However, we noted that increasing the height threshold from 3 m to 5 m decreased RMSD and AD but small CC were typically underestimated by FCI.

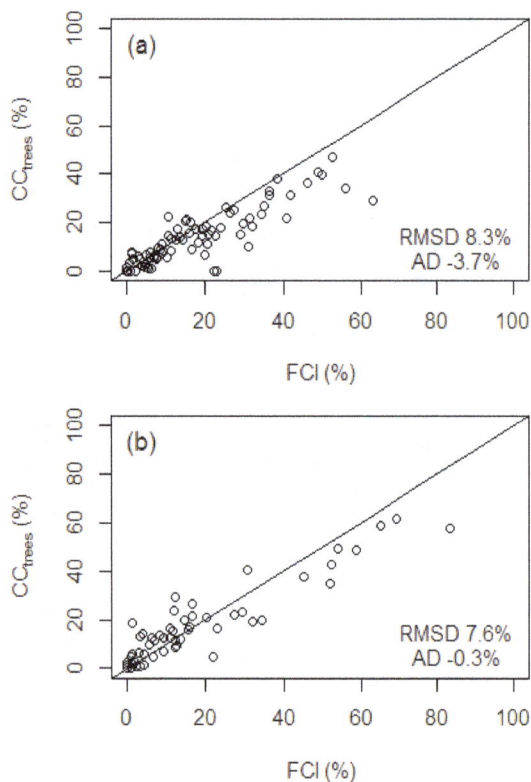

Figure 3. FCI from (a) ALS1 (5 m height threshold) and (b) ALS2 (4 m height threshold) against CC_trees.

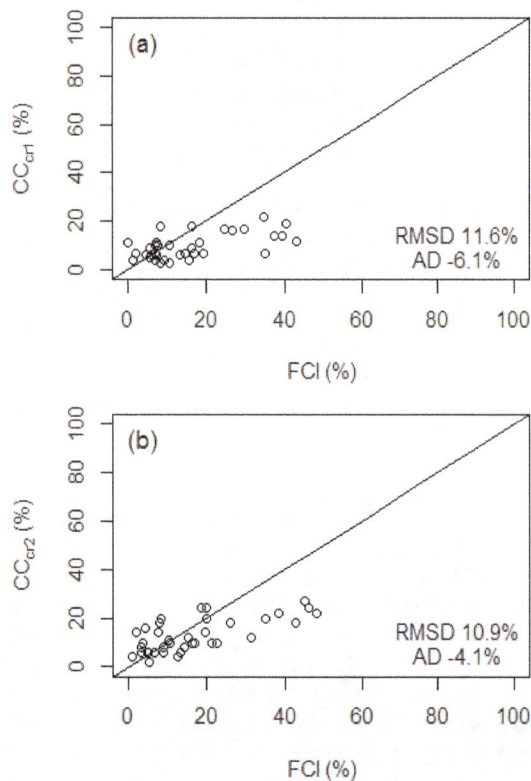

Figure 4. FCI against (a) CC_cr1 (CBAF = 100) and (b) CC_cr2 (CBAF = 200). Height threshold = 3 m.

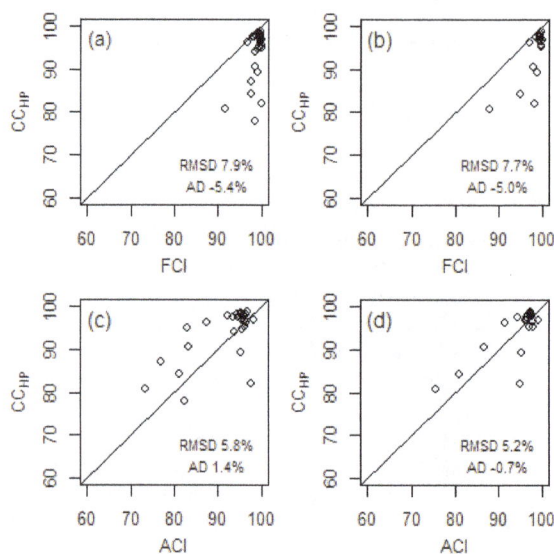

Figure. 5. Comparison of $CC_{HP}$ and (a) FCI from ALS1, (b) FCI from ALS3, (c) ACI from ALS1 and (d) ACI from ALS3.

Next, we compared $CC_{HP}$ and ALS cover indices. In contrast to the other field based CC estimates, $CC_{HP}$ were more closely related to ACI than FCI (Figure 5). As $CC_{HP}$ measurements were made in the indigenous forests and exotic plantations, $CC_{HP}$ was in general high. There was lots of variation in the lower range of the values, probably due to insufficient number of photo positions to estimate $CC_{HP}$ accurately in those plots. The larger 25 m plot radius and 3 m height threshold provided the best agreement.

Finally, we compared FCI and ACI in the overlapping plots of ALS1 and ALS3 scans using 3 m height threshold (Figure 6). Both indices showed good agreement across the complete range of FCI and ACI although FCI had somewhat smaller AD. Furthermore, in the plots with the greatest CC, ALS3 had larger values of ACI than ALS1.

## 4. DISCUSSION

Reliable CC maps would be useful for the land cover characterization of complex tropical landscapes. In this study, we compared several field and ALS based CC estimates across multiple tropical land cover types.

FCI based on single and first returns had good agreement with $CC_{Cajanus}$, which is often used as a reference method in the methodological comparisons (Korhonen et al., 2006; Stenberg et al., 2008). Good agreement between $CC_{Cajanus}$ and FCI is in line with the previous results from the boreal forests (Korhonen et al., 2011). As our data covered the complete range of CC and all the main land cover types and land uses, the results suggest that FCI provides a reliable mapping of CC for land cover characterization and forest area delineation also in the tropical areas.

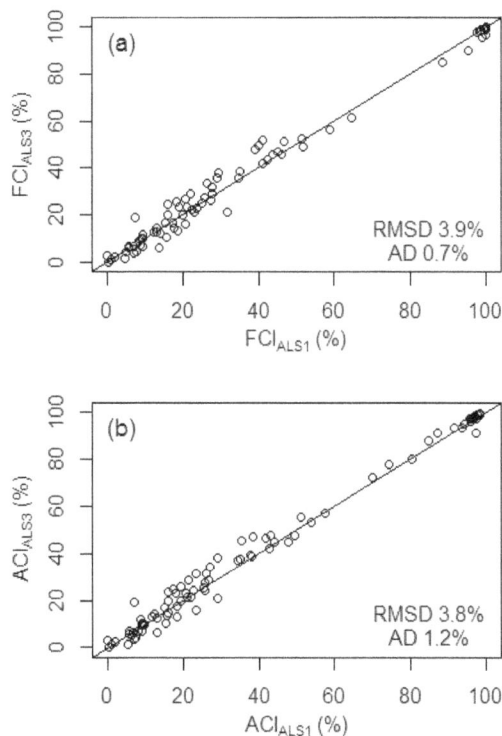

Figure 6. Comparison of (a) FCI and (b) ACI based on ALS1 and ALS3 data sets. Positive AD indicates that ALS1 shows on average smaller values.

The other field based methods showed more considerable average differences when compared to FCI. $CC_{trees}$ provided rather good agreement with FCI in relatively when CC < 20%. However, $CC_{trees}$ was smaller than FCI in the higher values. In the field inventory trees were included if DBH $\geq$ 10 cm. Therefore, the shrubs that had DBH < 10 cm and height greater than height threshold were included in ALS based CC but not in the field estimates. Furthermore, tree-like herbaceous plants, such as bananas, were not mapped in the field. This explains large differences in CC in some of the plots where bananas were abundant (i.e. large FCI but very small $CC_{trees}$). Also, the best height threshold for $CC_{trees}$ varied between the hills and the lowlands. The differing height-diameter relationship between the areas, and differences in the density of shrubs and small trees could affect the best height threshold.

Crown relascope has potential for the fast estimation of CC in the areas of low stem density, such as tropical woodlands and croplands, but the agreement with FCI was weak. In some of the plots, FCI was considerably higher, which is due to dense shrub cover in some of the plots similar to $CC_{trees}$. Furthermore, because the radii to include trees were rather large because of large crowns (i.e. mean CD), it was difficult to see if crowns were overlapping in the denser plots. Therefore, larger CBAF could be considered in the future studies. It should also be noted that fixed and variable radius plots are not directly comparable, even if the radius is set separately for each plot based on the CD.

We also demonstrated that $CC_{HP}$ has better agreement with ACI, which considers all returns types. The same has been observed with GF and LAI in the boreal forests (Korhonen et al., 2011) and in this study area (Heiskanen et al., in press). In the dense

forests, CC can be estimated accurately using relatively few photo positions (images) but more images are needed if CC is close to 50% (Korhonen and Heikkinen, 2009). In our comparison, this was visible as greater scatter in the lower $CC_{HP}$ values.

In general, ALS cover indices that include also intermediate and last returns, such as ACI, are more sensitive to changes in scanning parameters than those using only single and first returns, such as FCI (Korhonen and Morsdorf, 2014). Although FCI can also change between different scanners, it should be fairly reliable as long as the scan angle is < 15°. In this study, the differences between indices were small in the overlapping area although although ALS1 and ALS3 were acquired with different sensors and from different flying heights. Some differences were observed in ACI in the closed forest stands (Heiskanen et al., in press).

In conclusion, the good agreement between FCI and $CC_{Cajanus}$ highlights the potential of ALS for CC mapping in the tropical landscapes. ALS data provides CC estimates with unambiguous height definition. Hence, height based separation between trees and shrubs in the field should provide better correspondence between the field and ALS data than DBH based separation. Furthermore, it should be noted that tree-like herbaceous plants are included in ALS based indices and cause overestimation of CC when present.

## ACKNOWLEDGEMENTS

This work was supported by the Ministry for Foreign Affairs of Finland under Building Biocarbon and Rural Development in West Africa (BIODEV) project. We also thank Taita Research Station of the University of Helsinki for the logistical support, and Jessica Broas and Darius Kimuzi for their contribution in the field work.

## REFERENCES

Beckschäfer, P., Seidel, D., Kleinn, C., Xu, J., 2013. On the exposure of hemispherical photographs in forest. *iForest* 6, pp. 228-237.

Di Gregorio, A., 2005. Land cover classification system. Classification concepts and user manual. Software version 2. FAO, Rome.

FAO, 2010. Global forest resources assessment 2010. Main report. FAO Forestry Paper 163. FAO, Rome.

Gschwantner, T., Schadauer, K., Vidal, C., Lanz, A., Tomppo, E., di Cosmo, L., Robert, N., Englert Duursma, D., Lawrence, M., 2009. Common tree definitions for national forest inventories in Europe. *Silva Fenn.* 43(2), pp. 303-321.

Heiskanen, J., Korhonen, L., Hietanen, J., Pellikka, P. K. E., in press. Airborne LiDAR for estimating canopy gap fraction and leaf area index of tropical montane forests. *Int J. Remote Sens.*

Jennings, S. B., Brown, N. D., Sheil, D., 1999. Assessing forest canopies and understorey illumination: canopy closure, canopy cover and other measures. *Forestry* 72(1), pp. 59-74.

Jonckheere, I., Fleck, S., Nackaerts, K., Muys, B., Coppin, P., Weiss, M., Baret, F., 2004. Review of methods for in situ leaf area index determination: Part I. Theories, sensors and hemispherical photography. *Agr. Forest Meteorol.* 121(1-2), pp. 19-35.

Korhonen, L., Heikkinen, J. 2009. Automated analysis of in situ canopy images for the estimation of forest canopy cover. *Forest Sci.* 55(4), pp. 323-334.

Korhonen, L., Korhonen, K. T., Rautiainen, M., Stenberg, P., 2006. Estimation of forest canopy cover: a comparison of field measurement techniques. *Silva Fenn.* 40(4), pp. 577-588.

Korhonen, L., Korpela, I., Heiskanen, J., Maltamo, M., 2011. Airborne discrete-return LiDAR data in the estimation of vertical canopy cover, angular canopy closure and leaf area index. *Remote Sens. Environ.* 115(4), pp. 1065-1080.

Korhonen, L., Morsdorf, F., 2014. Estimation of canopy cover, gap fraction and leaf area index with airborne laser scanning. In: *Forestry applications of airborne laser scanning*, Springer, Dordrecht, pp. 397-417.

Morsdorf, F., Kötz, B., Meier, E., Itten, K. I., Allgöwer, B., 2006. Estimation of LAI and fractional cover from small footprint airborne laser scanning data based on gap fraction. *Remote Sens. Environ.* 104(1), pp. 50-61.

Nobis, M. Hunziker, U., 2005. Automatic thresholding for hemispherical canopy photographs based on edge detection. *Agr. Forest Meteorol.* 128(3-4), pp. 243-250.

Pellikka, P., Lötjönen, M., Siljander, M., Lens, L., 2009. Airborne remote sensing of spatiotemporal change (1955–2004) in indigenous and exotic forest cover in the Taita Hills, Kenya. *Int. J. Appl. Earth Obs.* 11(4), pp. 221-232.

Pellikka, P. K. E., Clark, B. J. F., Gonsamo Gosa, A., Himberg, N., Hurskainen, P., Maeda, E., Mwang´ombe, J., Omoro, L. M. A., Siljander, M., 2013. Agricultural Expansion and Its Consequences in the Taita Hills, Kenya. In: *Kenya: a Natural Outlook*, Elsevier, Amsterdam, pp. 165-179.

Schäfer, E., 2015. Tree species diversity estimation using airborne imaging spectroscopy. MSc thesis, University of Helsinki.

Solberg, S., Næsset, E., Hansen, K. H., Christiansen, E., 2006. Mapping defoliation during a severe insect attack on Scots pine using airborne laser scanning. *Remote Sens. Environ.* 102(3-4), pp. 364-376.

Stenberg, P., Korhonen, L., Rautiainen, M., 2008. A relascope for measuring canopy cover. *Can J. For. Res.* 38(9), pp. 2545-2550.

Zomer, R. J., Trabucco, A., Coe, R., Place, F., 2009. Trees on farm: analysis of global extent and geographical patterns of agroforestry. ICRAF Working Paper 89. World Agroforestry Centre, Nairobi, Kenya.

# 32

# MULTI SENSOR DATA INTEGRATION FOR AN ACCURATE 3D MODEL GENERATION

S. Chhatkuli [a, *], T. Satoh [a], K. Tachibana [a]


[a] PASCO CORPORATION, Research & Development HQ, 2-8-10 Higashiyama, Meguro-ku, Tokyo, Japan
(cshahb7460, tuoost7017, kainka9209) @pasco.co.jp

**IV/7 and V/4**

**KEY WORDS:** 3D model, 3D TIN, Data fusion, Point cloud

**ABSTRACT:**

The aim of this paper is to introduce a novel technique of data integration between two different data sets, i.e. laser scanned RGB point cloud and oblique imageries derived 3D model, to create a 3D model with more details and better accuracy. In general, aerial imageries are used to create a 3D city model. Aerial imageries produce an overall decent 3D city models and generally suit to generate 3D model of building roof and some non-complex terrain. However, the automatically generated 3D model, from aerial imageries, generally suffers from the lack of accuracy in deriving the 3D model of road under the bridges, details under tree canopy, isolated trees, etc. Moreover, the automatically generated 3D model from aerial imageries also suffers from undulated road surfaces, non-conforming building shapes, loss of minute details like street furniture, etc. in many cases. On the other hand, laser scanned data and images taken from mobile vehicle platform can produce more detailed 3D road model, street furniture model, 3D model of details under bridge, etc. However, laser scanned data and images from mobile vehicle are not suitable to acquire detailed 3D model of tall buildings, roof tops, and so forth. Our proposed approach to integrate multi sensor data compensated each other's weakness and helped to create a very detailed 3D model with better accuracy. Moreover, the additional details like isolated trees, street furniture, etc. which were missing in the original 3D model derived from aerial imageries could also be integrated in the final model automatically. During the process, the noise in the laser scanned data for example people, vehicles etc. on the road were also automatically removed. Hence, even though the two dataset were acquired in different time period the integrated data set or the final 3D model was generally noise free and without unnecessary details.

## 1. INTRODUCTION

3D city modelling has been gaining quite a lot of attention recently. There are numbers of commercial software (*for e.g. Street Factory*[TM], *Smart3DCapture*[®], *Agisoft PhotoScan, etc.*) available that can convert imageries into realistic 3D city models without much, users', intervention. An accurate 3D city model has several potential areas of application which ranges from simple 3D archiving to mapping, navigation, city planning, hydrodynamic simulation, etc. to name a few. Data sources from different sensors taken from different platforms have their own pros and cons. For example, airborne oblique imageries produce an overall decent 3D city models. However, an automatically generated 3D model, from aerial imageries, generally suffers from the lack of accuracy in deriving the 3D model of road under bridges, details under tree canopies, isolated trees, etc. Moreover, an automatically generated 3D model from aerial imageries also suffers from undulated road surfaces, non-conforming building shapes, loss of minute details like street furniture, etc. in many cases.

On the other hand, laser scanner data and images taken from mobile vehicle platform can produce more detailed 3D road model, street furniture model, 3D model of details under bridge, etc. However, laser scanner data and images from mobile vehicle are not suitable to acquire detailed 3D model of tall buildings, roof tops, etc.

Integration of multiple sensor data can add more information and reduce uncertainties to data processing (Gruen et al., 2013). Recently there has been an increase in interest to fuse multiple sensor data to create more detailed 3D models. Bastonero et al., (2014) demonstrated the fusion of point cloud generated from images acquired by remotely piloted aircraft system with the terrestrial point cloud generated from Terrestrial laser scanner. Gruen et al., (2013) presented a report on joint processing of UAV imagery and terrestrial mobile mapping system data for very high resolution city modelling. In most of these prior works, the multiple sensor data acquisition was conducted basically for the integration purpose.

Multi sensor data acquired during different time for different purposes could also serve the purpose of generating more detailed and geometrically more accurate 3D model if there is a novel way to integrate them.

In this paper, we have introduced a technique to integrate two different type of data sets: 3D model generated from aerial imageries and mobile vehicle laser scanned data (MMS point cloud) acquired at two different times. The data integration produced a final composite model with more details and better geometric accuracy, especially on the road surface, compared to the original 3D model generated from aerial imageries.

---

* Corresponding author.

## 2. TEST SITE AND DATA SOURCES

The data integration test was performed between the 3D textured TIN model and long range MMS RGB point cloud acquired in the Yokohama area in Japan. The 3D textured TIN model was created by using a commercial software Street Factory™ utilizing imageries taken by Leica RCD30 oblique camera during the month of January 2013. MMS RGB point cloud data was acquired by long range laser scanner system from Riegl called *RIEGL* VQ-450 during May 2013.

## 3. RESEARCH OBJECTIVE

As seen in Figure. 1, the 3D model automatically generated from the oblique imagery suffered from an undulated road surface and non-confirming shapes of buildings at many places. On the other hand, MMS point cloud could capture more detailed features compared to the aerial imagery derived 3D maps, especially those that are closer to the sensor position. However MMS point cloud data also suffered from occlusion and also could not acquire details of buildings roof and so on. Hence, the purpose of this research is to formulate a workflow to create a 3D model with non-undulated road surface and with more accurate feature shapes by utilizing information from MMS point cloud. In additional to that, the further objective is to integrate the additional features like trees, electric poles, etc. from the MMS point cloud, which were missing on the original 3D model, onto the newly derived 3D model.

## 4. DATA INTEGRATION PROCEDURE & RESULTS

Figure 2 shows the general workflow of the proposed data integration technique. A brief description of the data integration technique that we have devised to achieve our objective is explained below.

### 4.1  3D model re-meshing

To evaluate and to select the common features between the MMS point cloud and the 3D model, it is desirable to have a similar resolution between the vertices of 3D model or 3D maps generated from aerial imageries and MMS point cloud. Hence, the 3D model or 3D maps derived from imagery (which will be called *image model* hereafter) was re-meshed to a refined size of about 50 cm. This is the first step in the data integration process.

### 4.2  MMS point cloud processing

Outliers of MMS point cloud were removed by utilizing a statistical outlier filter. Moving Least Square (MLS) smoothing was performed on the remaining data to smooth the small unevenness or undulation which is inherent to laser scanner data. Then, eigenvalues ($\lambda\_1 > \lambda\_2 > \lambda\_3$) of the point cloud data were calculated. The eigenvalues were then normalized to extract shape descriptors D1, D2 and D3 where D1 = $\lambda\_1/(\lambda\_1 + \lambda\_2 + \lambda\_3)$, D2 = $\lambda\_2/(\lambda\_1 + \lambda\_2 + \lambda\_3)$, D3 = $\lambda\_3/(\lambda\_1 + \lambda\_2 + \lambda\_3)$. The point cloud with D1 >=0.85 were considered as a linearly distributed and classified as a pole like features. On the remaining data, MLS smoothing was performed again. Eigenvectors of the point data were

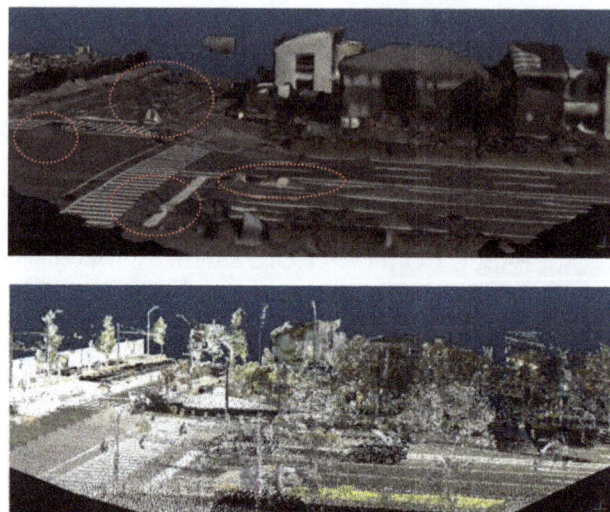

Figure 1. 3D model from aerial imagery (top) & MMS RGB point cloud (bottom) of the same area

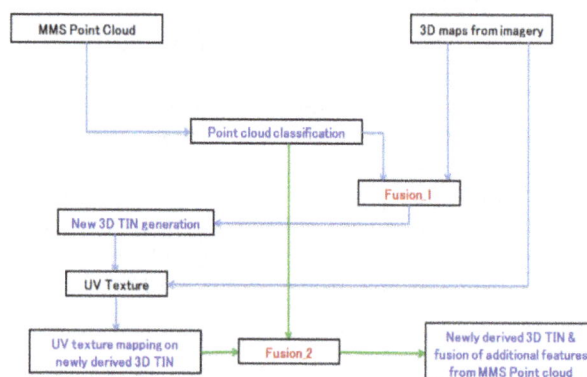

Figure 2. Data integration workflow

calculated by performing plane fitting. Curvature analysis was then performed utilizing the computed eigenvectors. In the next step, the absolute values of eigenvectors ($e_1$, $e_2$, $e_3$) were normalized. By evaluating the normalized eigenvectors $N_x$, $N_y$ and $N_z$ (where $N_x = abs\ (e_1)/\ abs\ (e_1) + abs\ (e_2) + abs\ (e_3)$, $N_y = abs\ (e_2)/\ abs\ (e_1) + abs\ (e_2) + abs\ (e_3)$, and $N_z = abs\ (e_3)/\ abs\ (e_1) + abs\ (e_2) + abs\ (e_3)$), maximum and minimum eigenvalue of curvature and the height information, the point cloud was then classified into (1) volumetric objects like trees, (2) planar flat objects like road pavement, side walk, and flat roof of a building, (3) planar vertical objects like walls, (4) curbs or road edges and (5) slopped objects in the respective order. The Point Cloud Library (PCL), which is an open source library for point cloud processing, was used to compute the parameters for point cloud classification (1). Moreover, for the classification, data clustering was performed by utilizing Jenks optimization method. Figure 3 shows the original MMS point cloud and the point cloud after classification utilizing the proposed technique.

In the next step, from the classified point clouds, MMS point cloud that represent road surface was selected. As seen in Figure. 4 (top) the road MMS point cloud could suffer from holes due to the occlusion created by people on roads, vehicles, etc. Hence to fill up the holes and to create a uniformly

# 33

# Precision Measurement for the grasp of Welding Deformation Amount of Time Series for Large-scale Industrial Products

Ryogo Abe [a, *], Kunihiro Hamada [b], Noritaka Hirata [b], Ryotaro Tamura [b], Nobuaki Nishi [c]

[a] Kokusai Kogyo Co., Ltd, Research and Development Division, Digital Sensing Group, 2-24-1 Harumi-cho, Fuchu City,Tokyo 183-0057 JAPAN - ryogo_abe@kk-grp.jp
[b] Hiroshima University
[c] Tsuneishi Shipbuilding Co., Ltd.

**Commission/WG**

KEY WORDS: Large-scale Industrial measurement, Weld Deformation, Precise Measurement, Ship Building

ABSTRACT:

As well as the BIM of quality management in the construction industry, demand for quality management of the manufacturing process of the member is higher in shipbuilding field. The time series of three-dimensional deformation of the each process, and are accurately be grasped strongly demanded. In this study, we focused on the shipbuilding field, will be examined three-dimensional measurement method. The shipyard, since a large equipment and components are intricately arranged in a limited space, the installation of the measuring equipment and the target is limited. There is also the element to be measured is moved in each process, the establishment of the reference point for time series comparison is necessary to devise. In this paper will be discussed method for measuring the welding deformation in time series by using a total station. In particular, by using a plurality of measurement data obtained from this approach and evaluated the amount of deformation of each process.

## 1. INTRODUCTION

In construction of structures (bridge or building, etc.) in the construction sector including BIM (Building Information Modeling), the quality management of the three-dimensional shape including the control of diremption from the design data in each in every work process is required. Similarly, also in the shipbuilding sector that is representative of large-scale industrial products, a quality management method of amount of deformation in each process represented by welding work is investigated. Unlike a general machine product, in shipbuilding sector, size of material in itself is large (some 10m - some 100m scale) and amount of deformation is also large accordingly. Therefore, a study has been pushed forward about measurement technique to meet the demanded precision of the quality control in the field of shipbuilding (Nomoto, et. al, 1997, Takeichi, et. al, 2000). Furthermore, after 2000, price reduction and the precision improvement of laser scanner have been realized and the examples that applied the laser scanner to the precision evaluation of deformation in the field of shipbuilding came out (Hiekata and Matsuo, 2012, Ono, 2012). But, in these studies, deformation before and after the single welding process has been compared, and there are few examples that measured the deformation in each real welding process in multiple processes.

In this paper, we have executed precise three-dimensional measurement for the hatch covers of which shape is comparatively simple among the ship members, and discuss about its quality management method. The hatch cover means an opening and closing type cover to be installed in the upper part of hatch to do putting in and out of the freight on the deck

of the ship. There are following three types as the representative measurement equipment for large-scale industrial products; a Stereo Image Measurement, a Total Station (TS), and a Laser Scanner (LS). TS is relatively short and it is possible to execute highly precise measurement even in narrow factory, we adopted TS as the measurement equipment.

## 2. TARGET AND ENVIROMENT

### 2.1 Measurement Target

Dimension of the hatch cover as the measurement object was 18m in height, 8m in width, and 0.8m in depth. In the initial process, the material is installed in opposite direction to completion, and the material is reversed and it is in the same direction as completion on the way(Fig.1). In the lower part (the top surface at the time of completion) before the inversion, a top plate has been installed from the first of the process. Inside the hatch cover, two and four longitudinal members are installed in parallel with longitudinal direction and in parallel with lateral direction of the member have been installed respectively.

We pasted a target seal on a member directly as a measurement point at the four corner points of member of outside surface of the hatch cover, the survey point with the longitudinal member, and the intermediate point between the survey points. By executing measurement at these measurement points, we tracked the relative shape deformation of the member throughout the process. We named each measurement surface as follows: Center: Center side of hull, Side: Outside of hull,

---

* Corresponding author. This is useful to know for communication with the appropriate person in cases with more than one author.

Aft: Rear side of hull, Fore: Forward side of hull, Upper: Upper part (Top plate part) of hull at the time of completion, and Deck: Lower part (Opening part) of hull at the time of completion. At the Center side and the Side side, 22 points in total for 11 points at each of upper and lower part, and at the Fore side and the Aft side, 14 points in total for 7 points at each of upper part and lower part, that is, 72 points in total, we set up the measurement points. Installing position of TS was set at the diagonal two places. We measured the dimension from the setting places at the Center/Aft side and Side/Fore side before inversion, and after inversion, we set the TS at the Center/Fore side and Side/Aft side.

## 2.2 Welding Process

We took five processes (i,…,v) from temporary welding to distortion removing welding as the measurement objects (Fig.2). Member was moved at each welding process and inverted between manual welding processes. Measurement by TS was executed before and after the each welding process and movement. Comparison of each process was executed between [A] before auto welding and [B] after auto welding (Comparison I) and [A] before auto welding and [F] after inverted welding (Comparison II), respectively.

## 2.3 Shipyard Environment

Since the hatch cover moves at each welding process, we executed measurement while changing the setting point of TS at each welding process (Fig.2). Since the factory inside was narrow and the wall was close to the both side, and there was other hatch covers in the anteroposterior direction, the place to set up TS was strictly limited (Fig.1).

## 3. MEASUREMENT METHOD

### 3.1 Measurement Parts

We used NET1200 (Sokkia's First class total station) for measurement. Since the target seal comes off by abrasion or welding at the edge part, we pasted the seal at the point approximately 50mm apart from the edge part. Because it is impossible to measure diagonal distance when the incident angle at the time of measurement becomes shallow, we measured the intersection point by setting the L-type target in a right angle against the seal target (Fig.3). As for measurement point that is impossible to collimate directly, we measured the target by off-setting horizontally and subtracted the offset value after coordinate conversion.

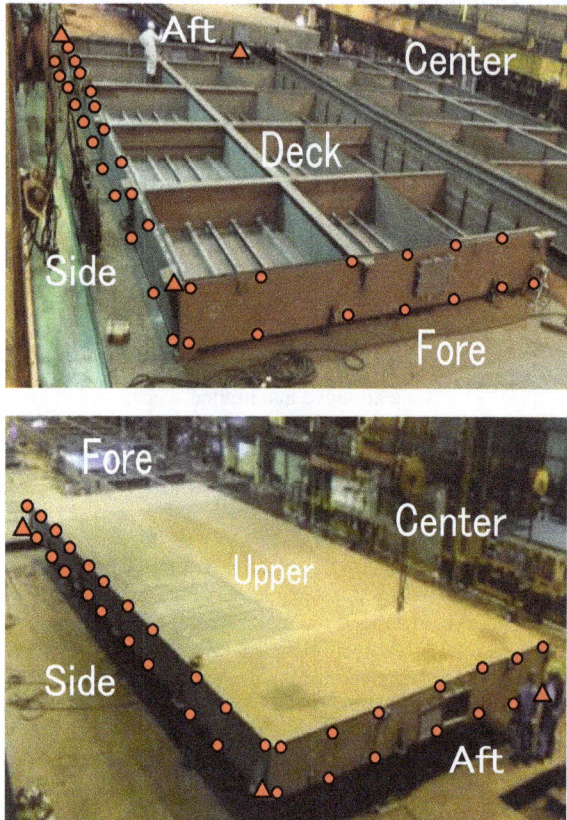

Figure 1.   The shipyard and Hatch Cover
●Measurement Points    ▲Reference Points

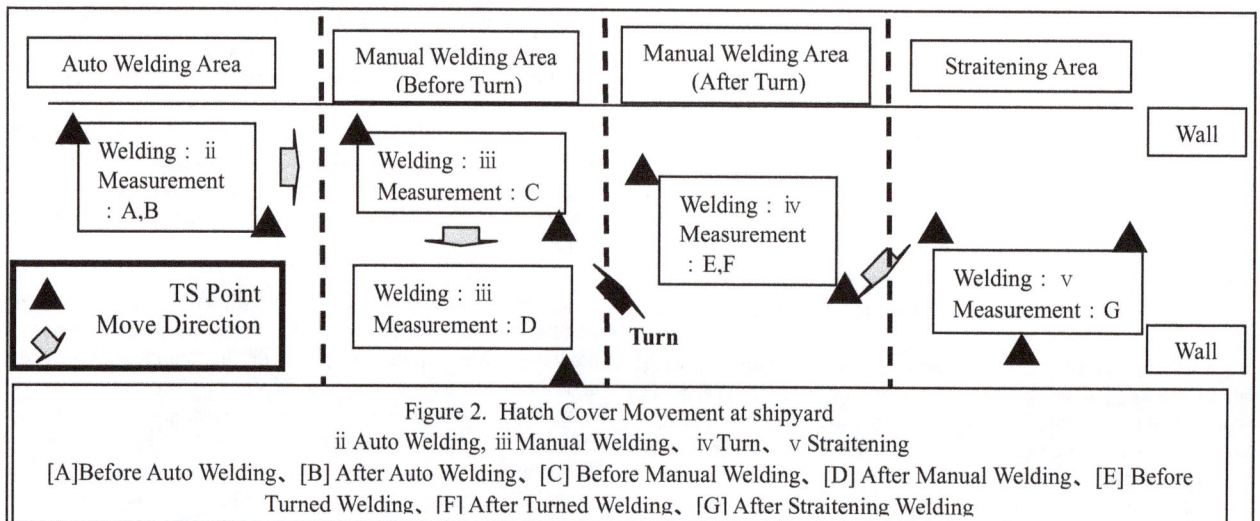

Figure 2.  Hatch Cover Movement at shipyard
ii Auto Welding, iii Manual Welding、iv Turn、v Straitening
[A]Before Auto Welding、[B] After Auto Welding、[C] Before Manual Welding、[D] After Manual Welding、[E] Before Turned Welding、[F] After Turned Welding、[G] After Straitening Welding

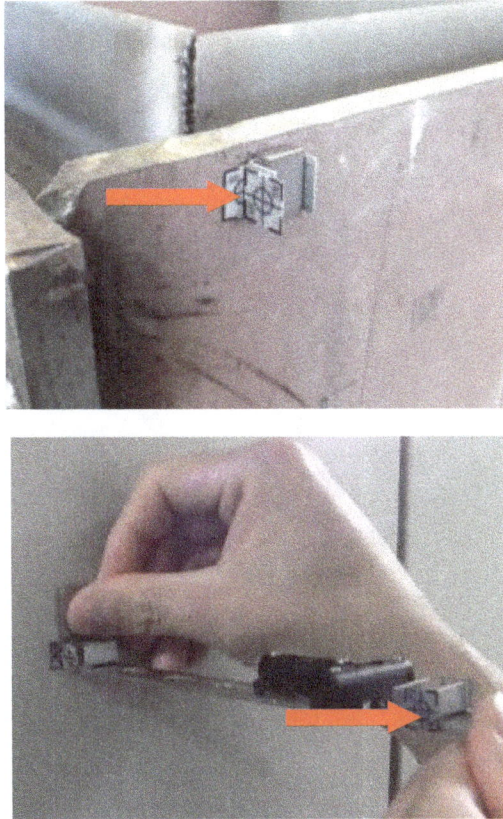

Figure 3. L-type target and Offset target

## 3.2 Coordinate Transform

Since the object moves at each process, we took the four corners as the reference points that are expected to have less deformation inside the hatch cover. And compared the deformation by applying a coordinate conversion on measured data. Taking the 3 points at the edge part as the conversion reference point for coordinate conversion, we executed rotating coordinate conversion (1).When we executed coordinate conversion taking the reference point at a certain side block (example: Block I (Center/Fore)), deformation was not represented correctly at the opposite side block (example: Block II (Side/Aft). This is because that, when amount of deformation at the conversion reference point is large, the amount of deformation at the reference point affects to the opposite side block and the relative deformation becomes unknown. Therefore, we extracted the relative deformation by executing coordinate conversions separately at the Center/Fore surface (Block I) and at the Side/Aft surface (Block II) (Fig.4).

$$\begin{pmatrix} X \\ Y \\ Z \end{pmatrix} = \begin{pmatrix} cos\gamma & sin\gamma & 0 \\ -sin\gamma & cos\gamma & 0 \\ 0 & 0 & 1 \end{pmatrix} \begin{pmatrix} cos\beta & 0 & -sin\beta \\ 0 & 1 & 0 \\ sin\beta & 0 & cos\beta \end{pmatrix} \begin{pmatrix} 1 & 0 & 0 \\ 0 & cos\alpha & sin\alpha \\ 0 & -sin\alpha & cos\alpha \end{pmatrix} \begin{pmatrix} x \\ y \\ z \end{pmatrix}$$

(1)

## 3.3 Diagram Representation

When representing the amount and direction of deformation before and after each process, representation that is possible to recognize the spatial position inside the member at each measurement point is required.

In order to make possible to understand spatially the deformation of each measurement point, we represented the

amount of deformation in vertical and lateral direction for horizontal direction with "Direction of vector" (Fig.5). For height direction, we represented the amount of deformation with "Size of circle" shown by diameter of circle, distinguish the upper (Upper) and lower (Deck) direction at the time of completion by color.

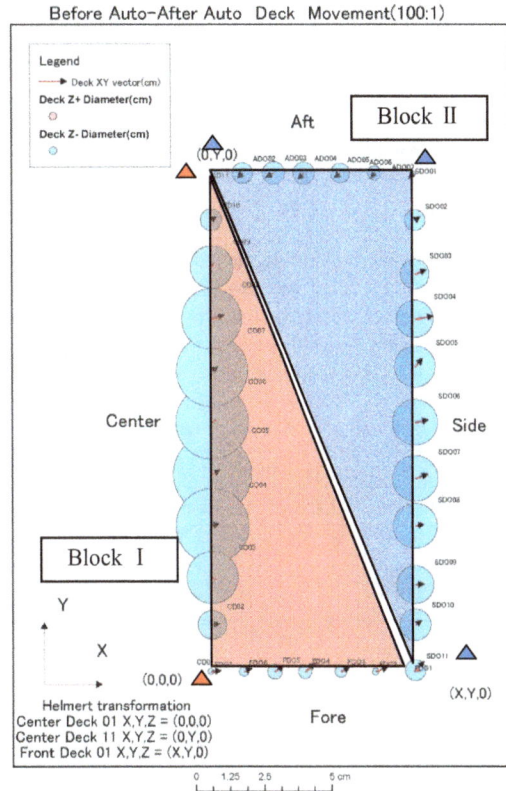

Figure 4. Coordinate Transform blocks

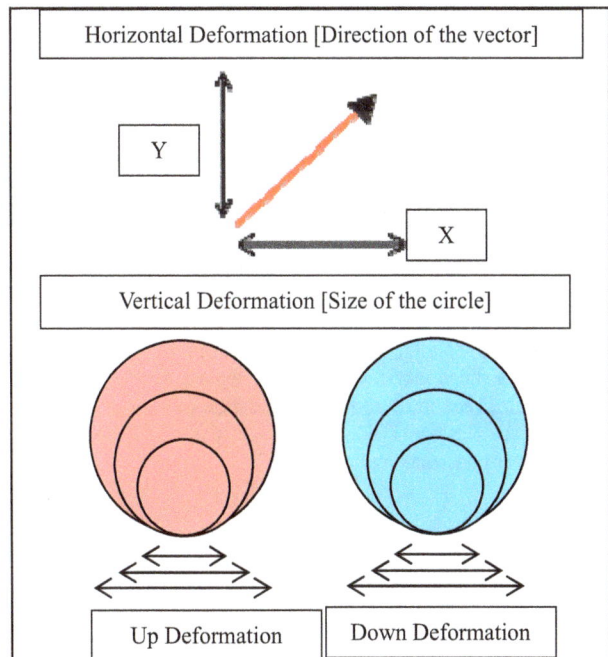

Figure 5. Representation of Horizontal Deformation and Vertical Deformation

## 4. RESULT AND DISCUSSION

### 4.1 Measurement state of measurement target

Since measurement time of one process was 1 - 1.5 hours limited to the time between processes or lunch time, efficient measurement was required. Therefore, we set up the TS at 2 points of four corners that is possible to measure 2 surfaces from single TS point. Although we could executed measurement in all of 7 processes, we could not measure deformation in 2 processes in which allowed measurement time was short ([C] before manual welding, [E] before inverted welding). By making the surface to be measured from TS to be the same one every time, we secured accuracy with a short measurement time.

We had to measure the members that move inside the narrow factory with few TS. Since we also could measure the target that incident angle is shallow or there is a shielding substance by using L-type target and offset target, we could obtain a good result that ratio of miss-measurement was about 5%(Table1).

Table1. Measurement method

| ALL | Measurement Method | | | |
|---|---|---|---|---|
| | Direct | L-type | Offset | Unmeasurable |
| **360 Points** | 40 Points | 285 Points | 24 Points | 11 Points |
| **100%** | 11% | 79% | 7% | 3% |

### 4.2 Deformation amount in each welding process

I. [A]Before Auto/[B]After Auto （Figure 6）
Auto welding is executed along the longitudinal member inside the hatch cover, and heat input is the largest among all the processes. At the Center side and the Side side, deformation in height direction (in-surface deformation) became greater to the center part. Also the Force side and Aft side, deformation in height direction was great. Furthermore, in horizontal direction, deformation directing from opening part to the center of hatch cover was great and a deformation outside surface such that top plate part becomes narrow was generated.

II. [A]Before Auto/[F]After Turn （Figure 7）
In order to calculate amount of work for removing distortion, we compared deformation before auto welding with deformation after inverted welding that is before distortion removing welding. In height direction, the state was that amount of deformation at the time of auto welding was remained. Especially, similar to I, deformation at the Center side that has been fabricated by a single plate was the largest. In horizontal direction, deformation that becomes short along the side surface was observed. It is thought that this is because the hatch cover constricts by heat input.

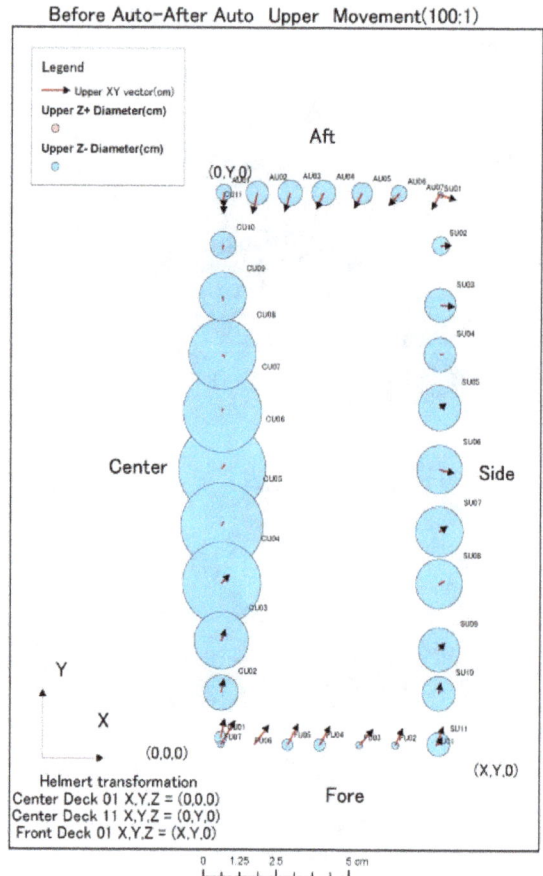

Figure 6. [A]Before Auto/[B]After Auto Comparison

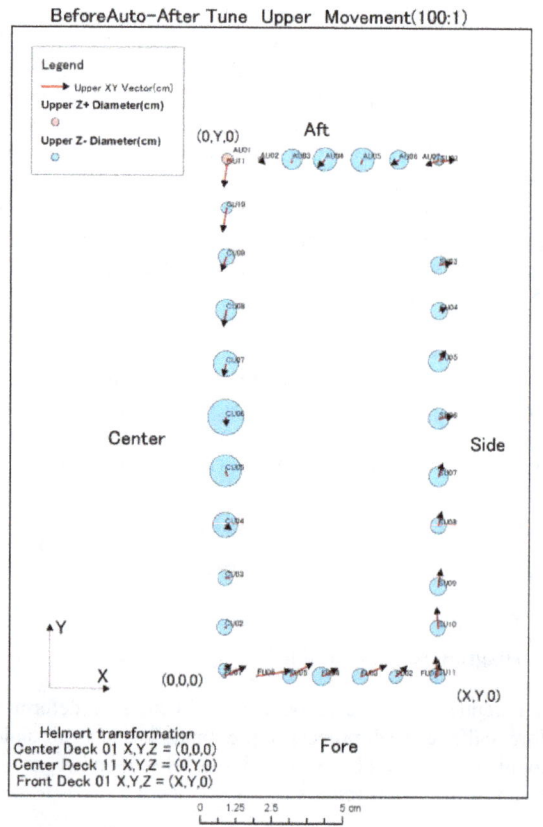

Figure 7. [A]Before Auto/[F]After Turn Comparison

## 5. CONCLUSIONS

In this paper, we could comprehend a deformation of the large industrial products in a small factory by using TS and target seal. Since a measurement speed was improved particularly with measuring precision, we could clarify a deformation in each process which was not frequently carried out up to the present. Furthermore, setting a measurement point at end part in component as a reference point of coordinate conversion was valid to comprehend a deformation tendency for hull component in each process. By executing a conversion by each block in particular, there was no influence on the deformation volume for a deviated measurement point when a measurement point as standard was largely deformed or inclined. It would be required to examine an application method for process improvement based on PDCA cycle in terms of deformation volume by each welding process selected for the working procedure. Specifically, it is necessary to examine a difference between design drawing (CAD) and final deliverable and also a reflecting method for design or work process.

## 6. REFERENCES

Nomoto, T, et al., 1997, Basic Studies on Accuracy Management System Based on Estimating of Weld Deformations, *Techno marine: bulletin of the Society of Naval Architects of Japan*, 181, pp. 249-260.

Takechi. S, et al., 2000, Studies on the Block Positioning Metrics System for the Hull Erection Stage, *Techno marine: bulletin of the Society of Naval Architects of Japan*, 188, pp. 399-408.

Hiekata K, and Matsuo, A., 2012, Application to shipbuilding industry of the three-dimensional measurement, *KANRIN*, 40, pp. 2-5.

Ono, T., 2012, Lecture 6 Bundle adjustment: Bundle adjustment and the camera calibration in precision industrial measurement, *Photogrammetric Engineering & Remote Sensing*, 51, pp. 387-396.

## 7. ACKNOWLEDGEMENTS

Measurement experiment field was kindly provided to IWAKITEC Co., Ltd.

# ON GROUND SURFACE EXTRACTION USING FULL-WAVEFORM AIRBORNE LASER SCANNER FOR CIM

K. Nakano [a,b] *, H. Chikatsu [b]

[a] Research and Development Center, AERO ASAHI CORPORATION,
3-1-1, Minamidai, Kawagoe, Saitama, 350-1165, Japan - kazuya-nakano@aeroasahi.co.jp
[b] Division of Architectural, Civil and Environmental Engineering, Tokyo Denki University,
Ishizaka, Hatoyama, Saitama, 350-0394, Japan - chikatsu@g.dendai.ac.jp

**Commission V, WG V/4**

**KEY WORDS:** Airborne laser scanner, Full-waveform data, Ground surface extraction, Evergreen broad-leaved trees, Virtual ground surface

**ABSTRACT:**

Satellite positioning systems such as GPS and GLONASS have created significant changes not only in terms of spatial information but also in the construction industry. It is possible to execute a suitable construction plan by using a computerized intelligent construction. Therefore, an accurate estimate of the amount of earthwork is important for operating heavy equipment, and measurement of ground surface with high accuracy is required. A full-waveform airborne laser scanner is expected to be capable of improving the accuracy of ground surface extraction for forested areas, in contrast to discrete airborne laser scanners, as technological innovation. For forested areas, fundamental studies for construction information management (CIM) were conducted to extract ground surface using full-waveform airborne laser scanners based on waveform information.

## 1. INTRODUCTION

Since the Great East Japan Earthquake Disaster, coastal residential areas have been shifting to higher elevations. However, there are no suitable places for relocation that is unaffected by tsunamis. Therefore, cutting and filling of earth must be performed in mountains for converting them to residential areas. In order to achieve planned construction in a section of a mountain, it is necessary to obtain accurate ground surface information before felling trees. Estimation of accurate soil volume is important for the planned work because it determines the cost and operating time of heavy equipment. However, it is difficult to accurately estimate soil volume because of the influence of dense vegetation in the mountain. Airborne laser scanners (ALS) have been widely used because of their effectiveness in extracting micro topography or ground surface in forested areas, which are not detected in photogrammetry, and many of its applications such as city modeling, DTM generation, electrical power line monitoring, and calculation of forest area volume have been proposed. The abilities of ALSs are significantly increasing through technological innovations. Many ALSs such as high-pulse-rate laser scanners and full-waveform ALSs were developed by improving the time resolution in pulse analyses. Full-waveform ALSs can extract waveform information to continuously record reflection information from one laser irradiation in a short time interval. Therefore, full-waveform ALSs are expected to be made capable of increasing the density and accuracy of point cloud of ground surface in forest areas in contrast to conventional discrete ALSs. However, it is considered that it would be very difficult to classify very weak reflection information. It would be difficult to separate noise passing through the gap between the trees from the laser irradiation reflected from the ground.

This paper describes a fundamental study on extracting ground surface from forested areas using a full-waveform ALS for construction information management (CIM).

## 2. FULL-WAVEFORM AIRBORNE LASER SCANNER

One of the features of the full-waveform ALS is that reflection information of laser can be continuously obtained as waveform. Figure 1 is a conceptual diagram showing the difference between conventional discrete airborne scanners and full-waveform airborne scanners.

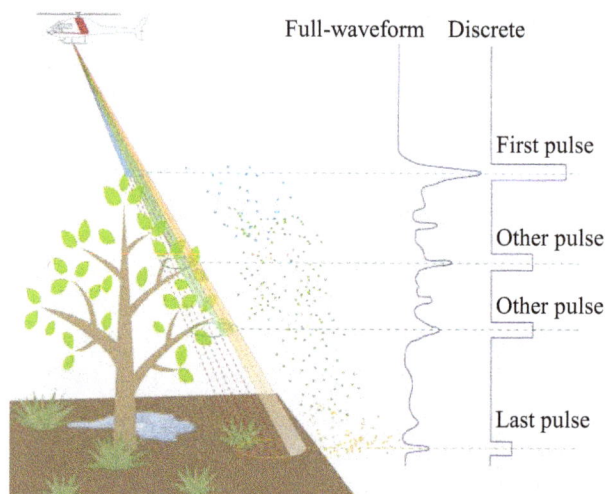

Figure 1. Airborne laser scanner conceptual diagram

---

* Corresponding author

Conventional discrete ALSs obtain three-dimensional coordinates and reflection intensity of discrete pulses as shown in the Figure 1; the first reflection from the crown of a tree is taken as the first pulse, the last reflection from the ground surface is taken as the last pulse, and all reflections between them are taken as other pulses.

The application of output from conventional ALSs to our original algorithm is difficult for processing in the analysis software, which was provided by the ALS manufacturer. On the other hand, the full-waveform ALS can analyze weak reflections for continuously recording reflection information with very short time resolutions such as one nanosecond. It is also possible to compute new three-dimensional coordinates using the continuous reflection information.

## 3. DATA ACQUISITION

In order to confirm the ability of the full-waveform ALS to extract ground surface in forested areas, data acquisition was performed at the Institute for Nature Study, which is located at Shirokanedai, Minato-ku, Tokyo. The area has a variety of vegetation such as evergreen broad-leaved and deciduous trees. Table 1 shows data specifications used in this fundamental study. Figure 2 shows a simple ortho-image that was created using images taken by a digital camera mounted on the ALS in order to confirm the situation of the vegetation. Radiometric correction was not performed for the simple ortho-image for realistic visibility. Data was acquired on October 20, and it can be seen from Figure 2 that the vitality of most of the trees have not been significantly reduced.

Figure 2. Ortho-image

| Domain | Item | Contents |
|---|---|---|
| Data acquisition | Date | October 20, 2012 |
| Platform | Helicopter | AS-350BA |
| Flight | Speed | 120 km/h |
| | Altitude | 950 m |
| Laser | Model | ALTM Orion M300 |
| | Scan rate | 100 kHz |
| | Scan angle | 30 degree (full) |
| | Number of sample of waveform | 288 |
| | Waveform digitization | 10 bits |
| | Planned density | 4 points/m$^2$ |
| Digital Camera | Focal length | 60 mm |
| | Sensor size | 9 x 9 µm |
| | Image size | 5440 x 4080 |
| | Colour band | R, G, B, IR |

Table 1. Data specifications

## 4. DATA INTERPRETATION

We extracted samples from the field in order to develop an analysis method for understanding the characteristics of the full-waveform ALS.

### 4.1 Riverside vegetation

Thick reeds become an issue in ground surface extraction in river areas. Therefore, their waveform was confirmed by extracting data of reeds distributed behind a wetland in the institute, as shown in Figure 3. Figure 4 shows the waveform of the reeds. In the figure, the horizontal axis is the time in nanoseconds, and the vertical axis represents digital value of the reflection intensity. Digital values of the reflection intensity can be converted into Volts using the digitizer offset value and digitizer gain value. The digital values were not converted in this study because the characteristics of the waveform remain unchanged even after unit conversion. Two peaks can be identified in the figure; the first peak is considered to be the top of the reed and the next peak is the ground surface. Since the three-dimensional coordinates of the two peaks are generated using the manufacturer's analysis software, special procedures are not required.

Figure 3. Wetland situation

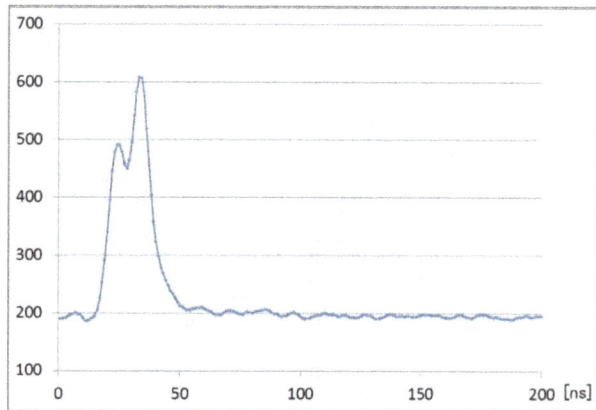

Figure 4. Waveform of reeds

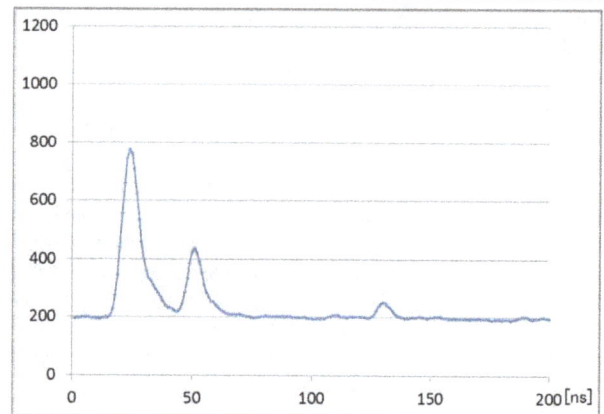

Figure 6. Waveform of Sudajii (three points)

## 4.2 Evergreen broad-leaved trees

Laser beam reaching the ground surface is significantly reduced because they are blocked by leaves and branches in areas with thick evergreen broad-leaved trees. We considered waveform information of Sudajii (Castanopsis sieboldii) as the target, as shown in the right center of Figure 5. Figure 6 shows the three peaks of beams that passed through small gaps of leaves and branches of Sudajii. The three points were generated using the manufacturer's analysis software, and the point of last pulse in the three points is the one that reached the ground surface. However, many of the laser beam reflections from Sudajii were configured as a single peak; this is confirmed by the waveforms in Figure 7 and 8.

It is difficult to extract ground surface when all reflections are concentrated in one crown of Sudajii, as shown in Figure 7. Several minor peaks occur after the major peak shown in Figure 8. The peak noise at 128 ns is likely from the ground surface. Classification of valid ground information from one irradiation is difficult, and the ground surface is obtained by analyzing a single pulse based on clear reflection information of the ground, as shown in Figure 7.

Figure 7. Waveform of Sudajii (single points)

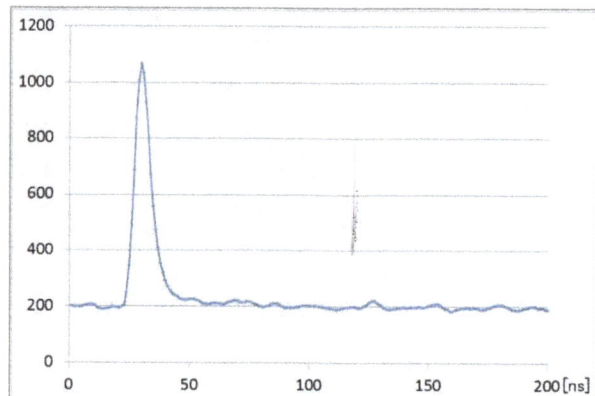

Figure 5. Figure placement and numbering

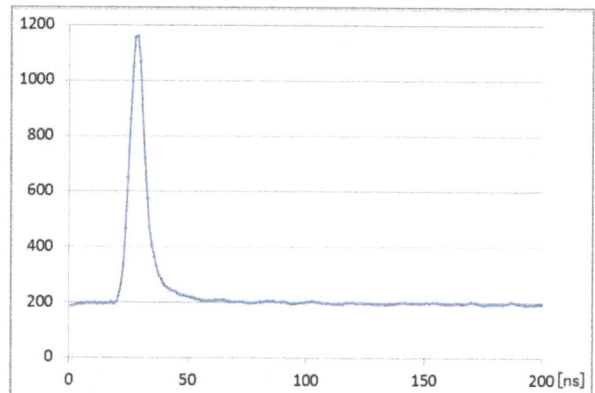

Figure 8. Waveform of Sudajii (single points with noise)

## 5. PROPOSED METHOD

We examined the waveform analysis method based on spatial distribution considering the possibility of ground surface extraction in evergreen broad-leaved trees. The proposed method aims to improve the acquisition rate and the accuracy of ground surface. An airborne laser survey was conducted using grid interpolation to compute three-dimensional coordinates of ground surface in difficult areas such as the area with Sudajii.

Figure 9 shows a flowchart of the processes involved in the proposed method. First, a three-dimensional point cloud with waveform information is generated by integrating the

GNSS/IMU and distance information using the manufacturer's analysis software. Then, a virtual ground surface, giving priority to the lowest point from the three-dimensional point cloud, is generated. Three-dimensional point clouds generated during the waveform analysis are extracted as peaks, assuming waveform as normal distribution, using the Expectation Maximization (EM) algorithm (Dempster, et al., 1977), which was frequently used in previous studies (Persson et al., 2005, Chhatkuli et al., 2012). Finally, ground-surface candidate points are computed using the reflection information and the virtual ground surface.

Point cloud data are managed in units of flight course to maintain a close relationship with flight trajectory of the platform. Japanese survey regulations specify a standard flight course sidelap of 30% lap for ALSs. However, we adopted a 50% lap by in-house standards, and adjacent courses are able to supplement the entire area for a course. Therefore, a single course has the potential for laser irradiation to reach the ground by passing through different paths of adjacent courses. Point cloud from adjacent courses was used for generating virtual ground surface.

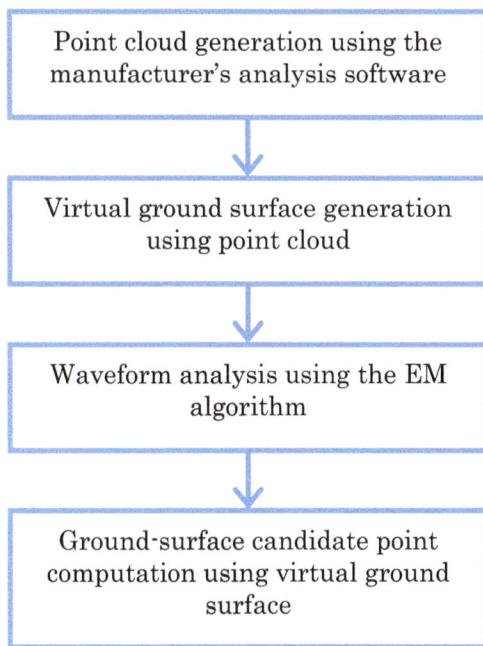

Point cloud generation using the manufacturer's analysis software

↓

Virtual ground surface generation using point cloud

↓

Waveform analysis using the EM algorithm

↓

Ground-surface candidate point computation using virtual ground surface

Figure 9. Flawchart of the proposed method

## 6. EVALUATION

The ground-surface candidate points were evaluated by referring to the measured results of the total station. Reference data were observed before the Great East Japan Earthquake. Tokyo area was moved by approximately 29 cm to the east and it subsided by approximately 3 cm. However, elevation data were obtained directly from the results because accurate parameters for elevations could not be obtained from the Geospatial Information Authority of Japan.

The proposed method was verified on a selected test area (40 m × 40 m) at the Institute for Nature Study as shown in Figure 2. The test area comprises a dense community of mainly Sudajii, and includes 10 checkpoints.

Figure 10 shows the virtual ground surface from a single course, and Figure 11 shows the virtual ground surface including the adjacent course.

A virtual ground surface of 2.5 m grid spacing was created using the low point priority of point cloud; in addition, a virtual ground surface of 20 m grid spacing was created using 2.5 m grid spacing. Spike noises in virtual ground surface from the single course are confirmed, as shown in Figure 10. In contrast, spike noises are reduced in virtual ground surface using adjacent courses.

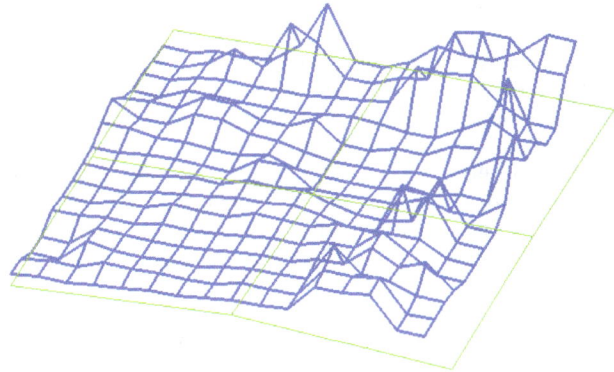

Figure 10. Virtual ground surface from a single course

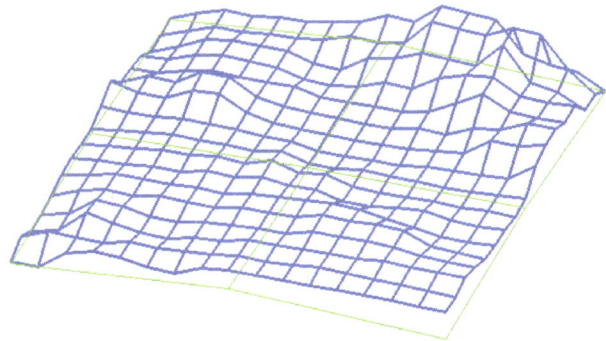

Figure 11. Virtual ground surface including the adjacent course

The accuracy of this method was evaluated from checkpoints for ground-surface candidate points computed using the virtual ground surface of the single and adjacent courses.

Table 2 shows RMS error for 10 checkpoints, which were computed using checkpoint height and interpolated height under the inverse distance weighted method using the horizontal position of checkpoint and ground surface candidate point coordinates.

The effect of the adjacent courses is confirmed by the fact that RMS error is reduced from 2.4 m to 0.6 m as shown in Table 2. As a reference, it shows the RMS error calculated from the lowest elevation in the range of 1 m around the checkpoints using the last pulse, which has been calculated using the manufacturer's analysis software. The test area consists of thick trees, and the RMS error of the last pulse is greater than eight meters.

| Case of computation | RMS error |
|---|---|
| Ground-surface candidate points using single course | ±2.445 m |
| Ground-surface candidate points using including adjacent courses | ±0.604m |
| Last pulse points from discrete point cloud | ±8.365 m |

Table 2. RMS error for 10 checkpoints

## 7.  CONCLUSION

A fundamental study on ground surface extraction for CIM was conducted using a full-waveform ALS. The proposed method was analyzed with weak reflection information from the ALS considering spatial distribution using virtual ground surface. The proposed method was able to improve the accuracy of the ground candidate points and virtual ground surface using point cloud of adjacent courses in the test area with thick evergreen broad-leaved trees such as Sudajii.

It can be concluded that the proposed method is useful for ground surface extraction in heavily forested areas, where it cannot be extracted by discrete ALSs.

However, there are several issues in that the accuracy of ground-surface candidate points is influenced by the virtual ground surface, and difficulties arise when all reflections are concentrated on only the crown of trees, as shown in Figure 7.

## ACKNOWLEDGEMENTS

Checkpoint data were obtained with the cooperation of the Institute for Nature Study. We would like to express our gratitude to this institute.

## REFERENCES

Dempster A, Laird N, Rubin D, 1977. Maximum likelihood from incomplete data via the EM algorithm. J Roy Stat Soc Series B, 39, pp.1-38.

Persson, Å., Söderman, U., Töpel, J., Alhberg, S., 2005. Visualization and analysis of full-waveform airborne laser scanner data. International Archives of Photogrammetry, Remote Sensing and Spatial Information Sciences 36 (Part3/W19), 103  108

Chhatkuli S, Mano K, Kogure T, Tachibana K, and Shimamura H, 2012. Full Waveform LIDAR Exploitation Technique and Its Evaluation in the Mixed Forest Hilly Region, Int. Arch. Photogramm. Remote Sens. Spatial Inf. Sci., XXXIX-B7, pp.505-509.

# Permissions

All chapters in this book were first published in TIAPRSSIS, by Copernicus Publications; hereby published with permission under the Creative Commons Attribution License or equivalent. Every chapter published in this book has been scrutinized by our experts. Their significance has been extensively debated. The topics covered herein carry significant findings which will fuel the growth of the discipline. They may even be implemented as practical applications or may be referred to as a beginning point for another development.

The contributors of this book come from diverse backgrounds, making this book a truly international effort. This book will bring forth new frontiers with its revolutionizing research information and detailed analysis of the nascent developments around the world.

We would like to thank all the contributing authors for lending their expertise to make the book truly unique. They have played a crucial role in the development of this book. Without their invaluable contributions this book wouldn't have been possible. They have made vital efforts to compile up to date information on the varied aspects of this subject to make this book a valuable addition to the collection of many professionals and students.

This book was conceptualized with the vision of imparting up-to-date information and advanced data in this field. To ensure the same, a matchless editorial board was set up. Every individual on the board went through rigorous rounds of assessment to prove their worth. After which they invested a large part of their time researching and compiling the most relevant data for our readers.

The editorial board has been involved in producing this book since its inception. They have spent rigorous hours researching and exploring the diverse topics which have resulted in the successful publishing of this book. They have passed on their knowledge of decades through this book. To expedite this challenging task, the publisher supported the team at every step. A small team of assistant editors was also appointed to further simplify the editing procedure and attain best results for the readers.

Apart from the editorial board, the designing team has also invested a significant amount of their time in understanding the subject and creating the most relevant covers. They scrutinized every image to scout for the most suitable representation of the subject and create an appropriate cover for the book.

The publishing team has been an ardent support to the editorial, designing and production team. Their endless efforts to recruit the best for this project, has resulted in the accomplishment of this book. They are a veteran in the field of academics and their pool of knowledge is as vast as their experience in printing. Their expertise and guidance has proved useful at every step. Their uncompromising quality standards have made this book an exceptional effort. Their encouragement from time to time has been an inspiration for everyone.

The publisher and the editorial board hope that this book will prove to be a valuable piece of knowledge for researchers, students, practitioners and scholars across the globe.

# List of Contributors

**R. Kaijaluoto**
Finnish Geospatial Research Institute (FGI) - Center of Excellence in Laser Scanning Research, FI-02430 Masala, Finland

**A. Kukko**
Finnish Geospatial Research Institute (FGI) - Center of Excellence in Laser Scanning Research, FI-02430 Masala, Finland

**J. Hyyppä**
Finnish Geospatial Research Institute (FGI) - Center of Excellence in Laser Scanning Research, FI-02430 Masala, Finland

**S. Kanai**
Graduate School of Information Science and Technology, Hokkaido University, Kita-ku, Sapporo 060-0814, Japan

**R. Hatakeyama**
Graduate School of Information Science and Technology, Hokkaido University, Kita-ku, Sapporo 060-0814, Japan

**H. Date**
Graduate School of Information Science and Technology, Hokkaido University, Kita-ku, Sapporo 060-0814, Japan

**Jie Xiong**
School of Electronic Information, Wuhan University, Wuhan, Hubei 430072, China

**Sidong Zhong**
School of Electronic Information, Wuhan University, Wuhan, Hubei 430072, China

**Lin Zheng**
School of Electronic Information, Wuhan University, Wuhan, Hubei 430072, China

**Tee-Ann Teo**
Dept. of Civil Engineering, National Chiao Tung University, Hsinchu, Taiwan 30010

**V. A. Knyaz**
St. Res. Institute of Aviation Systems (GosNIIAS), 125319, 7, Victorenko str., Moscow, Russia

**R. A. Guryanova**
Sechenov's First Moscow State Medical University, Moscow, Russia

**A. S. Guryanovb**
Sechenov's First Moscow State Medical University, Moscow, Russia

**I-C. Lee**
Center for Space and Remote Sensing Research, National Central University, Taiwan

**F. Tsai**
Center for Space and Remote Sensing Research, National Central University, Taiwan

**K. Nagara**
Dept. of Civil Engineering, University of Tokyo, 7-3-1 Hongo Bunkyo Tokyo 113-8656, Japan

**T. Fuse**
Dept. of Civil Engineering, University of Tokyo, 7-3-1 Hongo Bunkyo Tokyo 113-8656, Japan

**Boris V. Vishnyakov**
State Research Institute of Aviation Systems (FGUP GosNIIAS), Moscow, Russia

**Sergey V. Sidyakin**
State Research Institute of Aviation Systems (FGUP GosNIIAS), Moscow, Russia

**Yuri V. Vizilter**
State Research Institute of Aviation Systems (FGUP GosNIIAS), Moscow, Russia

**G. R. Petrosyan**
Institute for Informatics and Automation Problems of NAS RA, International Scientific - Educational Center of NAS RA, Armenia, Yerevan

**L. A. Ter-Vardanyan**
Institute for Informatics and Automation Problems of NAS RA, International Scientific - Educational Center of NAS RA, Armenia, Yerevan

**A.V. Gaboutchian**
Russian Federation, Moscow

**T. Anai**
General Technology Div., R&D Dept., TOPCON CORPORATION, 75-1, Hasunuma, Itabashi, Tokyo

**N. Kochi**
General Technology Div., R&D Dept., TOPCON CORPORATION, 75-1, Hasunuma, Itabashi, Tokyo
R&D Initiative, Chuo University, 1-13-27, Kasuga, Bunkyo-ku, Tokyo, Japan

**M. Yamada**
General Technology Div., R&D Dept., TOPCON CORPORATION, 75-1, Hasunuma, Itabashi, Tokyo

**T. Sasakia**
General Technology Div., R&D Dept., TOPCON CORPORATION, 75-1, Hasunuma, Itabashi, Tokyo

**H. Otani**
Smart Infrastructure Company, Technology Development Dept., TOPCON CORPORATION, 75-1, Hasunuma, Itabashi, Tokyo

**D. Sasaki**
Smart Infrastructure Company, Technology Development Dept., TOPCON CORPORATION, 75-1, Hasunuma, Itabashi, Tokyo

**S. Nishimura**
Keisoku Research Consultant Co.,Ltd.Creative design group Kasuga, Bunkyo-ku, Tokyo, Japan

**K. Kimoto**
Keisoku Research Consultant Co.,Ltd.Creative design group Kasuga, Bunkyo-ku, Tokyo, Japan

**N.Yasui**
Keisoku Research Consultant Co.,Ltd.Creative design group Kasuga, Bunkyo-ku, Tokyo, Japan

**D. Borini Alves**
University of Zaragoza (UNIZAR), Department of Geography and Spatial Management, and Aragon University Research Institute in Environmental Science (IUCA), Zaragoza, Spain

**F. Pérez-Cabello**
University of Zaragoza (UNIZAR), Department of Geography and Spatial Management, and Aragon University Research Institute in Environmental Science (IUCA), Zaragoza, Spain

**M. Rodrigues Mimbrero**
University of Zaragoza (UNIZAR), Department of Geography and Spatial Management, and Aragon University Research Institute in Environmental Science (IUCA), Zaragoza, Spain

**H. Adhikari**
University of Helsinki, Department of Geosciences and Geography, P.O. Box 68, FI-00014, Helsinki, Finland

**J. Heiskanen**
University of Helsinki, Department of Geosciences and Geography, P.O. Box 68, FI-00014, Helsinki, Finland

**E. E. Maeda**
University of Helsinki, Department of Geosciences and Geography, P.O. Box 68, FI-00014, Helsinki, Finland

**P. K. E. Pellikka**
University of Helsinki, Department of Geosciences and Geography, P.O. Box 68, FI-00014, Helsinki, Finland

**K. Zakšek**
University of Hamburg, CEN, Institute of Geophysics, Bundesstr. 55, 20146 Hamburg, Germany

**A. Gerst**
ESA, European Astronaut Centre, Linder Höhe, 51147 Köln, Germany

**J. von der Lieth**
University of Hamburg, CEN, Institute of Geophysics, Bundesstr. 55, 20146 Hamburg, Germany

**G. Ganci**
INGV, Sezione di Catania, Piazza Roma, 2, 95125 Catania, Italy

**M. Hort**
University of Hamburg, CEN, Institute of Geophysics, Bundesstr. 55, 20146 Hamburg, Germany

**P. Bholanath**
Guyana Forestry Commission, Planning and Developmnet Division, 1 Water Street, Kingston, Georgetown, Guyana

**K. Cort**
Guyana Forestry Commission, Manager, Forest Area Assessment Unit, 1 Water Street, Kingston, Georgetown, Guyana

**M. Nakagawa**
Dept. of Civil Engineering, Shibaura Institute of Technology, Tokyo, Japan

**T. Yamamoto**
Dept. of Civil Engineering, Shibaura Institute of Technology, Tokyo, Japan

**S. Tanaka**
Dept. of Civil Engineering, Shibaura Institute of Technology, Tokyo, Japan

**M. Shiozaki**
Nikon Trimble Co., Ltd., Tokyo, Japan

**T. Ohhashi**
Nikon Trimble Co., Ltd., Tokyo, Japan

**Fuan Tsai**
Center for Space and Remote Sensing Research National Central University, Zhong-li, Taoyuan 320 Taiwan

**Tzy-Shyuan Wub**
Department of Civil Engineering National Central University, Zhong-li, Taoyuan 320 Taiwan

**I-Chieh Leea**
Center for Space and Remote Sensing Research National Central University, Zhong-li, Taoyuan 320 Taiwan

**Huan Chang**
Department of Civil Engineering National Central University, Zhong-li, Taoyuan 320 Taiwan

**Addison Y. S. Suc**
Research Center for Advanced Science and Technology, National Central University, Zhong-li, Taoyuan 320 Taiwan

**D. Pagliari**
DICA-Dept. of Civil and Environmental Engineering, Politecnico di Milano, Milan, Italy

**N.E. Cazzaniga**
DICA-Dept. of Civil and Environmental Engineering, Politecnico di Milano, Milan, Italy

**L. Pinto**
DICA-Dept. of Civil and Environmental Engineering, Politecnico di Milano, Milan, Italy

**T.Fuse**
Dept. of Civil Engineering, University of Tokyo, Hongo 7-3-1, Bunkyo-ku, Tokyo, 113-8656,Japan

**R.Harada**
Dept. of Civil Engineering, University of Tokyo, Hongo 7-3-1, Bunkyo-ku, Tokyo, 113-8656,Japan

**J. Veitch-Michaelis**
Imaging Group, Mullard Space Science Laboratory (University College London, Holmbury St Mary, RH5 6NT, UK

**J-P. Muller**
Imaging Group, Mullard Space Science Laboratory (University College London, Holmbury St Mary, RH5 6NT, UK

**J. Storey**
IS Instruments Ltd, 220 Vale Road, Tonbridge, TN9 1SP, UK

**D. Walton**
Imaging Group, Mullard Space Science Laboratory (University College London, Holmbury St Mary, RH5 6NT, UK

**M. Foster**
IS Instruments Ltd, 220 Vale Road, Tonbridge, TN9 1SP, UK

**E. Zillmann**
BlackBridge, Dept. of Application Research, 10719 Berlin, Germany

**M. Schönerta**
BlackBridge, Dept. of Application Research, 10719 Berlin, Germany

**H. Lilienthal**
Julius-Kühn-Institute (JKI), Federal Research Centre for Cultivated Plants, 38116 Braunschweig, Germany

**B. Siegmann**
Institute for Geoinformatics and Remote Sensing, University of Osnabrueck, 49076 Osnabrueck, Germany

**T. Jarmer**
Institute for Geoinformatics and Remote Sensing, University of Osnabrueck, 49076 Osnabrueck, Germany

**P. Rosso**
BlackBridge, Dept. of Application Research, 10719 Berlin, Germany

**H. Weichelt**
BlackBridge, Dept. of Application Research, 10719 Berlin, Germany

**D. N. H Thanh**
Tula State University, Institute of Applied Mathematics and Computer Sciences, Lenin ave. 92 Tula city, Russian Federation

**S. D. Dvoenko**
Tula State University, Institute of Applied Mathematics and Computer Sciences, Lenin ave. 92 Tula city, Russian Federation

**Hideharu Yanagi**
Japan Association of Surveyors, 1-33-18, Hakusan, Bunkyo-ku, Tokyo 113-0001, Japan
School of Science and Engineering, Tokyo Denki University, Hatoyama, Hiki-gun, Saitama 350-0394, JAPAN

**Hirofumi Chikatsu**
Division of Architectural, Civil and Environmental Engineering, Tokyo Denki University, Hatoyama, Hiki-gun, Saitama 350-0394, JAPAN

**Yewei Huang**

**Haiting Wang**

**Kefei Zhan**

**Junqiao Zhao**

**Popo Gui**

**Tiantian Feng**

**M. Piper**
INSTAAR / CSDMS, University of Colorado Boulder, U.S.A

**T. Bahr**
Exelis VIS GmbH, 82205 Gilching, Germany

**Á . Marqués-Mateu**
Universitat Politécnica de Valéncia (UPV), Department of Cartographic Engineering, Geodesy, and Photogrammetry, 46022 Val`encia, Spain

**M. Balaguer-Puig**
Universitat Politécnica de Valéncia (UPV), Department of Cartographic Engineering, Geodesy, and Photogrammetry, 46022 Val`encia, Spain

**H. Moreno-Ramón**
Universitat Politécnica de Valéncia (UPV), Department of Plant Production, 46022 València, Spain

**S. Ibáñez-Asensio**
Universitat Politècnica de Valéncia (UPV), Department of Plant Production, 46022 València, Spain

**Zhenfeng Shao**
State key Laboratory for Information Engineering in Surveying, Mapping and Remote Sensing, Wuhan University, Wuhan, 430079, China

**Congmin Li**
State key Laboratory for Information Engineering in Surveying, Mapping and Remote Sensing, Wuhan University, Wuhan, 430079, China

**Sidong Zhong**
Wuhan University, Wuhan, 430079, China

**Bo Liu**
Geomatics Center of Guangxi, 5 Jianzheng Road, Nanning China, 530023

**Honggang Jiang**
Geomatics Center of Guangxi, 5 Jianzheng Road, Nanning China, 530023

**Xuehu Wend**
The Third Surveying and Mapping Engineering Institute of Sichuan, Chengdu Sichuan 610500

**L. Hojas-Gascón**
European Commission Joint Research Centre, Via E. Fermi 2749, 21027 Ispra (VA), Italy
CIFOR Center for International Forestry Research, Bogor Barat 16115, Indonesia

**A. Belward**
European Commission Joint Research Centre, Via E. Fermi 2749, 21027 Ispra (VA), Italy

**H. Eva**
European Commission Joint Research Centre, Via E. Fermi 2749, 21027 Ispra (VA), Italy

**G. Ceccherini**
European Commission Joint Research Centre, Via E. Fermi 2749, 21027 Ispra (VA), Italy

**O. Hagolle**
CESBIO / CNES, Toulouse, France

**J. Garcia**
Unidad de Investigación de Teledetección, Dr. Moliner 50, 46100 Valencia, Spain

**P. Ceruttid**
CIFOR Center for International Forestry Research, Bogor Barat 16115, Indonesia

**I. Harada**
Department of Informatics, Tokyo University of Information Sciences, 4-1 Onaridai Wakaba-ku, Chiba, 265-8501 Japan

**K. Hara**
Department of Informatics, Tokyo University of Information Sciences, 4-1 Onaridai Wakaba-ku, Chiba, 265-8501 Japan

**J. Park**
Department of Informatics, Tokyo University of Information Sciences, 4-1 Onaridai Wakaba-ku, Chiba, 265-8501 Japan

**I. Asanuma**
Department of Informatics, Tokyo University of Information Sciences, 4-1 Onaridai Wakaba-ku, Chiba, 265-8501 Japan

**M. Tomita**
Department of Informatics, Tokyo University of Information Sciences, 4-1 Onaridai Wakaba-ku, Chiba, 265-8501 Japan

**D. Hasegawa**
Department of Informatics, Tokyo University of Information Sciences, 4-1 Onaridai Wakaba-ku, Chiba, 265-8501 Japan

**K. Short**
Department of Informatics, Tokyo University of Information Sciences, 4-1 Onaridai Wakaba-ku, Chiba, 265-8501 Japan

**M. Fujihara**
Graduate School of Landscape Design and Management, University of Hyogo/Awaji Landscape Planning and Horticulture Academy, 954-2 Nojimatokiwa, Awaji, Hyogo Prefecture, 656-1726 Japan

**J. Haarpaintner**
Norut, P.O. Box 6434, Tromsø Science Park, N-9294
Tromsø, Norway

**C. Davids**
Norut, P.O. Box 6434, Tromsø Science Park, N-9294
Tromsø, Norway

**H. Hindberg**
Norut, P.O. Box 6434, Tromsø Science Park, N-9294
Tromsø, Norway

**E. Zahabu**
Sokoine University of Agriculture, Morogoro, United
Republic of Tanzania

**R. E. Malimbwi**
Sokoine University of Agriculture, Morogoro, United
Republic of Tanzania

**J. Heiskanen**
Department of Geosciences and Geography, University of
Helsinki, P.O. Box 68, FI-00014, Helsinki, Finland

**L. Korhonen**
Department of Forest Sciences, University of Helsinki,
P.O. Box 27, FI-00014, Helsinki, Finland

**J. Hietanen**
Department of Geosciences and Geography, University of
Helsinki, P.O. Box 68, FI-00014, Helsinki, Finland

**V. Heikinheimo**
Department of Geosciences and Geography, University of
Helsinki, P.O. Box 68, FI-00014, Helsinki, Finland

**E. Schäfer**
Department of Geosciences and Geography, University of
Helsinki, P.O. Box 68, FI-00014, Helsinki, Finland

**P. K. E. Pellikka**
Department of Geosciences and Geography, University of
Helsinki, P.O. Box 68, FI-00014, Helsinki, Finland

**S. Chhatkuli**
PASCO CORPORATION, Research & Development HQ,
2-8-10 Higashiyama, Meguro-ku, Tokyo, Japan

**T. Satoh**
PASCO CORPORATION, Research & Development HQ,
2-8-10 Higashiyama, Meguro-ku, Tokyo, Japan

**K. Tachibana**
PASCO CORPORATION, Research & Development HQ,
2-8-10 Higashiyama, Meguro-ku, Tokyo, Japan

**Ryogo Abe**
Kokusai Kogyo Co., Ltd, Research and Development
Division, Digital Sensing Group, 2-24-1 Harumi-cho,
Fuchu City,Tokyo 183-0057 JAPAN

**Kunihiro Hamada**
Hiroshima University

**Noritaka Hirata**
Hiroshima University

**Ryotaro Tamura**
Hiroshima University

**Nobuaki Nishi**
Tsuneishi Shipbuilding Co., Ltd

**K. Nakano**
Research and Development Center, AERO ASAHI
CORPORATION, 3-1-1, Minamidai, Kawagoe, Saitama,
350-1165, Japan
Division of Architectural, Civil and Environmental
Engineering, Tokyo Denki University, Ishizaka, Hatoyama,
Saitama, 350-0394, Japan

**H. Chikatsu**
Division of Architectural, Civil and Environmental
Engineering, Tokyo Denki University, Ishizaka, Hatoyama,
Saitama, 350-0394, Japan